SUICIDE TERROR

SUICIDE TERROR
Understanding and Confronting the Threat

EDITED BY

Ophir Falk

Henry Morgenstern

A JOHN WILEY & SONS, INC., PUBLICATION

Published by John Wiley & Sons, Inc., Hoboken, New Jersey
Published simultaneously in Canada

Library of Congress Cataloging-in-Publication Data:
Suicide terror : understanding and confronting the threat / Ophir Falk, Henry Morgenstern, editors.
p. cm.
Includes bibliographical references and index.
ISBN 978-0-470-08729-9 (cloth)
1. Terrorism–Psychological aspects. 2. Terrorism–Prevention. 3. Terrorists–Suicidal behavior. 4. Suicide bombers. 5. Suicide bombings. I. Morgenstern, Henry, 1951– II. Falk, Ophir, 1968–
HV6431.S83 2009
363.325—dc22

2008038591

Printed in the United States of America

10 9 8 7 6 5 4 3 2 1

CONTENTS

PREFACE

Sunday is the first day of the week in Israel. It has that "Monday morning" feeling, when everyone heads back to their routines. Both men and women serve in the Israel Defense Forces (IDF), two years for women and three for men from the age of 18.

These young men and women were making their way, sleepily, from the welcome relief of home and home comforts, back to their bases for another hectic week of service on the cool morning of Sunday, January 22, 1995. Many of them converged on a junction to the north of Tel Aviv and near the coastal city of Netanya, where buses would ferry them all over the country. The junction served as a meeting point for paratroop units that were reporting back to duty at 9 A.M. sharp. They frequently set their watches by the beeps of the 9 A.M. news; a news bulletin few Israelis miss. That morning the junction would have been swamped with men and women in uniform rushing to get back to their bases on time.

HaSharon Junction, commonly known as Beit Lid Junction, is not a scenic crossroads. It intersects Highway 4, which goes north and south, and Highway 57, which goes east and west. It was at that time really a large bus stop with several covered and uncovered stops for dozens of different bus lines that came in and out to pick up anyone there. On the southwest corner of the junction is the Ashmoret Prison—a civilian jail.

For weeks, a group of young Palestinian men allowed into Israel to work and earn money for their families had been doing reconnaissance on the junction, and it had nothing to do with which bus they should take.

Notwithstanding the effort that had been made to establish the Palestinian Authority as a result of Oslo agreements, along with the efforts of the Rabin Government to placate the demands of the then-head of the Palestinian Authority, Yassir Arafat, other factions within the Authority were more than ready to play the bad guys while Arafat feigned conciliation. The Palestine Islamic Jihad funded by people—such as Professor Sami al-Arian at the South Florida University in Tampa—and a network that spread across the United States—had other plans.

At 9:30 A.M., on that fateful morning a Palestinian named Anwar Soukar feigned intense stomach pains and dropped to his knees. As soldiers gathered around him to help, Soukar reached into his bag and detonated the first bomb.

Bodies were instantly transformed into bleeding projectiles of disconnected appendages as a result of the blast wave—many to be found as shreds embedded in the surrounding trees, on fences, and under bus benches. The packed scene of devastation was now the scene of hundreds running toward the blast to rescue whomever they could. As first responders began descending on the scene in great numbers to tend to those wounded with a chance of survival, another member of the Palestinian Islamic, Jihad, Salaah Shaaker, detonated a bomb that he wore on his chest so that the blast would go out and kill and maim as many first responders as possible.

The massacre caused the deaths of 21, and more than 69 were injured. It was not Israel's first suicide bombing; there had already been others, but it was the first double bombing; a suicide terror mission designed to instill fear and hopelessness in the population. Almost all the victims were paratroopers from the same brave units that had once freed Jerusalem—now helplessly slaughtered—and it underlined the vulnerability of first responders to this type of attack.

At nearby Ashmoret Prison, which held the founder of another bloody terrorist group known as Hamas, Sheik Ahmed Yassin was whisked away; the prison officials believed that this might be the first volley in an attempt to free him. In the Palestinian Authority, the engineer or master bomb maker Ihyyah Ayyash, would have been celebrating his handy work.

For the Israeli Government's Prime Minister, Yitzhak Rabin, who told his cabinet a few days later that suicide terror was a strategic threat to the existence of the State of Israel, this was an attack that marked a turning point. Benjamin Netanyahu, who would be Prime Minister of Israel after Rabin, writing back in 1986, claimed that suicide terror was a strategic threat to the world (see the case studies in Chapter 4).

Indeed, in the 1980s and 1990s, international terror or the Global Jihad prepared itself and conducted many bloody attacks using the tactic or weapon of suicide terror, culminating in the bloodiest day of them all, 9/11/2001. Since then there have been thousands of attacks, especially in Iraq.

This book provides the professional first responder and student of Homeland Security with an understanding of suicide terror as a tactic and also as a strategic tool used by terrorists worldwide. We have based the text on diligently researched findings, aimed at constructing a full picture of the challenge posed by the threat.

The Introduction sets the international context of the development of this weapon. Chapter 1 shows how the Global Jihad justifies the use of indiscriminate murder, the sources for the justification by radicals, and where this may take us. Chapter 2 looks at the Israeli experience in the eyes of those who most contributed to confronting this strategic threat to Israel's existence and, in that sense, makes a unique contribution to understanding the Israeli difficulties in dealing with such an effective mode of terror. Nothing would be complete in the realm of terrorism that is detailed here in 3 chapters without a close look at the Iraqi experience that U.S. forces have encountered—without precedence in the history of terrorism. Chapter 4 looks at the wide reach of suicide terror and its internationalism. Some of the probable scenarios that pose the greatest risk and some ideas on mitigation are detailed in Chapter 5. Chapter 6 addresses U.S. law enforcement's challenge in dealing with the threat of this weapon. No study would be complete without a look at the medical response necessary to save lives and provide insights into the results of attacks as provided by Chapter 7.

It is our hope that we create a better understanding—and since knowledge is power—enabling our first responder community and homeland security professionals to be more ready for the challenges that this terrorist weapon of choice poses.

Through better understanding, we hope to make a modest contribution to avoiding the death and destruction that suicide terror wreaks on its victims.

OPHIR FALK
HENRY MORGENSTERN

Tel Aviv, Israel
Miami Florida
April 2009

ACKNOWLEDGMENTS

Writing about suicide terrorism is much easier than confronting it. We acknowledge and appreciate that this book was made possible by those that led the confrontation against terrorism and some of who kindly shared with us from their in-depth insight and experience. The interviewees to whom we owe our deep gratitude for their time and efforts are (in alphabetical order): Uri Bar Lev, Israeli Police Commander of the Southern District, and former Commander of Elite Counterterrorism Units; Avi Dichter, Head of the ISA 2000–2005, and Internal Security Minister since 2006; Zohar Dvir, Yamam commander 2002–2005; Giora Eiland, General (retired) Head of the National Security Council 2003–2006; Yitzchak Eitan, Genral (retired), Commander of the Central Command 2000–2002; Dr. Boaz Ganor, Head of the International Counterterrorism Center (ICT); H.D. Lieutenant Colonel, Head of IDF Negotiation Unit; Yossi Kuperwasser, Brigadier General (retired), Head of the IDF's Intelligence Research Department 2001–2006; Lior Lotan, Colonel (retired) former Head of IDF Negotiations Unit; Dan Meridor, former Minister of Justice, and the minister in charge of the intelligence community 2001–2003; Benjamin Netanyahu, Prime Minister of Israel 1996–1999; Ilan Paz, Brigadier General (retired), brigade commander during Intifada and head of the civil administration in Judea and Samaria, 2001–2005; Hagai Peleg, Commander of Special Israeli Police Unit—Yamam, 1999–2001; Moshe Yaalon, IDF Chief of Staff 2002–2005; Danny Yatom, Head of the "Mossad" 1996–1998 and Aharon Zeevi-Farkash, Head of IDF Intelligence 2002–2006.

We owe a debt of gratitude to a man whose voice has been warning of this scourge for some time. Benjamin Netanyahu, who, as this book goes to print, is charged with forming the next Israeli government, has been a life-long fighter against terrorism. Mr. Netanyahu knows more than most, of the very painful consequences of terror attacks and has been warning the West about the threat and consequences of terrorism as far back as the 1970s. He has been a strong source of inspiration to the editors and many others besides, for which we now take the opportunity to thank him.

Finally, we have the hope that this book makes a very modest contribution to understanding this threat and that it may be a contributing factor in leaving the world a safer place for our children. For the authors, and for most of humanity, our children are the ultimate inspiration.

O. F.
H. M.

CONTRIBUTORS

Dr. Leonard A. Cole is an adjunct professor of Political Science at Rutgers University, Newark, New Jersey and an expert on bioterrorism and terror medicine. Trained in the health sciences and public policy, he holds a Ph.D. in Political Science from Columbia University, New York, and a DDS from the University of Pennsylvania School of Dental Medicine in Philadelphia. Cole is a fellow of the Phi Beta Kappa Society and recipient of grants and fellowships from the Andrew Mellon Foundation, the National Endowment for the Humanities, and the Rockefeller Foundation. He has written for professional journals as well as *The New York Times*, *The Washington Post*, *Los Angeles Times*, and *Scientific American*. Cole has appeared frequently on network and public television and is the author of seven books including the award-winning *The Anthrax Letters: A Medical Detective Story* and *Terror: How Israel Has Coped and What America Can Learn*.

Willam Cooper is the Chief Executive Officer for Leading Beyond Tradition in Mukilteo, Washington and a retired police officer with extensive background in law enforcement and corporate security. He holds an MBA and second Master degree in Public Administration and is a graduate of the FBI National Academy and Northwest Law Enforcement Executive Command College. Cooper is the author of *First Responders Guide to Terrorist Attacks* and is a frequent public speaker and teacher, specializing in anti-terrorism and international terrorism.

Ophir Falk is a partner at the Naveh, Kantor Even-Har law firm, Ramat Gan, Israel, and a research fellow at the Institute for Counter-Terrorism, Herzilya, Israel where he has published numerous articles in the field. Mr. Falk holds an MBA and has more than a decade of experience in various security capacities. Falk was a risk consultant for the 2004 Athens Olympic Games, where he took part in risk assessments for Olympic venues and critical national infrastructure.

Dr. Ofer Israeli teaches courses on International Relations Theory, Complexity of International Relations, and Foreign Policy Decision-Making at the University of Haifa, Tel Aviv University, and at the Israel Defense Forces (IDF) Academy, The Colleges for Strategic and Tactical Commanders, Israel. He is scheduled to begin post-doctorate research at Georgetown University, Washington D.C., in 2010.

Hadas Kroitoru is a researcher at the Institute for Counter-Terrorism, Herzliya, Israel. She specializes in global terrorist networks and risk assessments, with a particular interest in Turkey. Ms. Kroitoru is a former journalist with articles published in newspapers and magazines in the United States and throughout the Middle East.

Amir Kulick is a research fellow at the Institute for National Security Studies and earned a Master degree in Middle Eastern Studies from Tel Aviv University, Tel Aviv, Israel. Between 1994 and 2006, he served as an officer in the Israeli Defense Force (IDF). Kulick's research focuses on Palestinian terror organizations and Hezbollah terror activity. In addition, he works as an international security consultant on intelligence evaluation and threat assessments.

Henry Morgenstern is the president of Security Solutions International, Miami Beach, Florida . His company has trained more than 700 federal, state, and local agencies to effectively confront the threat of terror and regularly takes groups of first responders to Israel to study homeland security. Mr. Morgenstern is a widely published author on the subject of suicide terror and has offered expert commentary on terror-related issues for NBC, ABC, CBS, Fox, numerous radio stations, and Web broadcasts.

Yaron Schwartz is a security analyst focusing on terrorism, Middle East affairs, and homeland security. He has advised various government entities and private sector companies, including the 2004 Athens Olympic Committee (ATHOC). Mr. Schwartz is the former director of the Institute for Counter-Terrorism's (ICT) U.S. office. He also served in the Israeli Foreign Ministry and the Israeli Security Agency.

Dr. Shmuel Shapira is professor of Medical Administration at the Hebrew University Faculty of Medicine, currently the Deputy Director General of the Hadassah Medical Organization, and Director of the Hebrew University Hadassah School of Public Health Jerusalem, Israel. Shapira received his medical degree from the Hadassah-Hebrew University School of Medicine. Professor Shapira is a Lieutenant Colonel (Res.) in the Israel Defense Forces (IDF) and has served as the IDF Head of Trauma Branch. An authority on terror, trauma and emergency medicine, he instructs medical students, physicians, EMS, medical leaders, and rescue teams on terror medicine, management of mass casualty events, and advanced trauma life support. Dr. Shapira is co-editor of the textbook *Essentials of Terror Medicine* and has written many other trauma, terror medicine, and mass casualty management publications.

Shabtai Shoval is head of Suspect Detection Systems, a homeland security company in Tel Aviv, Israel. His involvement in combating terrorist began in 1979 when he joined the Israeli Defense Forces, serving in an elite anti-terrorism combat unit. In 1982, after completing his three years tour of duty including participation in the 1982 Israel-Lebanon war, he studied political science at Tel Aviv University while continuing involvement in classified intelligence issues. In 1982, in Washington D.C., Shoval served as an Israeli Embassy staff member for media-related issues and as advisor to Mr. Benjamin Netanyahu, then deputy Ambassador General. Shoval has held high- profile marketing and public relations positions, including spokesperson for the City of Tel Aviv; a senior position in Gitam-BBDO, a leading advertising agency; and president of the ICTA (Israel Cable TV Association) where he was responsible for broadcasting content, political lobbying, marketing, and business activities for CATV companies. In 1992, Benjamin Netanyahu asked Shoval to join the Prime Minister's office as a special advisor for background on the Madrid Talks regarding the future of the Middle East. In 1995, Shoval joined the telecommunication investment company Telrad Holdings, as Vice President for Digital TV, and in 1999 he joined Comverse Inc., a leading high-tech company (NASDQ CMVT), and successfully managed their Digital TV division.

INTRODUCTION: OVERVIEW AND HISTORICAL ACCOUNT OF THE WEAPON

Ophir Falk

For evil to triumph, the good need only to remain silent.
—Edmund Burke
Irish orator, philosopher, and politician (1729–1797)

BACKGROUND

The fall of the Berlin Wall in November 1989 marked the end of the Cold War and raised two major, yet contrasting, views of the paradigm to come. The first view was advocated by Dr. Francis Fukuyama in his brow-raising article "The End of History?"[1] In this article, Fukuyama asserted that "history," in terms of major human conflicts, had come to an end with the collapse of Soviet Communism. The new world order, he argued, and convinced many, would be immune from significant ideological wars, and future conflicts would be limited to mere hodgepodges of localized nuisances that posed no substantial threat to Western civilization and its way of life.

The second view was subsequently articulated by the late Harvard professor Samuel Huntington in an article entitled "The Errors of Endism"[2] and, afterward,

Suicide Terror: Understanding and Confronting the Threat. Edited by Falk and Morgenstern

in the more widely read and famously controversial article "The Clash of Civilizations?"[3] In his classic analysis in the second article, Huntington argued, *inter alia*, that ethnically volatile regions previously held as stable satellite entities of the Soviet Union would gradually erupt and that "Islam has bloody borders."[4]

In the West, the fall of the Berlin Wall and with it, the Iron Curtain, was seen as the end to a 50-year-long silent war and an event that brought about the liberalization of peoples. In stark contrast, influential Islamic extremist clerics and fighters, including the "Afghan alumni," viewed these happenings as a direct corollary—indeed, a climax—to their successful struggle against the Soviet invasion of Afghanistan.

In light of the events that have occurred in the past decade and a half, nearly all observers have come to the view that "history" did not end with the collapse of Soviet Communism. In fact, while the liquidation of the Soviet Union removed the ideological impetus of Communist domination, it also released the tight grip that the Kremlin had on the ambitions of many satellite republics, peoples, and frivolous dictators. Furthermore, the downfall of the Soviet superpower unleashed the specter of nuclear know-how and materiel that could leak to almost anyone willing and able to pay for it. In addition, the great ideological and political void created by the decline of Communism helped pave the way for the accelerated pace at which militant and fundamentalist Islam has gained ground in many parts of the Middle East and beyond—regions that had previously toyed with Communism as their manifesto.[5]

It may still be premature to conclude which thesis, Fukuyama's or Huntington's, was more accurate; the authors would prefer to leave that assessment to historians in the decades to come. Nevertheless, those future historians may indeed construe that the growth in a new form of warfare, *suicide terrorism*, attests to the fact that "history" did not end in 1989, but rather set our generation into a clash of cultures that served as a backdrop to World War III.

With the end of the Cold War and the advent of low-intensity conflict and asymmetrical warfare, terrorism—particularly suicide terrorism coupled with the potential for use of weapons of mass destruction (WMD)—has become an imminent threat to global peace and security. Yet, terrorism by means of suicide attacks did not begin with the fall of the Berlin Wall. Thus, in order to understand and confront this phenomenon, it is first necessary to define "suicide terrorism" and identify its origins.

DEFINING THE THREAT

For many counterterrorism practitioners, the distinction between terror in general and suicide terror in particular may seem a fine one at best, and a bureaucratic technicality at worst. This approach is wrong. After all, defining the threat is of paramount importance in attempting to successfully confront it. In other words, *if a problem is not defined, it cannot be solved*. Therefore, the first rule in understanding and confronting suicide terrorism is to *define the threat*.

Less than two weeks after 9/11, President Bush addressed a joint session of congress and the American people and declared that "on September the 11th, enemies of freedom committed an act of war against our country" and that America will lead a war on terror.[6] But is terrorism "war"? In the context of international law, the relations between nation-states can be classified in terms of two general statuses: war and peace. The internationally accepted definition of "war" is "[t]he widespread use of force, between sovereign states, by means of their military."[7] Using this definition, can the "War on Terror" be considered a true "war"? At face value, most terrorist attacks are seemingly instigated or at least carried out by nonstate actors. Hence, terrorism does not ostensibly fall under the classic definition of "war." So, then, should terrorism simply be considered a criminal offense? Before clearly categorizing terrorism as either an act of war or a criminal act, a clear definition of the threat should be established.

In 1988, a U.S. Army study found that there were 109 distinct official definitions of terrorism in common use in the federal government. The study's authors enumerated each of these definitions and determined that they consisted of combinations of 22 different elements ascribed to terrorism.[8] Eleven years later, terrorism expert Walter Laqueur also counted over 100 definitions.[9] The discrepancies between these authors' taxonomies do not arise only out of variances in the definitions proposed by various states or international organizations. Rather, there turn out to be different meanings given to the term even among various agencies of the U.S. government. For example, the State Department defines "terrorism" as "[p]remeditated, politically motivated violence perpetrated against noncombatant targets by subnational groups or clandestine agents, usually intended to influence an audience."[10] Similarly, the U.S. *National Strategy for Combating Terrorism* defines "terrorism" as "premeditated, politically motivated violence perpetrated against noncombatant targets by subnational groups or clandestine agents."[11] The U.S. Department of Defense (DoD), on the other hand, uses a somewhat different definition.

Despite these differences, there are a number of similarities between the many definitions of "terrorism." For one, each definition is ambivalent on the scope of the term "noncombatant." More importantly, however, all of these definitions exclude the concept of "state-sponsored" or "state" terrorism. What many would consider as "state terrorism" has existed since at least the French Revolution, when the word "terror" was first used to describe the Jacobin "Reign of Terror" in 1789,[12] and it has claimed far more victims than terrorism perpetrated by non-state actors.[13] The lethality of the actions carried out by nonstate groups such as al-Qaeda, the Tamil Tigers, and Hamas are modest compared to the consequences of state terrorism in Stalin's Russia, Mao's China, Pol Pot's Cambodia, Saddam Hussein's Iraq, or Mahmoud Ahmadinejad's Iran. By excluding state terrorism, these definitions give rogue states the benefit of the moral doubt.

As previously mentioned, further obfuscating the meaning of "terrorism" is the vague notion of "noncombatants." Where is this line to be drawn? Would the deadly suicide attack on the USS *Cole* on October 12, 2000 (see Figure I.1) be considered an act of terror? This attack, which was carried out by al-Qaeda

Figure I.1. Photograph of the USS *Cole* being removed from the high seas on October 31, 2000. (Source: http://www.lockport-ny.com/images/USS_Cole_Hole.jpg)

operatives in Yemen's Port of Aden, killed 17 American sailors and injured an additional 39.[14] The operational concept of the attack was planned by Abdullah al-Nashiri, who first became acquainted with Osama bin Laden during the period in which the two men fought together against Soviet troops in Afghanistan. Carried out by two suicide attackers, the plan included fitting a boat (a service vessel) with 500 kilograms of explosives and then detonating it alongside the American destroyer.[15] Osama bin Laden was so pleased and impressed by the outcome of the attack on the USS *Cole* that he ordered the production of a short film on the attack for media exposure and as a recruiting tool.[16] It is also believed that al-Nashiri was later responsible for planning a similar attack on the French oil tanker *Limburg* in the Gulf of Aden in 2002.[17]

While the attack on the USS *Cole* served as an instrumental vehicle in proving al-Qaeda's operational credibility, provided the group with media exposure, and served as a recruitment enhancer, can the attack be considered an example of "premeditated, politically motivated violence perpetrated against *noncombatants*"? Not if one considers U.S. sailors aboard an American destroyer in Yemen to be *combatants*.

Another example further illustrates the sort of confusion that often surrounds use of the word "noncombatant." On Wednesday, July 12, 2006, at approximately 9:00 A.M., an internationally recognized terrorist organization known as Hezbollah attacked a number of targets on Israel's northern border with Lebanon. During the attack, two Israeli soldiers were abducted, three were killed, and another two were wounded.[18] At the time of the attack, the soldiers were carrying out a run-of-the-mill defensive patrol in an effort to secure their country's border. This attack was the backdrop to an escalation in violence between Israel and Hezbollah throughout the summer of 2006, which was ultimately labeled as the

"Second Lebanon War." The direct consequences of that "war" included over 150 Israeli fatalities and hundreds of fatalities on the Hezbollah and Lebanese side, not to mention damage to property and infrastructure; the direct and indirect costs associated with these damages totaled millions of dollars. So, was Hezbollah's attack of July 12, 2006, an act of terrorism? Again, if one considers the attacked infantry soldiers to be *combatants*, the answer is no.

The DoD defines terrorism as the "calculated use of unlawful violence to inculcate fear, intended to coerce or intimidate governments or societies in pursuit of goals that are generally political, religious, or ideological."[19] This definition is seen by many as subjective and therefore ineffective in obtaining a wide enough range of international consensus to lead to broad acceptance. Furthermore, while the *intended* use of violence or *threat* to apply force at times results in as much disruption as an actual attack, it is not addressed in the DoD definition.

Thus, certain *ad hoc* acts of terrorism seem difficult to identify as such due to the subjective elements detailed in the previous paragraphs. Logic would dictate that terrorist organizations should be more easily categorized in light of their systematic dissemination of ideology and the sequenced attacks that they carry out. However, even allied states that share similar values, such as the United States and United Kingdom, can and have disagreed over what sorts of actions should be categorized as terrorism. For example, for many years, some branches of the United States government refused to label members of the Irish Republic Army (IRA) as terrorists, despite the fact that the IRA was using terrorist methods against one of the United States' closest allies (Great Britain) and, more to the point, despite the fact that this ally had itself branded the attacks as instances of terrorism.[20] This dissonance was highlighted by the case of *Quinn v. Robinson*, a U.S. trial involving the extradition of William Joseph Quinn, a confessed member of the IRA.[21]

Edward Peck, former U.S. Chief of Mission in Iraq and ambassador to Mauritania, expressed the following opinion concerning the value of defining terrorism:

> In 1985, when I was the Deputy Director of the Reagan White House Task Force on Terrorism ... they asked us to come up with a definition of terrorism that could be used throughout the government. We produced about six, and in each and every case, they were rejected because careful reading would indicate that our own country had been involved in some of those activities. ... After the task force concluded its work, Congress got into it, and you can Google into U.S. Code Title 18, Section 2331,[22] and read the U.S. definition of terrorism. And one of them in here says—one of the terms, "international terrorism," means "activities that," I quote, "appear to be intended to affect the conduct of a government by mass destruction, assassination, or kidnapping. ..." Yes, well, certainly, you can think of a number of countries that have been involved in such activities. Ours is one of them. Israel is another. And so, the terrorist, of course, is in the eye of the beholder.[23]

Mr. Peck's thesis is dangerous because it inadvertently legitimizes terrorism by casting it as subjective and suggests that a terrorist act is "in the eye of the beholder." The statement that "one man's terrorist is another man's freedom fighter" has become not only a cliché, but also one of the most difficult impediments in coping with terrorism.[24] Can anyone of sound conscious and clear mind confuse a suicide attacker that intentionally blows up a bus full of civilians with a "freedom fighter"? In the eyes of *which* beholder can people who hijack civilian aircrafts for the purpose of flying them into metropolitan skyscrapers be considered "fighters for freedom"?

Attackers of this ilk and the organizations they represent *must* be viewed as terrorists and treated accordingly by the entire international community. A clear definition accepted by the international community and its institutions must be put in place and jointly acted upon.

Definition Criteria

The two main drafting methods used by legislative bodies to define terrorism are the *specific* and *general* approaches. The specific approach identifies unambiguous activities, such as hijacking and hostage taking or the introduction of biological material, as terrorism, without seeking to define a general category of terrorism per se. On the other hand, the general approach (examples of which were given earlier in this chapter) seeks to arrive at a general definition of terrorism by reference to criteria such as motivation, mode of operation, type of target, and so on.[25]

There are benefits and disadvantages to both methods. The specific approach, used by countries such as New Zealand, Canada, South Africa, and Australia,[26] keeps subjective observations to a minimum but also limits acts of terrorism to an ostensibly closed list. The general approach, adopted by countries including the United States and the United Kingdom,[27] may cover the entire scope of the phenomenon, yet it is hindered by the "eye of the beholder" effect and thus too often remains ineffective. Common criteria in the general approach include "the use of violence," "psychological impact and fear," "political goals," "deliberate targeting of noncombatants," and "unlawfulness," among others.

This is not to say that the results of the two approaches do not feature some similarities. After close examination of the different definitions used or proscribed, one may notice that these definitions most commonly include the following elements: violence and/or force (appeared in 83.5% of the definitions); political motivation (65%); fear and/or an emphasis on terror (51%); use of threats (47%); psychological effects and anticipated reactions (41.5%); discrepancy between the targets and the victims (37.5%); intentional, planned, systematic, and/or organized action (32%); and methods of combat, strategy, and/or tactics (30.5%).[28]

Proposed Definitions

In this volume, the authors' proposed definitions of *terrorism, terrorist organizations or states*, and *suicide terrorist attack* are based on both the general and the specific approach. These definitions encompass the entire scope of the phenomenon as the authors see it. The authors have also provided specific examples for the sake of clarification. These examples are not a closed list; rather, they serve merely as illustrations and can be amended with the advent of new forms of attack not experienced to date and that are currently less conceivable.

The proposed definitions herein are intended to serve as working tools for lawyers, academics, legislators, journalists, security experts, and, perhaps most importantly, law enforcement agencies. These definitions are provided in the following sections.

Terrorism. For the purposes of this text, "terrorism" is defined as follows:

> A premeditated attack, or a threat to attack civilians or civilian targets, in an attempt to attain political, ideological, and/or religious aims[29]; examples include but are not limited to, intentionally shooting and/or bombing civilians and/or civilian targets; exploding bombs within and/or hijacking means of civilian transportation.

This definition features four components:

- *Account of Action.* First, the definition states that terrorism involves "*a premeditated attack or threat of attack.*" Note the use of "attack" rather than "violence," which is the word more commonly employed in such definitions. The rationale behind preferring "attack" to "violence" is that "violence" may exclude terrorist actions against electronic systems; such actions can cause mass disturbance and even destruction, but they do not necessarily include the use of violence.
- *Type of Target.* Next, the definition declares that terrorism involves action against "*civilians or civilian targets.*" The decision to use the word "civilian" rather than "noncombatant" was neither simple nor trivial. While "noncombatant" has a much broader scope and may therefore seem justifiable, it leaves too much room for interpretation, moral relativism, and manipulation.[30] For example, working from the "noncombatant" point of view, one possible alternate wording of this portion of the definition could read as follows: "… civilians, or any other persons not taking an active part in hostilities in a situation of armed conflict …"[31]; however, this expansion remains vague, and it would not receive global acknowledgment.[32] With use of the word "civilian," the definition can obtain a much wider international consensus.

In an effort to learn from the different modes of operation used by suicide attackers, this book does not limit its analyses to attacks against civilians or noncombatants. This is primarily intended for law enforcement personnel that need to confront the threat on ground.

- *Motivation.* Third, the suggested definition states that terrorism involves "*an attempt to attain political, ideological, and/or religious aims.*" There are some commentators who attest that the motivation behind an attack is irrelevant, thereby implying that terrorism should be considered as such based solely on the outcome of an attack, regardless of just or unjust motives. Nevertheless, this component is included in the definition in order to differentiate terrorist actions from common criminal acts.

- *Examples.* Finally, the definition concludes with several examples of terrorist acts that include but are not limited to: "*intentionally shooting and/or bombing civilians and/or civilian targets; exploding bombs within and/or hijacking civilian transportation.*" These examples are provided in an effort to illustrate actual attacks and therefore clarify the definition's purpose. The examples are obviously not a closed list and may be amended.

Terrorist Organization or State. Within this text, a "terrorist organization or state" is defined as any organization or state that carries out or knowingly facilitates terrorism as defined in the previous section.

Suicide Terrorism. Finally, for the purposes of this book, "suicide terrorism" is defined as follows:

> A politically, religiously, and/or ideologically motivated attack perpetrated by one or more individuals who intentionally cause their own death while harming or attempting to harm civilians and/or civilian targets.[33]

Historical Account. If the first rule in understanding and confronting the threat of suicide terrorism is to define the threat, the second rule is this: *One should study history and the mistakes made in the past in an effort to avoid similar errors in the future.* After all, in the words of George Santayana, "Those who cannot remember the past are condemned to repeat it."[34]

The first recorded suicide attack may have been the biblical account of Samson, who intentionally brought about his own death along with scores of Philistine inquisitors.[35] As a military tactic, suicide attacks became widely known and used during World War II, due to the actions of Japanese kamikaze pilots who aimed to cause maximum damage by flying their explosive-laden aircraft into Allied ships and other military targets. The backdrop to those attacks occurred on December 7, 1941, in Pearl Harbor, where 19 American warships were sunk by Japanese pilots, killing over 2400 servicemen and serving as an impetus for U.S. entrance into World War II the following day.[36]

This book focuses on suicide as a mode of terror rather than a military tactic. The authors do not purport to provide a deductive history lesson that details the chronological development of terror and suicide terror attacks. Rather, they note a number of key events and processes that triggered the dawn of terrorism and suicide attacks, focusing on those elements that led such attacks to become terrorists' primary weapon of choice.

What is commonly regarded as the first suicide terror attack in modern times took place in Lebanon on April 18, 1983 (Figure I.2).[37] In this attack, the perpetrators deployed a reconnaissance vehicle to observe the physical security of the U.S. embassy in Beirut. Subsequently, the car drove a few blocks and flashed

An aerial view of the American Embassy as heavy cranes continue to remove rubble from the upper floors on 21 April, 1983, following the terrorist bombing three days earlier.

Photo courtesy of Claude Salhani/U.S. Marines in Lebanon 1982–1984 History and Museums Division, Headquarters, U.S.M.C., Washington, D.C.

Figure I.2. Aftermath of the April 1983 bombing of the U.S. Embassy in Beirut.

its lights to a truck awaiting the signal. As the truck, laden with explosives, sped toward its destination, embassy staff—including the entire intelligence division—had no idea their world was about to end. Within minutes, 63 people were dead, and hundreds more were injured.[38]

The success of that strike encouraged enemies of the U.S. peace mission in Lebanon to execute far deadlier attacks approximately six months later. In the subsequent attacks on October 23, 1983, instead of a 2000-pound bomb like the one used in the embassy strike, the terrorists deployed a 12,000-pound charge against the U.S. Marine headquarters in Beirut, while almost simultaneously attacking French barracks in that city (Figure I.3).[39] Further detail on these ground-breaking incidents is provided in Chapter 4.

Both of the October 23 attacks were perpetrated by Hezbollah ("The Party of God"), a group of Iranian-funded Shiite terrorists. At the time, Hezbollah was a rather small organization, but these actions helped elevate the group to a place of international notoriety.

The Beirut attacks were successful in serving as a catalyst for the withdrawal of American, British, French, and Italian peacekeepers from Lebanon. They set a precedent for future attacks, both in Lebanon and throughout the world. They were also the first preplanned suicide attacks carried out in modern times by a non-state terrorist organization. Further analysis concerning the Hezbollah, its leaders and the precedence they set will be detailed in Chapter 4.

ISRAEL. Beyond Lebanon, another Middle Eastern state that has suffered significantly from suicide attacks is Israel. Following the signing of the Oslo Accords on September 13, 1993 (in which the Israeli government and the Palestine Liberation Organization agreed to the creation of the Palestinian Authority,

Figure I.3. Explosion of the U.S. Marine Corps barracks in Beirut, Lebanon on October 23, 1983.

among other things), several Palestinian terrorist groups—especially Hamas, the Palestinian Islamic Jihad (PIJ), and Fatah—unleashed unprecedented waves of suicide attacks against Israeli civilians.[40]

The particular attack that represented a turning point in Israel's perspective vis-à-vis suicide terrorism took place on January 22, 1995, at the Bet Lid juncture north of Tel Aviv. Although this was not the first suicide terror attack in Israel, it was the first in which two bombers participated in the same incident, exploding their charges almost simultaneously. The first bomb targeted a crowded area at the juncture, while the second was meant to maximize damage once rescue personnel arrived on the scene. As a result of this coordinated attack, 21 people were killed and 69 were wounded. Subsequently, Israeli Prime Minister and Defense Minister Yitzhak Rabin said, in closed quarters, that terror had become a *strategic* threat to Israel's existence.[41] In fact, more than 175 suicide attacks were launched in Israel between 1993 and 2005 (Figure I.4).[42] While these attacks only represented about 1% of terrorist incidents in the country at the time, they were by far the most lethal means employed by terrorist organizations—responsible for almost 50% of terrorism-related fatalities in Israel.[43]

A Global Phenomenon. A common misconception is that suicide terrorism is a tactic that has been monopolized by terrorist activists that subscribe to a radical and militant form of Islam. While the vast majority of suicide attacks worldwide have been carried out by individuals that identify as Muslim, this mode of attack has actually also been widely and effectively used by non-Muslim groups in many places throughout the world. One leading example is in Sri

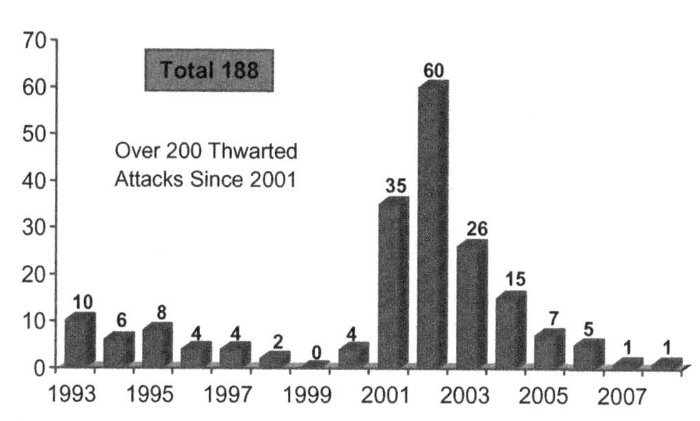

Figure I.4. Palestinian suicide terrorism. (Source: Israeli Security Agency Report.)

Lanka. For the purposes of this volume, however, the authors have decided to highlight the use of the suicide tactic by those individuals representing the most imminent threat to the United States and the West—radical Islamic terrorists. The basis for this assessment is that the majority of suicide attacks have been carried out by Muslims and because no non-Muslim has ever committed this sort of attack on American soil or against American targets abroad.

Islam was considered to be the world's most open, enlightened, creative, and powerful religion and culture for many centuries. In medieval times, Islam represented the greatest military power on earth, and its armies simultaneously invaded Europe, Africa, India, and China. Followers of Islam were also the world's foremost economic power, and they accomplished perhaps the greatest human achievements to date in the arts, philosophy, and the sciences.[44] Then, beginning with the Renaissance and continuing into the Enlightenment, the West began to win victory after victory, first on the battlefield, then in the marketplace, and eventually in multiple aspects of daily life.[45]

With Saladin's victories over the Crusaders between 1187 and 1189, Islam's military achievements reached their peak. However, Islam's influence gradually diminished, especially after the Battle of Lepanto in 1571 and the subsequent rejection of Ottoman armies from Vienna in 1683.[46]

In truth, the Islamic quest to return to prominence in terms of religious dominance and might was never really relinquished by Islamic fundamentalists and militants. These ambitions were merely suppressed until the twentieth century and the emergence of the Global Jihad movement, beginning with the foundation of the Muslim Brotherhood in 1928.[47] Over time, such efforts expanded and evolved, leading to the creation of organizations such as al-Qaeda, established in 1988.

The German invasion of Poland in September 1939 is considered by most historians as the official beginning of World War II. There is no clear date of the beginning of what now seems to be World War III, but 1979 can definitely be considered a watershed year. In that year, militant sects of both Sunnis and Shiites dramatically reemerged on the world stage. The Soviet Union's invasion of Afghanistan gave birth to al-Qaeda, while the Shiite religious revolution in Iran brought down the Shah and led to the formation of the first-ever Shiite Islamic republic. Both al-Qaeda and Iranian radicals championed the resurrection of an Islamic empire that would dominate the world and correct what, in their view, was an accident of history that enabled the rise of the West. At the time, many Americans saw the Iranian revolution simply as the backdrop to an inconvenient hostage situation at the U.S. embassy in Tehran. Meanwhile, the Soviet invasion of Afghanistan seemed to be a good reason to boycott the 1980 Olympics in Moscow and refrain from ratifying the SALT agreements, but beyond those considerations, few Americans gave the situation much thought. On the other hand, militant Muslims throughout the Middle East and beyond saw the invasion as cause for a holy war—Jihad.

Thus, an almost immediate reaction to the 1979 Soviet invasion of Afghanistan was the influx of thousands of young Muslims from all over the world into Afghanistan to join the fight against the Soviets. The reinforced "Afghan" force included thousands of Saudis, Yemenis, Egyptians, and Algerians; hundreds of Tunisians, Iraqis, and Libyans; and scores of Jordanians. After the Soviets withdrew from Afghanistan and victory was in hand, these recruits went on to serve as the "base" (or, in Arabic, "al-Qaeda") to a network of Jihadists that would reach up to 50 different nations around the world.[48]

The rivalry between Shiites (who account for about 10–15% of the Islamic population) and Sunnis (who account for most of the rest of this population) dates back to Muhammad's death in 632; at that time, the Shiites supported the successorship of Ali, while the Sunni supported that of Abu Bakr.[49] Over the years and even in the last century, the conflict between Muslims has given rise to far more bloodshed than the Israel–Arab conflict.[50] It probably caused more casualties than the Israel–Arab conflict and the "War on Terror" combined.

Though they are rival sects within Islam, both extreme Shiites and extreme Sunnis serve as key players in the Global Jihad movement. These militants simultaneously compete and cooperate with one another. Both seek to destroy perceived infidels and establish their leadership and supremacy within the Muslim world—for example, al-Qaeda by way of the September 11, 2001 attacks and subsequent strikes; Iran via its sponsorship of Hezbollah and its promise to develop nuclear weapons and use them to wipe out Israel; and Hamas via its rocketing of Israeli cities. Despite their differences, they do agree on one important point: The new empire should be an *Islamic* realm, cleansed of infidel presence or power. This is why Sunni and Shiite groups often cooperate with one another in their efforts against the common enemy, as Hamas and Hezbollah have done against Israel and other Western targets.

Many experts and outside observers dismiss militant Islamic zeal for conquest as simply an expression of frustration that will pass with time. However, if left unmatched, this militant ideology could continue to grow. As officials and experts commonly argue, if Iran succeeds in developing nuclear weapons, an extremist group may gain access to such weapons through a state sponsoring terrorism, an unprecedented development in terms of modern terrorism. Since this possibility must be thwarted, the third major rule in understanding and confronting the threat of suicide terrorism can be stated as follows: *Regard threats at their full value and act accordingly.*

The Bush administration's view of Iran, other rogue states, and their terrorist allies is clear-cut: "Iran aggressively pursues [weapons of mass destruction] and exports terror," Bush declared, "while an unelected few repress the Iranian people's hope for freedom. ... States like these, and their terrorist allies, constitute an axis of evil, arming to threaten the peace of the world."[51] Further analysis on the Jihad and its possible ramifications on the United States and the West is presented in detail by Henry Morgenstern in Chapter 1.

Why Suicide Terrorism? According to Alberto Abadie, Professor of Public Policy at Harvard University, doubt should be "cast ... on the widely held belief that terrorism stems from poverty."[52] Abadie's research reaffirms an earlier analysis conducted by the National Bureau of Economic Research, which found very little, if any, correlation between terrorism and poverty.[53] The exact rationales for suicide terrorism—and contrary to common belief, there are rationales—can be diverse, ranging from hate, to incitement, to revenge, to the ease of executing attacks. At their core, however, each of these rationales is mainly motivated by religious, national, political, or ideological zeal.

Suicide terrorism is a complex individual, organizational, and psychosocial phenomenon that has been studied by a number of scholars who have reached various findings.[54] Nevertheless, in the opinion of the authors of this book, the underlying reason behind the choice of suicide terrorism as the preferred mode of operation of the world's most dangerous terrorist organizations is primarily the fact that *it works*. Suicide terrorism is by far the most effective means of terrorist attack. Therefore, the fourth major rule in understanding and confronting the threat of suicide terrorism can be summarized as follows: *The potential suicide attacker is highly motivated by rationale intent, and he or she should never be underestimated.* In fact, potency is a catalyst for additional attacks, and it has helped make suicide terrorism one of the most widely used terrorist tools, as illustrated in Figure I.5.

The effectiveness of suicide terrorism is derived from the ability of the perpetrator, more often than not, to choose the precise time and place of attack. Most attackers select a time and location that allows for maximum casualties and damage to the innocent and therefore provides very public exposure for the terrorists' cause. Aside from the grave physical consequences of many suicide attacks, the devastating blow to public morale that stems from the reputed inability to counter the threat has forced governments to change policy. Bringing about such changes is, in fact, an underlying strategy behind the use of this tactic.

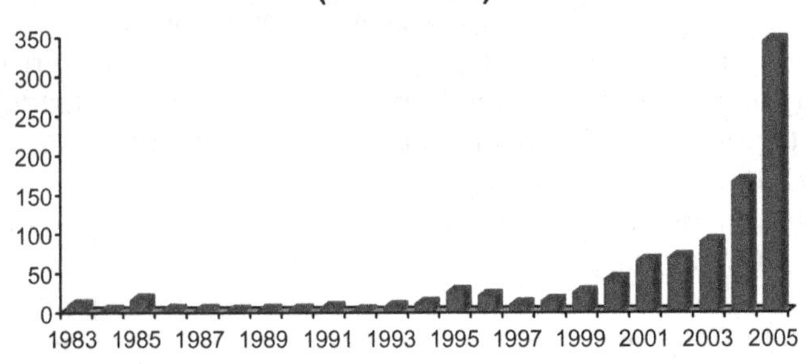

Figure I.5. Number of suicide attacks per year (1973–2005). (Source: Based on Bruce Hoffman of the RAND Institute and Ami Pedhazur of the University of Texas.)

Suicide terrorism has therefore become the ultimate "smart bomb" for terrorist organizations. Islamic organizations that pose a global threat (e.g., al-Qaeda), as well as those that are primarily active against Israel (e.g., Hamas), previously viewed this tactic as contradictory to the Islamic faith[55]; nevertheless, these groups have adopted suicide terrorism as a means of attack primarily due to its effectiveness. Additional catalysts for suicide terrorism include its "cost-effectiveness," relatively minor operational logistics, and need for little fallback planning. (After all, suicide attacks do not require escape plans, because if they are successful, there is no assailant to capture or fear of information leaking out during subsequent interrogation of the perpetrator.) Such "benefits" are described in greater detail in the sections that follow.

Lethality. On average, suicide attacks inflict four times more fatalities than more conventional terrorist attacks[56] and 26 times more casualties (Figure I.6).[57]

Cost Effectiveness. On average, it costs well-organized terrorist organizations like Hamas as little as $150 to launch a suicide attack.[58] This cost lies in direct opposition to the huge damages (in terms of both life and treasure) such an attack can produce. The Bank of Israel estimates the damage to the Israeli economy from just one year of Intifada (uprising) in, this case, the year 2002, at 3.8% of the GDP. By the year 2000, an accumulated three-year loss to the Israeli economy was in the range of 90 billion shekels—an amount roughly equal to $22 billion, or one-fifth of one year's gross domestic product.[59]

Other examples further illustrate the cost-effectiveness of suicide terrorism. For instance, al-Qaeda did not need a very large sum to finance the most devastating terrorist attack in history. The 9/11 plotters spent somewhere between $400,000 and $500,000 to plan and conduct their attack,[60] which ultimately resulted in the murder of nearly 3000 civilians and caused billions of dollars in direct damage—not a bad return on terrorism investment. Al-Qaeda leader Ayman al-Zawahiri described the situation as follows: "The method of

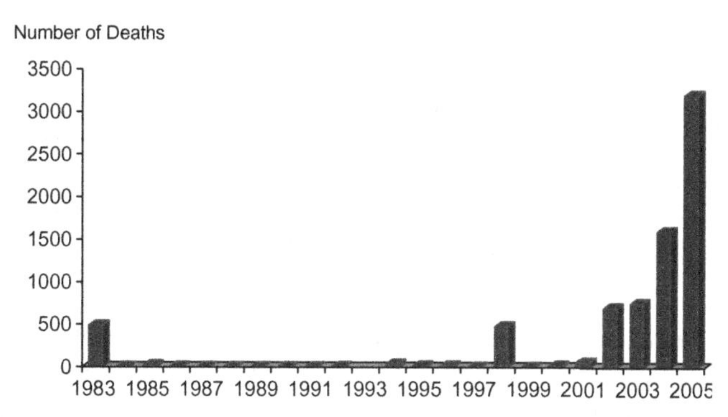

Figure I.6. Frequency and lethality of suicide attacks. (Source: Bruce Hoffman, RAND Chronology of Terrorist Incidents excluding 9/11 attacks.)

martyrdom operation [is] the most successful way of inflicting damage against the opponent and the least costly to the mujahidin in terms of casualties."[61]

The issue of financing attacks and, more importantly, the economic damage that successful attacks can inflict on an economy are key factors in a terrorist's choice of targets. Therefore, the fifth major rule in understanding and confronting the threat of suicide terrorism is as follows: *Follow the money trail in order to assess what may constitute key potential targets.*

The Individual's Perspective. One of the main differences between a terrorist attack and a criminal offense is that the latter is primarily motivated by individual gain, whereas the former is usually based on a far-reaching ideology. Nevertheless, terrorism in general, along with suicide terrorism in particular, also provides incentives for the individual terrorist. It seems mind-boggling that anyone could see personal gain from killing himself or herself, but this belief is an integral and essential part of suicide terrorism. These perpetrators are often stimulated from optimism rather than disparity.[62] A number of motives that lead individuals to carry out suicide terror attacks are as follows:

- *Religiously Motivated Incentives.* Interpretation of Koran writings and fatwas preached by Islamic spiritual leaders on a daily basis declare that suicide attackers are martyrs who will be rewarded in heaven with 72 virgins. If that is not enough, perpetrators are also guaranteed 70 virgins for their relatives.[63]
- *Self-Image.* A substantial portion of the attackers were presumably of low self-esteem and were susceptible to influence. The attacks were often viewed by the attackers as an act that would leverage their image.
- *National Patriotism.* Many, if not most, recent suicide attackers have been motivated by a sense of national pride, as well as a belief that their attacks will further the cause of their nation.[64]

- *Financial Support for Families.* Although it is not the main motivational factor for suicide terrorism, the financial aid that a perpetrator's family often receives is another element that must be considered. Because many suicide attackers come from disadvantaged socioeconomic backgrounds, these funds may help rescue their families from poverty. Much of this aid, especially in the case of Palestinian attackers, came from the former Iraqi regime of Saddam Hussein.[65]

The Organization's Advantage. The mass casualties inflicted by a suicide attack and subsequent media coverage of this event highlight the cause of the attacker's organization. Both immediately and, more commonly, over the long run, such attacks may coerce the targeted government's policy toward the organization and its goals.

In terms of logistics and planning, terrorist organizations are not hindered by the need to set escape routes for suicide attackers, nor must they worry about information being leaked during interrogation.

Terrorist organizations thus view suicide terrorism as a tool that helps them realize their strategic goals. These goals may range from ending what they view as occupation of their land to strengthening their extremist ideology in an attempt to turn it into a dominant one.

INTERNATIONALIZATION OF SUICIDE TERRORISM. In today's grim reality, the threat of suicide attacks has become a nightmare of many civilians and policymakers the world over. These attacks have evolved into a preferred mode of operation for terrorists, with staggering consequences. The risk suicide terrorism presents to both the international community and targeted countries is paramount. Parts of the world that have experienced suicide terrorism include India, Turkey, the United Kingdom, Indonesia, Pakistan, Afghanistan, Chechnya, Iraq, Israel, Saudi Arabia, Kenya, Tanzania, and of course the United States. Al-Qaeda and its affiliates are currently serving as catalysts for the phenomenon.[66]

In an age where terrorist organizations are decentralized and involve a network of operatives spread across most of the globe, there is a clear need for an internationally accepted definition of the terms "terrorism" and "terrorist organization." Hence, the sixth major rule in understanding and confronting the threat of suicide terrorism is to *arrive at an internationally recognized consensus regarding the need to confront suicide terrorism.*

AL-QAEDA. As previously mentioned, al-Qaeda ("the base") was established by Osama bin Laden in 1988, toward the end of the Soviet occupation of Afghanistan. This organization stemmed out of a "services office" (Maktab al-Khidamat) run by bin Laden, whose purpose was to absorb, place, and manage the thousands of volunteers who came to Afghanistan between 1979 and 1989 to fight alongside the local mujahidin against the Soviet invaders. Both during and after the Soviet occupation, Afghanistan served as recruiting grounds for Muslims from all over the world, and it was there that al-Qaeda's form of internationalization was shaped.[67]

Ironically, al-Qaeda has been directly responsible for only seven suicide attacks. The first attacks were on the American embassies in Kenya and Tanzania in August 1998, and they occurred after nearly five years of planning. Following these successes in Africa, al-Qaeda then attacked the USS *Cole* in October 2000. The history-changing attacks of September 11, 2001 came next, although they were preceded by a proxy attack in Afghanistan two days earlier. Then, on December 23, 2001, an al-Qaeda operative named Richard Reid attempted to blow himself up aboard an American Airlines flight from Paris to Miami by means of a shoe bomb. Later, on April 11, 2002, an al-Qaeda operator named Nizar Nawar detonated a bomb, killing himself and 21 civilians in a synagogue in Djerba, Tunisia. And the most recent suicide attack directly carried out by an al-Qaeda operative once again occurred in Kenya, this time on November 28, 2002, when a car bomb was driven into the Paradise Hotel, killing 13 people and injuring about 80 more.[68]

Thus, al-Qaeda's main source of strength is not its ability to carry out attacks, but rather its ability to network. The organization's skill in connecting with and motivating other like-minded organizations has resulted in major suicide attacks in Indonesia, Morocco, Saudi Arabia, Turkey, Spain, England, Chechnya, and Iraq; other nations are sure to join this list in the years to come.

SRI LANKA. Sri Lanka has been plagued by a separatist conflict involving its 200,000-person-strong Tamil minority. This conflict is spearheaded by a terrorist organization known as the Liberation Tigers of Tamil Eelam (LTTE or Tamil Tigers), which is notorious for its use of suicide bombings.

The LTTE is composed mainly of Hindus and Marxists that have been waging a terrorist campaign against the Sri Lankan government. Since 1987, the LTTE has carried out approximately 200 suicide attacks against civilian and political targets, with political assassinations being a primary objective. LTTE operatives pioneered the use of concealed suicide bomb vests, which are now used by many other organizations worldwide; moreover, they have exhibited naval suicide attack capabilities superior to those of all other terrorist organizations. The LTTE first deployed suicide bombers in 1991, when a female operative assassinated former Indian Prime Minister Rajiv Gandhi using a prototype suicide vest. The group killed Sri Lankan defense minister Ranjan Wijeratne later that same year. Sri Lankan president Ranasinghe Premadasa, the most senior Sri Lankan official assassinated by the group, was killed in 1993.[69]

The Tamil Tigers' capabilities are almost unrivaled and should therefore be a subject of study. However, the 1991 assassination of Gandhi is the only terrorist attack carried out by LTTE operatives outside the borders of Sri Lanka, a fact that seems to indicate that this group likely poses a low risk in terms of the threat to the U.S. homeland.

INDIA. Since its independence in 1947, India has faced waves of insurgency and terrorism, largely rooted in tensions with Pakistani-militant groups over control of Kashmir, a disputed area bordering Pakistan and India. Great Britain dismantled its Indian mandate in 1947, and since partition of the subcontinent, India and Pakistan have faced ongoing tension rooted in differing religious beliefs, history, and long-running dispute over the state of Jammu and Kashmir. Tensions have escalated into violence, several wars and terrorism.

One of the first modern-day suicide attacks in India was carried out by the LTTE in 1991, with the suicide-attack assassination of former Prime Minister Rajiv Gandhi. Another reported incident of a suicide attack took place in April 2000, when a 14-year-old suicide bomber targeted India's central army base in downtown Srinager, in Indian-administered Kashmir (IAK).[70] Soon after, in December 2001, India experienced an extremely devastating attack, when a suicide bomber targeted the Indian parliament, leading to a tense standoff between the two nuclear powers.[71] In both cases, Pakistani and Kashmiri militant groups seeking Kashmir's integration with Pakistan were presumed responsible.

In addition to facing nationalistic motivated terrorism, India has also faced religious-based terrorism from different sectarian groups. Despite the different militant and terrorist group's active in India, the tactic of suicide terrorism has primarily been adopted by Pakistan's Islamic Jihad groups operating in Jammu and Kashmir and in other parts in India. While terrorist attacks and bombings are all too common in India, compared to other regions of the world, the specific tactic of suicide terrorism has been used in a limited number of attacks launched in the country.

CHECHNYA. Suicide terrorism is a new mode of attack in the Chechen conflict. Although they were adopted in only 2000,[72] suicide attacks became a favored tactic. Thus, between June 2000 and 2005, approximately 800 people were killed in 25 different suicide attacks carried out by over 100 Chechens.

The conflict in Chechnya can be divided into two major periods of violence, lasting between 1994 and 1996 and from 1999 until today. This conflict has primarily been depicted as a struggle for independence from Russia. However, the second stage of the conflict has not been limited to nationalism, and it has revealed trends of Islamic extremism within the Chechen campaign.

In fact, two unprecedented attacks carried out by the Chechens in recent years were distinctly similar to al-Qaeda-type operations in that they required diligent planning, training, and dedication to self-sacrifice. The first incident took place at a Moscow theater in October 2002, while the second took place at a school in Beslan in September 2004. Both incidents involved multiple perpetrators and resulted in numerous casualties.

The first attack, which occurred at a Moscow theater, was carried out by about 40 Chechen attackers equipped with belt-bombs and other weapons. These attackers took hundreds of Russian theater-goers hostage, thereby precipitating an unsuccessful rescue attempt that left 129 hostages dead.

The second incident took place from September 1 to September 3, 2004, in northern Ossetia in a town called Beslan. Here, 32 terrorists carrying explosives and a variety of weapons, including belt-bombs, stormed a school, taking hundreds of children and adults hostage, resulting in the deaths of over 300 people.

Further analysis of these incidents will be detailed in Chapter 4.

EUROPE. Europe is another area of the world that has not been immune to attack. Major attacks have been thwarted in France, Germany, Holland, and elsewhere.[73] However, two countries that were less fortunate were Spain and Great Britain.

Spain. On the eve of controversial national elections in Spain on March 11, 2004, ten explosive devices were detonated within a short period at several train stations in southern Madrid. The police also discovered and diffused another three devices hidden in backpacks. The attacks on the Spanish trains were not suicide bombings; rather, they involved the use of explosive devices detonated remotely by means of cellular telephones. However, the operatives nonetheless exhibited a philosophy of self-sacrifice when they blew themselves up shortly after the attacks, when police found them and surrounded their hideout. Additionally, an explosive belt was later found in the attackers' hideout, indicating the potential for at least one suicide attack. The attacks that did occur claimed 191 fatalities, and approximately 1400 people were injured.[74]

One of the main issues surrounding Spain's 2004 elections was Prime Minister Jose Aznar's decision to send 1300 Spanish troops to Iraq as part of the coalition force. The Madrid attacks provided their terrorist organizers with direct and almost instant results, influencing the elections such that Aznar's underdog opponent was voted into office. Shortly thereafter, the new prime minister withdrew Spanish forces from Iraq.[75]

Great Britain. For the British, July 7, 2005 is a date that will go down in infamy, as four suicide bombers attacked London's public transportation system, the first act of suicide terrorism in a country with decades of experience combating local terrorist groups. The bombings carried out by local Islamic terrorists, left 52 commuters and the four suicide bombers dead, and injured more than 770.

With ongoing demographic changes taking place in European capitals, as well as the ease with which local recruits are found in Europe, the threat level is likely to escalate.

INDONESIA. Indonesia is the country with the world's highest number of Muslim inhabitants; these individuals constitute almost 90% of the country's population of 245 million.[76] However, Indonesian society is considered secular and tolerant; moreover, the nation's Muslims practice a moderate form of Islam, which is exactly what infuriates radical Islamic organizations like al-Qaeda and Jemaah Islamiyah, both of which operate in that country. A series of explosions rocked the country on December 24, 2000, part of an extensive campaign planned by al-Qaeda and Jemaah Islamiyah. The attack targeted churches in Jakarta and eight other places, leaving 18 people dead.

Later, on the night of October 12, 2002, terrorists carried out a number of additional attacks in Indonesia.[77] The targets selected by the perpetrators, members of a terrorist network operated by Jamaah Islamiya in affiliation with al-Qaeda, were two popular nightclubs in Bali on the Kuta coast, a heavily tourist area.[78] The attacks were conducted by two suicide terrorists: Jimmy, who wore an explosive belt on his waist and detonated it in a bar, and Iqbal, who detonated a TNT-laden car bomb outside a nightclub. The two men operated in coordination with each other. They detonated themselves nearly simultaneously, causing the clubs' roofs to cave in and igniting a giant fire, which in turn caused widespread damage in the surrounding area. The attack killed a total of 202 people, including 88 Australian tourists.[79]

IRAQ. Since May 2003, the aftermath of the U.S.-led military campaign to oust Saddam Hussein has been marked by the highest amount of suicide attacks in any one region to date, with frequency and a devastating level of lethality.

The military campaign against Saddam Hussein's government was followed by insurgency against the United States and its allies, as well as boundless sectarian violence. Resistance and a seemingly systematic uprising against coalition forces and between different Iraqi sects were encouraged by various parties of interest. For example, Osama bin Laden and his deputy encouraged Iraqis to carry out suicide attacks in order to strike at the occupiers and foreigners within the country:

> Use bombs wisely, not in forests and on hills. ... The enemy is scared primarily by fighting in the street in cities. ... We emphasize the importance of suicide operations against the enemy.[80]

Between May 2005 and March 2005, about 160 suicide attacks were carried out in Iraq by some 200 suicide operatives.[81] Further analysis on suicide terrorism in Iraq is detailed by Mr. Yaron Schwartz in Chapter 3.

In terms of terror, there is sound apprehensiveness that Iraq may become a rejuvenated model of Afghanistan following the Soviet invasion in 1979 and withdrawal in 1989 if U.S. and coalition troops withdraw in a similar manner.

How Is It Done? Most suicide attacks are carried out by means of improvised explosive devices (IEDs), which are primarily employed in four different ways: the human-borne suicide (HBIED), also known as the suicide bodysuit; the vehicle-borne suicide (VBIED), including use of trucks, cars, and motorcycles; the marine-borne suicide (MBIED), including use of watercraft and scuba divers; and the aerial-borne suicide (ABIED), including use of gliders, mini-helicopters, and airplanes. All these modes have been used in the past.[82]

The suicide attack phenomenon is multifaceted: Once the suicide bomber is psychologically set for the mission, he or she may be equipped with an explosive belt, a handbag, or a backpack (Figure I.7); he or she can then drive a car or a truck or use an aircraft or marine vessel to ram the desired target.

Figure I.7. Suicide vest, which is jacket containing plastic explosives with two pull cords: one to arm the device, and the other to detonate it. Such a device would be worn under the outer garments of a suicide bomber. (Source: Rohan Gunaratna, Suicide Terror A Global Threat, *Janes Security News*, Oct. 20, 2000.)

A number of studies have analyzed the state of mind experienced by suicide bombers when they approach their selected targets.[83] Most of these studies point out that a suicide attack is not an impulsive act or decision: It is the climax of a process, led by the people who plot the attack (but will rarely themselves go or send someone dear to them), which includes mental and religious preparation, technical training, and when possible, filming of a tape to be released to the media after the attack is executed. In many of the attacks, the perpetrators spent their final hours praying, either with their mission commanders or in a mosque, "to purify themselves" before their purported rendezvous with their maker.

Since the phenomenon of suicide attacks has spread worldwide, many people who are oppressed or perceived to be oppressed have begun to consider use of this method, which is considered by many to be "legitimate warfare." Parades of potential suicide bombers are now held in countries where terror organizations flaunt their cause in public. On the other hand, al-Qaeda, currently believed to be the world's most lethal terrorist organization, usually does not make public displays before taking action, instead preferring extensive planning and a silent mode of action in the period before an attack.

The international network that al-Qaeda has created is more difficult to deal with than other terrorist groups, largely because al-Qaeda recruits activists in many Western countries and has affiliates around the globe. Based on past experience and what seems like a future trend, al-Qaeda and organizations of its ilk present the most imminent threat to the West in general and, perhaps, to the

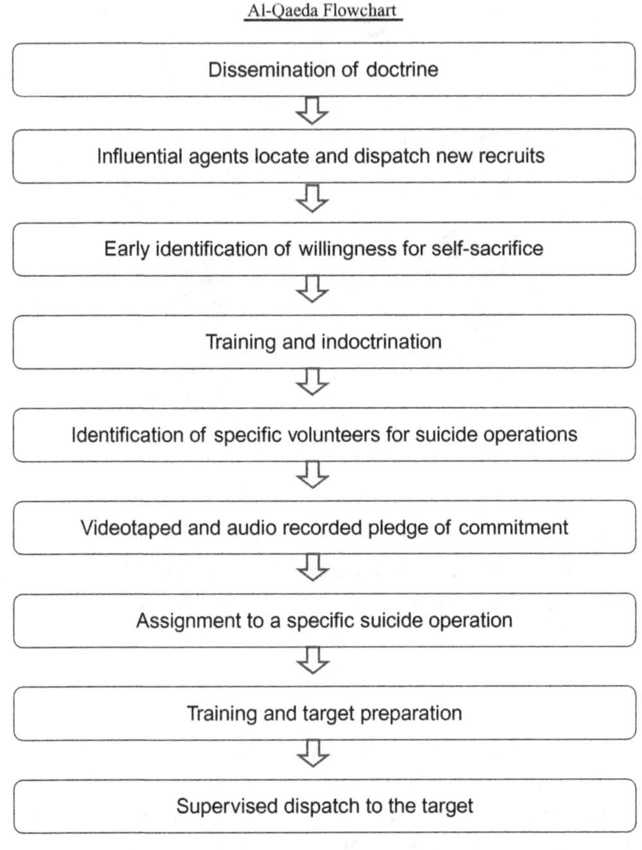

Figure I.8. Al-Qaeda flowchart.

United States in particular. Therefore, it is important to be acquainted with these groups' terror methodology as detailed by Yoram Schweitzer (see Figure I.8).

Profiling. In the past, determining the physical, biological, and socioeconomic characteristics of potential suicide attackers was often possible; today, that is no longer the case. Suicide bombers vary in age and economic status, and perhaps most interestingly, many bombers are largely secular in their outlook (though most are still Muslim). This lack of common traits may stem from the fact that suicide attacks are no longer limited to religious fanatics—they are now also employed by nationalist groups and other zealots.

Nevertheless, even if it has become much more difficult to characterize suicide bombers according to age, gender, economic status, and religious zeal, their mode of operation can and should be analyzed in an attempt to mitigate vulnerabilities.

Response. Ninety percent of the people who die in a suicide attack are killed almost immediately upon impact, but many of the remaining 10% can be saved.[84] In order to save part of those 10%, medical attention must be provided to wounded individuals within 10 minutes; after these individuals are quickly stabilized, they must be moved to the hospital immediately. However, haste provides opportunity for a second, more devastating attack—one that kills valuable emergency management personnel.

Thus, law enforcement should synchronize their response as follows:

- Immediately cordon off the attack scene and push back the crowd.
- Quickly search for possible secondary explosive devices or bombers.
- Get emergency crews on scene.

Further analysis of the medical response and management to suicide attacks is detailed by Professor Shmuel C. Shapira and Dr. Leonard Cole in Chapter 7.

Analyses of Data

In the following pages, a breakdown of most suicide attack incidents in a ten-year time period (1994–2004) is provided. This is done in an effort to detect common denominators, as well as to suggest an operational solution.

In this analysis, we attempt to address subjects of interest that may have operational significance. Within these issues, a breakdown of the types of methods applied, the times of the attacks, and the types of targets attacked is laid out.

The data spans from 1994 to 2004, at which time suicide attacks became the primary mode of operation for some of the world's most dangerous terrorist organizations. Some of the information concerning specific incidents is incomplete, often because sufficient information was not available from open sources. In addition, there may be incidents that are not cited due to inadvertent and unintentional omission; however, the authors believe that said omissions did not affect their analysis of trends.

Target Distribution. In the 10-year time period, the most popular targets of suicide bombers were public, crowd-concentrated areas. Such places included (but were not limited to) shopping centers, open markets, cafés, schools, and even religious quarters. Open and accessible areas were targeted more often than closed ones.

The second most popular target was various means of public transportation, together with military targets. This can be distributed into subsections: Most of the attacks against military targets were executed against armies that were in a different region and attackers found it too difficult to attack the civilians of the relevant nation. That was the case when Israel was in south Lebanon; and it continues to be the case today in Chechnya with the Russian army, as well as in Iraq with American and British forces.

Public transportation, mainly buses and trains, were targeted because of the high concentration of people in designated areas—many of which are closed because of air conditioning, thus maximizing the blast effect. Another reason that transportation is so often targeted is the lack of security checks at the entrance to bus and train facilities; such checks would necessitate a huge increase in government spending and cause significant delays.

Throughout the years surveyed, there is consistency in the choice of targets, although since 2003, attacks against military targets have increased. This can be explained by the U.S.-led occupation of Iraq, which is the main parameter to have changed since 2002.

Method of Attack. Between 1994 and 2004, the two primary methods of suicide attack were the human-carried charge and the vehicle bomb. The use of car bombs has become more common with the rise of such attacks against coalition forces in Iraq.

The use of car or truck bombs requires a more-developed infrastructure (and lack of security measures), due to the quantity of explosives needed and the accessibility of the vehicle to the chosen target.

With regard to pedestrian attacks, the information concerning many of these attacks is incomplete, because limited evidence is left after the explosions and only a modest number of lab tests were conducted. But based on the available data, it can be clearly asserted that the explosive belt was most frequently used device in such attacks. It can also be construed that the reason for use of such belts is their ease of concealment under a garment, whether it is a coat, shirt, or other item of clothing. On the other hand, the load capacity of such a charge is limited (in volume as well as weight) in comparison to explosive-laden vehicles.

Due to their volume constraints, many explosive belts or vests are loaded with screws, bolts, nuts, nails, steel balls bearings, and shrapnel to increase the potential for death and injury.

Most suicide bombing methods require an electric detonation device with an "on–off" switch activated by the actual bomber. Nonetheless, there were incidents in which the dispatcher of the bomber planted a remote-controlled detonating device to make sure that the suicide bomber did not falter at the last minute.

Of all vehicles used in suicide attacks, cars and trucks were most common. Incidents also have occurred in which planes, bicycles, motorcycles, carts, and even horses and donkeys were used as weapons.

Time of the Attack. Most suicide attacks in the 10-year time frame were executed in the morning hours, while almost no attacks occurred in the late-night hours.

When analyzing this subject, one must remember that the target chosen usually determines the time of the attack. Should the target be a bus, it is most likely to be attacked in the early morning hours, when many people are rushing to work and other places. On the other hand, when attacking a café or a restaurant, the evening hours are typically more "appropriate." Similarly, when the target is a military barrack, the late-night hours are generally preferred, because most troops are asleep and guards are less alert at this time.

From the case studies evaluated and other relevant material that will be detailed in this book, we can make five key observations:

- Nearly all suicide terrorist attacks are part of an organized campaign rather than a spontaneous action carried out by individuals.
- Attacks are part of the overall strategic objective of an organization.
- Most suicide attacks have been perpetrated by Islamist organizations.
- The tactical use of female attackers has become more common and may become a trend.
- Suicide terrorism is used because it works! The casualty rate is relatively high, the media exposure is obvious, and the underlying political objectives are many times enhanced.

OVERVIEW

This book highlights the U.S. and Israeli experience in dealing with the weapon of suicide terror based on diligently researched material. It also draws conclusions concerning future use of the weapon in the United States. The book is written from a practitioners' viewpoint and it provides policymakers, first responders, and students of homeland security with knowledge of the weapon. Descriptions and reenactments based on eyewitness testimony create the look and feel of actual incidents. Utilizing extensive case studies, the book additionally points out how best to prevent and confront this dangerous threat.

Based on extensive interviews with U.S. and Israeli policymakers and first responders (i.e., law enforcement, paramedics, military personnel, members of counterterrorism units, and others), this book aims to speak in the language of those responsible for understanding, preventing, and confronting the threat. Utilizing a variety of well-known and qualified Israelis and Americans who have dealt with these phenomena and synthesizing their experience, the authors aim at crafting a definitive study of the subject.

Furthermore, the book's explanation of the history of the weapon and its use by Islamic fundamentalists will be of interest to both professionals and general readers. Unlike other books that often approach this subject matter from specific political agenda, the aim here is to explain the appeal of the weapon to the groups most likely to use it against the United States, as well as to explain the mindset that makes deployment of this weapon possible. To that end, the book looks at the way terror groups are organized, considers these groups' modus operandi, and describes the mechanics of suicide attacks, such as recruitment of candidates.

This book makes extensive use of data that include details regarding most of the pertinent attacks over the last decade, various case studies, and statistical analyses of terror attacks. The information in the appendices should serve as a valuable resource for students and policymakers alike.

ENDNOTES

1. Francis Fukuyama, *The End of History? National Interest*, Summer 1989.
2. Samuel Huntington, *The Errors of Endism, National Interest*, Fall 1990.
3. Huntington, *The Clash of Civilizations? Foreign Affairs* **72**(3), 22–49, 1993.
4. *Ibid.*
5. Benjamin Netanyahu, *Fighting Terrorism: How Democracies Can Defeat Domestic and International Terrorists*. New York: Farrar, Straus, and Giroux, 2001, p. 129.
6. "Address to a Joint Session of Congress and the American People," by George W. Bush, September 20, 2001 http://www.whitehouse.gov/news/releases/2001/09/20010920-8.html
7. Yoram Dinstein, *Laws of War*. Tel Aviv: Schockon, 1983, p. 14; Ingrid Detter, *The Law of War*, 2nd ed. Cambridge: Cambridge University Press, 2000, pp. 5–9.
8. Alex P. Schmid et al., *Political Terrorism: A New Guide to Actors, Authors, Concepts, Data Bases, Theories, and Literature*. New Brunswick, NJ: Transaction Books, 1988, pp. 5–6.
9. Walter Laqueur, *The New Terrorism: Fanaticism and the Arms of Mass Destruction*. New York: Oxford University Press, 1999, p. 6.
10. Brian Whitaker, *The Definition of Terrorism, Guardian Unlimited*, May 7, 2001.
11. *National Strategy for Combating Terrorism*. Washington, DC: The White House, 2003, p. 1.
12. John F. Murphy, *Defining International Terrorism: A Way Out of the Quagmire, Israel Yearbook on Human Rights* **19**, 13–14, 1989.
13. Jeffrey Record, *Bounding the Global War on Terrorism*. Carlisle Barracks, PA: Strategic Studies Institute, U.S. Army War College, 2003, p. 8.
14. Bodies of U.S. Sailors Flown Home, *BBC News*, October 15, 2000, http://news.bbc.co.uk/2/hi/middle_east/972901.stm (accessed June 5, 2008).
15. Yoram Schweitzer and Sari Goldstein Ferber, *Al-Qaeda and the Internationalization of Suicide Terrorism*, Memorandum No. 78, Jaffe Center for Strategic Studies, Tel Aviv University, November 2005.
16. *Ibid.*
17. *Ibid.*
18. Israel Ministry of Foreign Affairs, IDF Spokesman: Hizbullah Attack on Northern Border and IDF Response, news release, July 12, 2006, http://www.mfa.gov.il/MFA/Terrorism+Obstacle+to+Peace/Terrorism+from+Lebanon+Hizbullah/Hizbullah+attack+on+northern+border+and+IDF+response+12-Jul-2006.htm (accessed June 5, 2008).
19. *Department of Defense Dictionary of Military and Associated Terms*. Washington, DC: United States Department of Defense, 2001, p. 428.
20. The British Terrorism Act of 2000 interprets terrorism as follows: "[T]he use or threat of action where (a) the action falls within subsection (2), (b) the use or threat is designed to influence the government or to intimidate the public or a section of the public, and (c) the use or threat is made for the purpose of advancing a political, religious or ideological cause. (2) Action falls within this subsection if it (a) involves serious violence against a person, (b) involves serious damage to property, (c)

endangers a person's life, other than that of the person committing the action, (d) creates a serious risk to the health or safety of the public or a section of the public, or (e) is designed seriously to interfere with or seriously to disrupt an electronic system. (3) The use or threat of action falling within subsection (2) which involves the use of firearms or explosives is terrorism whether or not subsection (1)(b) is satisfied."

21. *Quinn v. Robinson*, 783 F.2d. 776 (9th Cir. 1986).
22. Terrorism is defined in the U.S. Code as follows: "[A]ctivities that involve violent ... [or life-threatening acts] ... that are a violation of the criminal laws of the United States or of any [s]tate and ... appear to be intended (i) to intimidate or coerce a civilian population; (ii) to influence the policy of a government by intimidation or coercion; or (iii) to affect the conduct of a government by mass destruction, assassination, or kidnapping; and ... [if domestic] ... (C) occur primarily within the territorial jurisdiction of the United States. ..." This definition was also adopted in the USA Patriot Act of 2002.
23. Democracy Now, "Hezbollah Leader Hassan Nasrallah Talks with Former U.S. Diplomats on Israel, Prisoners, and Hezbollah's Founding," July 28, 2006, http://www.democracynow.org/article.pl?sid=06/07/28/1440244 (accessed June 5, 2008).
24. Boaz Ganor, *Defining Terrorism: Is One Man's Terrorist Another Man's Freedom Fighter?* International Institute for Counter-Terrorism, http://www.ict.org.il/index.php?sid=119&lang=en&act=page&id=5547&str=Defining%20Terrorism (accessed June 5, 2008).
25. For further discussion of legal definitions of terrorism, see: Ben Golder and George Williams, *What Is "Terrorism"? Problems of Legal Definition, UNSW Law Journal* **27**(2), 2004. http://www.gtcentre.unsw.edu.au/Publications/docs/pubs/terrorismDefinitions.pdf.
26. *Ibid.*
27. *Ibid.*
28. Schmidt et al., *Political Terrorism*, 5; as referred to in Ganor, "Defining Terrorism."
29. The definition proscribed in Ganor's "Defining Terrorism" served as the primary basis for the general part of this definition.
30. See the discussion in Ganor, of the USS Cole attack.
31. Article 2(1) b of *International Convention for the Suppression of the Financing of Terrorism.*
32. See the discussion of the abduction of Israeli soldiers on pages 6 and 7.
33. This book concentrates on understanding and confronting suicide terrorism. Incidents in which the probability of the perpetrator's ultimate death is high but not a precondition for the success of their mission are beyond the scope of this book.
34. George Santayana, cited in *The New Dictionary of Cultural Literacy*, 3rd ed., 2002. http://www.bartleby.com/59/3/thosewhocann.html
35. *Judges* 16:30.
36. "Pearl Harbor Raid: 7 December 1941," Naval Historical Center, U.S. Department of the Navy, http://www.history.navy.mil/photos/events/wwii-pac/pearlhbr/pearlhbr.htm (accessed June 5, 2008).

37. There are researchers who note an April 1981 suicide attack against the Iraqi embassy in Lebanon, allegedly carried out by a Lebanese Shiite attacker and operated from Iran, but there is not enough information on that incident to verify that it was indeed an example of suicide terrorism.
38. Henry Morgenstern, The Terrorist Weapon of Choice, www.nationalhomelandsecurityknowledgebase.com/research/International_Articles/VBIED_Terrorist_Weapon_of_Choice.html (last visited March 12, 2009).
39. Yoram Schweitzer and Sari Goldstein Ferber, *Al-Qaeda and the Internationalization of Suicide Terrorism*, Memorandum No. 78, Jaffe Center for Strategic Studies, Tel Aviv University, November 2005.
40. For a detailed analysis of the ramifications of the Oslo Accords, see "The Oslo Process: Fate or Folly?" by Ophir Falk and Yaron Schwartz, published on April 28, 2002, and available online at http://www.ict.org.il/articles/articledet.cfm?articleid=434.
41. Author's interview with Karmi Gilon (former head of Israel Security Agency), April 28, 1996.
42. Data presented by former Israel Security Agency (ISA) executive Ehud Ilan at the Fifth International Conference on Terrorism, held in Herzliya, Israel, on September 12, 2005.
43. *Ibid.*
44. Bernard Lewis, *What Went Wrong?* London: Weidenfeld and Nicolson, 2002, p. 6.
45. *Ibid.*
46. Benjamin Netanyahu, in forthcoming book *Israeli Tiger*, 2009.
47. Lorenzo Vidino, The Muslim Brotherhood's Conquest of Europe, *Middle East Quarterly*, Winter 2005 http://www.meforum.org/article/687#_ftn1
48. Attack on Terrorism—Inside Al-Qaeda, *Financial Times*, November 30, 2001. http://specials.ft.com/attackonterrorism/FT3RXH0CKUC.html
49. Febe Armanios, *Islam: Sunnis and Shiites*, CRS Report for Congress February 23, 2004. http://www.fas.org/irp/crs/RS21745.pdf
50. Benjamin Netanyahu, *A Place Among the Nations*. New York: Bantam Books, 1993, p. 106.
51. George W. Bush, State of the Union Address, Washington, DC, January 29, 2002.
52. Alberto Abadie, "Poverty, Political Freedom, and the Roots of Terrorism," October 2004, http://ksghome.harvard.edu/~aabadie/povterr.pdf
53. http://www.nber.org/digest/sep02/w9074.html
54. Anne Speckhard, Understanding Suicide Terrorism: Countering Human Bombs and Their Senders, *The Atlantic Council*, July 2005. In that study, Speckhard notes the following relevant studies: Nichole Argo, *The Banality of Evil: Understanding and Defusing Today's Human Bombs*, 2003. Mia Bloom, *Dying to Kill: The Global Phenomenon of Suicide Terror*, 2005. Mohammed Hafez, *Manufacturing Human Bombs: Strategy, Culture, and Conflict in the Making of Palestinian Suicide Bombers*, unpublished research paper, 2004. Ariel Merari, *Suicide Terrorism*, unpublished manuscript, 2003, p. 12. Assaf Moghadam, Palestinian Suicide Terrorism in the Second *Intifada*: Motivations and Organizational Aspects, *Studies in Conflict and Terrorism* 26.2, February/March 2003. Assaf Moghadam, *Suicide Bombings in the Israeli-Palestinian Conflict: A Conceptual Framework*. Project for the Research of Islamist

Movements, Inter-Disciplinary Center, Herzliya, Israel, 2002, p. 15, http://www.e-prism.org/ Anne Speckhard, *Soldiers for God: A Study of the Suicide Terrorists in the Moscow Hostage-Taking Siege*. Unpublished research proposal. J. Post, E. Sprinzak, and L. Denny, *The Terrorists in Their Own Words: Interviews with 35 Incarcerated Middle Eastern Terrorists, Terrorism and Political Violence* **15**(1), 171–184, 2003. Oliver McTernan, ed., *The Roots of Terrorism: Contemporary Trends and Traditional Analysis*. Brussels: NATO Science Series 2004.

55. Rohan Gunaratna, Suicide Terrorism: A Global Threat, *Jane's Security News*, October 20, 2000, http://www.janes.com/security/international_security/news/usscole/jir001020_1_n.shtml (accessed June 5, 2008).
56. Bruce Hoffman, *The Logic of Suicide Terrorism, Atlantic Monthly*, June 2003.
57. Data presented by Bruce Hoffman at the Third International Counter-Terrorism Conference, held in Herzliya, Israel, on September 11, 2003.
58. Scott Atran, *Genesis and Future of Suicide Terrorism*, July 1, 2003.
59. Sever Plotzker, "*The Landscape of the Israeli Economy, Society and Policy after 1200 Days of Intifada*," The Brookings Institution, November 13, 2003.
60. *The 9/11 Commission Report*, 169
61. http://www.fas.org/irp/world/para/ayman_bk.html
62. Author's interview on July 4, 2008 with Yochanan Alon, a former special operations officer in the IDF and the Israel Police Force. Mr. Alon led an in-depth research on Palestinian suicide terrorism that was completed in 2008. In the context of that study, Mr. Alon interviewed scores of close relatives of suicide attackers and a number of inmates that were arrested after failed suicide attacks.
63. *Ibid.*
64. *Ibid.*
65. Ibid.
66. Yoram Schweitzer and Sari Goldstein Ferber, *Al-Qaeda and the Internationalization of Suicide Terrorism*, Memorandum No. 78, Jaffe Center for Strategic Studies, Tel Aviv University, November 2005, p. 66.
67. *The Globalization of Terror*, Schweitzer and Shay, p. 55.
68. Schweitzer and Ferber, *Al-Qaeda and the Internationalization of Suicide Terrorism*, Memorandum No. 78, Jaffe Center for Strategic Studies, Tel Aviv University, November 2005, p. 80.
69. Jonathan Lyons, *Suicide Bombers: Weapon of Choice for Sri Lanka Rebels*, Reuters, August 20, 2006.
70. M. Burgess, In the Spotlight: Jaish-e-Mohammed (JEM) Terrorism Project. *Center for Defense Information*. April 8, 2002. Online http://www.cdi.org/terrorism/jem-pr.cfm; "Major Islamist Terrorist Attacks in India in the Post-9/11 Period," *South Asian Terrorism Portal*, online; Swami, P. (2001, October 18–31). *An Audacious Strike*. Retrieved December 10, 2008, from Frontline: http://www.hindunnet.com/fline/f11821/18210200.htm
71. "Parliament Suicide Attack Stuns India," *BBC News*, December 13, 2001. http://news.bbc.co.uk/2/hi/south_asia/1708853.stm
72. Schweitzer and Ferber, *Al-Qaeda and the Internationalization of Suicide Terrorism*, Memorandum No. 78, Jaffe Center for Strategic Studies, Tel Aviv University, November 2005, p. 80.

73. Matthew Levitt, "Islamic Extremism in Europe," Hearing before the Subcommittee on Europe and Emerging Threats, of the Committee on International Relations, House of Representatives, April 27, 2005 http://commdocs.house.gov/committees/intlrel/hfa20917.000/hfa20917_0f.htm
74. Yoram Schweitzer and Sari Goldstein Ferber, "*Al-Qaeda and the Internationalization of Suicide Terrorism*," Memorandum No. 78, November 2005, p. 74.
75. http://news.bbc.co.uk/2/hi/europe/3637523.stm
76. https://cia.gov/cia/publications/factbook/geos/id.html
77. http://news.bbc.co.uk/2/hi/in_depth/asia_pacific/2002/bali/
78. Schweitzer and Ferber, *Al-Qaeda and the Internationalization of Suicide Terrorism*, Memorandum No. 78, November 2005, p. 66.
79. Maria Ressa, *Seeds of Terror*. New York: Free Press, 2003, p. 168.
80. Bin Laden Tape: Full Text, www.bbc.com, February 12, 2003, as cited in Schweitzer and Ferber, *Al-Qaeda and the Internationalization of Suicide Terrorism*, p. 78.
81. Schweitzer and Ferber, *Al-Qaeda and the Internationalization of Suicide Terrorism*, Memorandum No. 78, November 2005, p. 80.
82. Rohan Gunaratna, Suicide Terrorism: A Global Threat, *Jane's Security News*, October 2000.
83. Anne Speckhard, *Understanding Suicide Terrorism: Countering Human Bombs and Their Senders*, Atlantic Council, July 2005.
84. Bruce Hoffman, in context of presentation he gave at the International Counter Terrorism Conference in Herzlyia in September 2007. The data was based on his research at the Rand Institute.

1

THE GLOBAL JIHAD

Henry Morgenstern

INTRODUCTION—WHERE IS THE JIHAD TODAY?

In the years since the tragic events of 9/11 the United States and the International community has landed heavy blows against the world-wide Islamist movement, or what we will call the Global Jihad. By destroying the Taliban State in Afghanistan, crippling the insurgency in Iraq, and making it more difficult for International financial support for the Jihad—along with a general increase in security measures, as well as advances in Intelligence gathering, not to mention killing and capturing senior al-Qaeda figures—the United States and the International community seems to be prevailing. But is this really the case?

The Global Jihad is a many-headed Hydra and, in the finest tradition of asymmetrical conflicts, is almost infinitely capable of change. More importantly, the same fanatical zeal that caused 19 Jihadists to bring about the deaths of nearly 3000 Americans has not diminished and may have increased and spread. Indeed, there may be unintended consequences of our success against the physical bastions of the movement. If the movement can recruit effectively on an international basis, arm its adherents with know-how about bomb-making

Suicide Terror: Understanding and Confronting the Threat. Edited by Falk and Morgenstern

techniques, and use the laws of democracies in its favor, it may be even more dangerous than it was before and may also be more capable of using the weapon of Suicide Terror and in more devastating ways than we have thus far seen despite our countermeasures—mostly based on the reaction to 9/11.

In today's post 9/11 era, from a Dorm room at Georgia Tech near Atlanta, two U.S. citizens were able to contact a group in Canada named "Toronto 18" who are alleged to have planned attacks against targets in Canada. Twenty-one-year-old Syed Haris Ahmed and his friend Ehsanul Sadequee admitted that they had taken films of Capitol Hill and other Washington landmarks to help plan terror attacks and had uploaded these to websites. They joined an International cast of characters that only begins with the Toronto 18, a group that included 14 adults and 14 youths. (See Appendix 2 for details.)

The videos they shot were found on the computer of one Younis Tsouli, better known on line as Irhabi007 or Terrorist007, who is now serving 10 years in the United Kingdom. According to a Senate Committee on Homeland Security and Governmental affairs,[1] Tsoulis had helped the late Abu Musa al-Zarqawi, head of al-Qaeda Iraq, and his spiritual mentor, Abu Muhammed Maqdisi, disseminate the writings of both, along with ghastly films of the beheadings of innocent Western journalists and U.S. workers in Iraq. Tsoulis also helped the progenitors of the Global Jihad: Abu Qatada; Abu Hamza al-Masri from the infamous Finsbury Park Mosque in London; Abdullah Azzam, the mentor of Osama bin Laden, and his early sources of inspiration; and Sayyid Qutb of the Muslim Brotherhood. These are all hallowed names for Jihadists over the world. The links go even deeper: Irhabi007 was known to have the trust of Dr. Ayman al-Zawahiri, al-Qaeda's leading theoretician. On the versatile 22-year-old's computers were also found the declaration of 45 doctors that wanted to fire bomb London.

This has caused an incredible proliferation of Internet activity and organization—certainly enough to firmly threaten our cities and institutions.[2] Anyone doubting that the virtual can become real should look carefully at the London Bombings of July 7, 2005, where four born-in-Britain terrorists caused havoc and destruction on a scale unknown in London since World War II utilizing Suicide bombing, inspired by just this type of Internet Jihad. The London attack, like others in Europe, is a disturbing precursor—almost a warning that home-grown terrorism and the use of Suicide Bombers in our own cities may not be long in coming.[3]

The Global Jihad has changed its method of operations, but the reasons for the Global Jihad are the same; they have not changed much since the 1970s and may not have changed much since the beginning of the seventh century.

DO WE NEED TO KNOW WHY?

Why should the context of Suicide Terror attacks interest First Responders, concerned as they should be with detecting, preventing, or responding to acts of terror? Are the "Why's" of Terror attacks at all relevant to carrying out the mission of protecting the U.S. Homeland?

In 1989, in Southern Florida, at the University of South Florida's Tampa campus, on a narrow, dead-end, 130th Street, a sign was affixed to the last house on the block memorializing Izz al-Din al-Quassam, the name of an early Jihadist who was killed by the British in Yabrod, Palestine in 1935.[4]

To anyone that understands the International Global Jihad, this name appearing on a building is a sign. The name has become a symbol of heroism, resistance, occupation, and invasion to an entire generation of terrorists. Indeed a whole modern terror brigade is named after Izz al-Din al-Quassam, a brigade with a fearsome record of killing and maiming innocent Israelis (see Chapter 2, page 94).

Seeing this on a University Building, being used as a Mosque by then Professor Sami Al-Arian of the University of South Florida (served a sentence for activities in support of terrorism), should have meant something.[5] Sami al-Arian's background should have indicated a problem; his whole life was a series of Jihadi networks. He was at the North Carolina Agricultural and Technical State University with Khalid Shaykh Mohammed (the operational head of the 9/11 attacks) and Mazen al-Najjar, later deported for supporting terrorism. All three were known as the *Mullahs* at North Carolina, and al-Arian and al-Najjar were members of Palestinian Islamic Jihad. These terror networks meant that al-Arian was associated with Khalil Shikaki, brother of the leader of the Palestinian Islamic Jihad (PIJ) Fathi Shikaki, and that he was closely associated with Ramadan Abdullah Shallah, former adjunct professor of Middle Eastern Studies at the University of Southern Florida who later turned up as the General Secretary of the Palestinian Islamic Jihad (PIJ).

We also have to ask if understanding the meaning of that sign might have had some tangible and important intangible results. If we had better understood, might we have curtailed the career of an important front-man, organizer, and financier for some of the most vicious terror groups in the world? We also have to ask how many radicalized members of the international Global Jihad are still in Tampa and how many have spread to other cities within the United States and other Western countries since 1989—ready to become active at any time? We have to ask how much money was raised between that sign appearing in 1989 and how much made its way to the coffers of al-Qaeda, possibly contributing to the events of 9/11?

There appear to be real measurable gains in having an understanding of an ideology and its goals. But there are also important intangible advantages.

These events, occurring here in the United States in 1989, before the first Iraq War, a dozen years before the 9/11 attacks, and much before the First Gulf War let alone the invasion of Afghanistan and Iraq, should, at the very least, help inform us that the Global Jihad began in another time and must be using the weapon of suicide terror for its own reasons—removed from whatever has happened in Iraq and Afghanistan recently. As we have already seen in the introduction to this book, suicide bombing became a tactical weapon of the

Global Jihad in Lebanon in 1983. By understanding the history, we understand that this is not a land dispute as some have judged.[6]

Without understanding why suicide terror is used by terror groups, there is no hope that we will achieve the goal of detecting and preventing these attacks; therefore we will be reacting to attacks, not even responding. To understand why these attacks are used, we need to understand what the Global Jihad is really about. Without this, First Responders will be denied any chance of achieving the goal of keeping the United States safe from attack. We have to develop public policy that confronts these acts and then influences the procedures and planning that regulate First Response accordingly.

If we do not have an accurate understanding but believe what the adversary wants us to believe, we will make costly policy and procedure errors. Believing, for example, that the prime motivation of terrorists is to liberate their land, not have foreigners on their shores, not be humiliated by our Western culture, and so on, brings us to the conclusion that this is a Foreign Policy/Political issue that can be settled by "talking to them." Terrorists *do* want to achieve all these goals, but they have much greater ambitions that make the achievements of these goals interim successes in a much longer struggle with a very clear final result. As Sami al-Arian put it, speaking in Arabic to his followers before he was unmasked as the terrorist instead of the friendly, harmless professor: "The War of the 'ummah' against the Kufar (the infidel) … won't happen in a single attack but will be a long struggle."[7]

A historical analogy can make this very clear. In the pre-World War II period, the appeasement movement in the United Kingdom, led by the then Prime Minister, Neville Chamberlain, interpreted Hitler's expansionist goals as a "land Issue" over the future of the Sudetenland, then a disputed part of Czechoslovakia. Although Hitler had been crystal clear in his own *Mein Kampf (in Arabic, this book's title is Jihadi—or my Jihad)* written almost 20 years before, about what he ultimately wanted to achieve, the British Prime Minister found it convenient to base policy on a mistaken understanding of the context of Hitler's Czechoslovakian actions. As a consequence, Britain did not re-arm quickly enough, did not make any preparations for countering Hitler's policy of extreme rearmament, and so was unprepared when Hitler attacked Poland and put the rest of his policy of conquest on full display. It is estimated that more than 60 million people lost their lives during the 6 years of World War II in Europe and Asia. A little understanding of context can go a long way.[8]

The Global Jihad is much more complex than Hitler's European ambitions because it is not being waged by one dictator but being waged, as we will see, by an amorphous, transnational force using al-Qaeda and others in much the same way as we understand a *brand* but with a radical twist: an idea, or a label under which they fight us. This should not obscure their real goals and ideology.[9]

Understanding the Global Jihad, and the origins and goals of the Jihadist cause, will therefore give First Responders and policy makers the basic understanding they need to *begin* confronting this threat.

THE MEANING OF THE JIHAD FOR THE JIHADISTS (*MUJAHEDEEN*)

The following is a fatwa issued by Osama bin Laden on February 23 1998, concerning Jihad against the Jews and Crusaders:

> The ruling to kill the Americans and their Allies—civilians and military—is an *individual duty for every Muslim who can do it in any country in which it is possible to do it.* ... We—with Allah's help—call on every Muslim who believes in Allah and wishes to be rewarded to comply with *Allah's order to kill Americans and plunder their money wherever and whenever they find it.* We also call on the Muslim ulema (scholars), leaders, youths, and soldiers to launch a raid on Satan's U.S. troops and the devil's supporters allying with them, and to displace those who are behind them so they may learn a lesson.

In the aftermath of 9/11, analysts in the West sought to understand how 19 individuals could slaughter nearly 3000 Americans in a hideous suicide terror attack that was unprecedented in scale and brutality and, most of all, clearly directed at innocent noncombatants with complete premeditation and with killing *per se* being one of its goals. As it became clear that al-Qaeda was responsible, the *fatwa* issued in 1998 declaring war against Crusaders and Jews took on new meaning.

"We have the right to kill 4 million Americans—two million of them children—and to exile twice as many and wound and cripple hundreds of thousands" said al-Qaeda spokesman Sulaiman Abu Ghaith.[10] Abu Ghaith's statement, proclaiming a veritable genocide against Americans, escalates the horror of 9/11 by stating numbers that signal unconventional attacks or worse. Where does this "right" to kill indiscriminately come from? This is especially important because these statements by Jihadists are backed up by the most careful religious exegesis.

THE ORIGINS AND EVOLUTION OF THE GLOBAL JIHAD

The concept of Jihad is embedded in the Quran throughout. Some of the critical verses are:

> Fight those who believe not in Allah nor the Last Day, nor hold forbidden that which hath been forbidden by Allah and his Messenger, nor acknowledge the religion of truth (even if they are) of the People of the Book, until they pay the Jizya with willing submission and feel themselves subdued. Quran 9:29

> Fight and Slay the Pagans wherever you find them, and seize them, beleaguer them and lie in wait for them in every stratagem (of war); but if they repent, and establish regular prayers and practice charity, then open the way for them: for Allah is oft-forgiving, Most Merciful. Quran 9:5

> Fighting is prescribed for you, and ye dislike it. But it is possible that ye dislike a thing which is good for you, and that ye love which is bad for you. But Allah knows and ye know not. Quran 2:216

Based on Islamic *Fiqh, or Islamic Jurisprudence*, Jihad has been defined in legal terms:

> Jihad is fighting anybody who stands in the way of spreading Islam or fighting anyone who refuses to enter into Islam. Based on Surah 8:39
>
> O you who believe! Fight the infidels who dwell around you and let them see how ruthless you can be. Know that Allah is with the righteous. Quran 9:123.
>
> The punishment of those who wage war against Allah and His Messenger, and strive with might and main for mischief through the land is: execution or crucifixion, or the cutting off of the hands and feet from opposite sides, or exile from the land. Quran 5:33

To go with many verses that speak in this vein, there are also at least 114 verses in the Quran that speak of love, peace, and forgiveness, especially the Surah named the Heifer (Surah 2:62, 109).[11] As with any faith, there are always matters of interpretation, but our concern *here is not a religious one. It is clear that all Jihadists are Muslim, but it is never asserted that all Muslims are Jihadists.* Rather we are concerned with the religious justification of Jihad against the "The Crusaders and Jews" by the radicals' own exegesis.

The interpreters use a variety of practices: *qiyas* or analogy and *nasikh* or overriding to reconcile difference in verses; and along with the many Hadith (sayings and examples), this forms the Sunna. This determines all the Usl al-Fiqh (Islamic jurisprudence) and its outcome, as well as Islamic law (Sharia) that regulates all of life for a believer. This is mentioned because this is just how the practitioners of Jihad make their arguments for issues like Offensive Jihad; the duty of all Muslims to undertake this; *Ishtihad* or martyrdom; Suicide Bombings; the killing of innocent noncombatants; the killing of Muslims during these attacks; and the declarations of an Islamic hegemony in the world. Which interpretation *do* the Jihadists (which is their own name for themselves), "Mujahedin," take to be correct, and how have they interpreted the Quran to serve their cause?

Only Jihadists waging the war can answer the following questions: Is this a clash of civilizations or a religious war comparable to the crusades? Is it being driven by hatred of the West and everything we represent, or is it about a new Caliphate and a new world order, subject to *Sharia* or Islamic law, a love for a new Islamic World Order?

OSAMA BIN LADEN AND DR. AYMAN AL-ZAWAHIRI—THE AL QAEDA VIEW OF THE JIHAD

One of the best sources for understanding the thinking of Osama bin Laden, and hence of his followers, is his answer to Saudi intellectuals and religious leaders in his essay "Moderate Islam is a prostration to the West."[12]

Osama bin Laden's letter, written in Arabic, is the expression of his outrage at a declaration by Saudi intellectuals that was written in answer to U.S. intellectuals, entitled "Why do we fight"? The Saudi response was entitled "How can we coexist" and was written by intellectuals and religious figures. Bin Laden's answer was clear and straightforward, unlike his Media releases which are full of *taqiyaa* or prudential dissimulation—a well-practiced tactic in the Holy War and religiously mandated for any Muslim to say anything as long as in his heart there is faith.[13]

Addressing the writers of what he sees as a complete prostration, he says:

> It's best you prostrate yourself in secret. ... There are only three choices in Islam; either willing submission or payment of the jizya, thereby physical submission to the authority of Islam or the Sword—for it is not right to let him (an infidel) live. The matter is summed up for every person alive: either submit, or live under the suzerainty of Islam or die. This it behooves the Saudi signatories to clarify this matter to the West—otherwise they will be like those who believe in part of the book while rejecting the rest.

He continues:

> The West is Hostile to us on account of Loyalty and Enmity and Offensive Jihad. So how can the writers of the declaration address those infidels who attack our faith by word and deed with such trivial matters that have nothing to do with the *heart of the conflict*. What the West desires is that we abandon (the doctrine of) loyalty and enmity and abandon offensive Jihad. This is the essence of their request and desire of us. *Do the intellectuals think its actually possible for Muslims to abandon these two commandments simply to coexist with the West?*

Bin Laden is clear in his opinion; for him, there is no moderate Islam and the choices for anyone in the conflict are clear. The apologists, as he sees them, are counting the amount of angels on a pin because it is simply impossible for the true believer to conduct himself in any way other than what is written in the Doctrine of Loyalty and Enmity (*walaa wa baraa*), which will be explained fully.

According to bin Laden, the debate really revolves around the word of the Messenger:

"I have been sent in the final hours with the sword so that none is worshipped but Allah alone, partnerless."

For bin Laden, it is clear that Radical Islam's war with the West is not limited to political grievances (whether justified or not) but is timeless and deeply rooted in his faith.

Osama's thinking harks back to the writings of Ibn Tamiyya, who wrote during the Mongol invasions 1263–1328 and whose writings were extremely influential to the Hanbali School and also to the later Abdul al-Wahhab, founder of the Wahabbi sect in Islam.[14] What we have is an emulation of tradition (*Taqleed*) and a harkening back to the first three generations of Islam, the Golden Age, which is essential to fight against the Far Enemy (*al-Abou al-Baeed*), the United States, today.

More evidence of the religious and historical inevitability of a conflict with the nonbelievers can be found in Dr. Ayman al-Zawahiri, who is the most prolific of the al-Qaeda leadership on the subject.

Here is Ayman al-Zawahiri speaking in an interview on the Internet.

He answers one questioner[15]: "I respond to it by stressing that we invite all people to Islam, and we invite the Muslims and their organizations to unite around the word of Tawhid [Islamic monotheism—i.e., there is no god but Allah], and from the requirements of this word is that they work to help Islam by ruling by his Shari'ah and not making it equal to any other rule, and that they confront the invaders usurping the homelands of the Muslims and neither recognize nor respect any obligation or agreement which gives up even a hand span of them, and that they work to dethrone and remove the corrupt, corrupting, hireling rulers who dominate their homelands."

Moreover, Zawahiri is clear on the way the enemy should be handled and points to the doctrine of Loyalty and Enmity to provide methodical proof from the Hadith (examples or sayings) and the Quran about the relationship with unbelievers. Loyalty and Enmity is an all-encompassing doctrine in Islam that outlines what is acceptable and what is completely unacceptable or unclean for a Muslim. It is as though anything outside Islam must be regarded as unclean and with Enmity. Any relaxing of this doctrine (pluralism for example) is a prime reason, according to Zawahiri, for the problems Muslims have endured.

And so he elaborates that the Almighty has informed us that the infidels despise the Muslims,[16] and that part of the duty of all Muslims is to hate the infidel and renounce their love because Muslims are forbidden from showing affection to those that oppose Allah and his Messenger. "Allah only forbids you those who made war upon you on account of your religion, and drove you forth from your homes and supported (others) in your expulsion—these do not befriend. And whoever befriends them is unjust [60:8:9]. According to the writings of Ahmad, there is no greater duty than to repulse the invading enemy."

Both leaders of al-Qaeda parallel their harkening back to a world that is free of "Innovation," which is what all heretical acts come from—in Arabic *Bidaa*—and reject completely any Western ideas such as Democracy, which they regard as abhorrent because it is man-made and not made by Allah.

Zawahiri continues: "In the Hadith there is evidence for not deposing a sultan even if he transgresses. The jurors have agreed unanimously on the necessity of obeying the victorious Sultan and waging jihad with him. Obeying him is better than fighting him. However, should the Sultan become an infidel, then he is not to be obeyed and it is obligatory to wage jihad against him.[17] Exchanging the Islamic Shaaria with something else is infidelity, especially in the despicable manner we see in the lands of Islam, according to al-Zawahiri.[18] If such a one persists in this (implementing laws borrowed from the Europeans (this is from Hamid al-Fiqi) and does not return to governing according to what Allah has revealed, he is without any doubt an apostate infidel."

The bottom line regarding democracies is that the right to make laws is given to someone other than Allah most high. *So whoever is agreed to this is an infidel—for he taken gods in place of Allah. Legislation is the exclusive right of the most high.*[19]

Al-Zawahiri justifies himself by quoting Sayyid Qutb from "In the Shade of the Koran." Qutb is probably the foremost theorist of reverting to purer Islamic belief in an aggressive way and was sentenced to death by the Egyptian Government in 1966 but not before his writings influenced a whole generation of Jihadists. "It is an issue (Democracy) of *Jahiliyya* (the period of pre-Islamic Chaos)" according to Zawahiri's quotation. We will encounter a lot more about Qutb in the next section, which discusses how much support among Muslims the Global Jihad really has achieved.

Part Two of the treatise on Jihad Martyrdom and the Killing of innocents (probably written before 9/11) begins with the heading "The permissibility of bombarding infidels when Muslims and others who are not permitted to be killed are dispersed among them" and continues on in that vein. In this, Zawahiri tries to justify suicide terror as a legitimate weapon, despite the many injunctions in Islam that ban the taking of your own life. Here is the exegetic birth of a tactic first employed by Shia Muslims and now exonerated by al-Zawahiri.

Zawahiri begins by quoting the Muslim prophet's famous statement about how war is deceit. He must (and he knows this) justify not only the use of Martyrdom Operations (as suicide bombings are known among the Jihadists) but also the killing of Muslims, women, children, and *dhimmmis* (infidels that have sworn fealty and prostrated themselves to the Muslim faith if not converted), something much trickier than other justifications. Not surprisingly, using a stretch of even his peculiar logic, he manages to find stories and analogies that justify his way of thinking. This is obvious in the following passage:

> *This is killing of an individual whose individual status is unknown, because clarification of status is in regard to the one under our control only, and these are not under our control, so the obligatory defensive Jihad is not suspended in order to determine their status. And Shaykh al-Islam Ibn Taymiyyah (may Allah have mercy on him) dealt with this matter in detail in his fatwas regarding the Mongols in the 28th volume of Majmuu'a al-Fataawa, so refer to it.*

Bin Laden and Zawahiri are not the only theoreticians of the Jihad, but also the foremost on the Sunni side. However, on the Shia side, Sheik Hassan Nasrallah of the Hezbollah (Party of God) Lebanon is increasingly a force to be reckoned with and estimated by many to be much shrewder and actually winning the battle for the hearts and minds of even some Sunni Muslims over bin Laden and Zawahiri.

Neither of these men are particularly fond of Nasrallah: Hassan Nasrallah welcomes the international Crusader forces which occupied Lebanon and came between its people and the Jihad in Palestine, and Rafsanjani states that we don't aim to remove Israel, and Iran is a member of the United Nations with Israel, and the United Nations charter obligates all members to respect the unity and safety of the other members territories and sovereignty.[20]

The leader of the Hezbollah, born in Lebanon and educated in Iran, certainly knows how to compromise in order to maintain power and the finance so critical to his movement. It is estimated that he has received between $100 and $200 million dollars from Iran and other backers in the Arab and Persian worlds. However, it should always be remembered that the modern era of suicide terror began with Amal (later became Hezbollah) in Lebanon and the deadly suicide terror attacks against the U.S. Embassy and later the U.S. Marine Barracks.

For example, on the subject of 9/11 and speaking to a reporter from the Washington Post, Nasrallah—who has been called the most intelligent and therefore the most dangerous of the International Terrorists[21]—took a shot at his Sunni rivals: "What do the people who worked in those two [World Trade Center] towers, along with thousands of employees, women and men, have to do with war that is taking place in the Middle East? Or the war that Mr. George Bush may wage on people in the Islamic world. Therefore we condemned this act—and any similar act we condemn. I said nothing about the Pentagon, meaning we remain silent. We neither favored nor opposed that act. Well, of course, the method of Osama bin Laden, and the fashion of bin Laden, we do not endorse them. And many of the operations that they have carried out, we condemned them very clearly."

However, here in a speech in Arabic and not an interview going out specifically to the readers of the Washington Post, the self-described "freedom fighter" that dissimulates his respect for innocent life sounds very different. "Let the entire world hear me. Our hostility to the Great Satan (America) is absolute. I conclude my speech with the slogan that will continue to reverberate so that nobody will think that we have weakened. Regardless of how the world has changed after the 11th of September, Death to America will remain our reverberating and powerful Slogan" (Al-Manar September 27, 2002).

Here is his statement to the United Press International on November 4, 2001 just after 9/11: "It is our pride that the Great Satan (America) and the head of despotism, corruption and arrogance in modern times considers us an enemy that is listed in the terrorism list. I say to every member of Hezbollah be happy and proud that your party has been placed on the list of terrorist organizations as the US views it." In a rally in the Bekaa Valley in Lebanon he continues: "Matrydom operations (suicide bombings) *should be exported outside Palestine. I encourage Palestinians to take suicide bombings worldwide. Don't be shy about it.*" The message may be sometimes overlaid, but the end justifies the means and the end is the destruction of the United States of America, the Big Satan, and its ally, the little Satan, Israel, which means that he and his rivals have the same goals. Regarding the enemy of my enemy to be my friend, as the saying goes in Arabic, would be naïve in Nasrallah's case. The present-day theorists of the Global Jihad, both Sunnite and Shia, have the following approach to a Holy War with the West and have justified their actions by taking the following views:

The resurgence in Islam will come about when the Muslims return to the values and practices of the first generation of the religion's founding. This "pure Islam" makes clear certain precepts:

1. This purer and renewed state will be achieved by strict adherence to the Quran and Hadith as the Jihadis see them.

2. Muslims should follow the "purer" Jihadi interpretation of the Quran and Hadith and emphasize the Jihad—bringing it back to its former glory and actively restoring it from being a forgotten obligation.
3. Citing of original sources shall determine the nature of the conflict, and these determine that there is no quarter given to any infidel that does not subjugate himself to Islam.
4. This means that suicide terror is also justified against anyone, even Muslims or Dhimmis (subjugated infidels).
5. Any compromise with the infidel is tantamount to apostasy and warrants death.
6. There shall be no compromise in the establishment of Sharia law over the earth, and the rule of the Caliphate must be as it is described in the Koran.

SUPPORT FOR THE GLOBAL JIHAD

Is the Global Jihad the work of a small group of Guerilla fighters with similar fundamentalist views, or is there wider support for the Global Jihad throughout the Muslim world? This question is not a political one; it assumes renewed importance because the Global Jihad has "gone freelance."[22]

The view that the Jihad today is not the work of hierarchically denominated networks such as the al-Qaeda of September 11, 2001 is widely supported by the way the Jihad has gone to the Internet. Some have called this *al-Qaeda Social* opposed to *al-Qaeda Central* and see radicalization continuing in a leaderless way. The process is more of an example or idea according to this view than actual leadership. It can lead to radicalization on a scale never before known.[23]

A 2004 Pew Research survey revealed in 2004 that Osama bin Laden is viewed favorably by large percentages in Pakistan (65%), Jordan (55%), and Morocco (45%). In the same study, 31% said that suicide terror attacks against American and other Westerners in Iraq are justifiable.[24] A leading London Newspaper showed that 6% of British Muslims fully supported the July 2005 Bombings in the London Underground.[25]

Apart from the collective support, there is the phenomenon of the Lone Wolf who becomes incited by the Jihadi goals. Mohammed Reza Taheri-Azar, an Iranian-born American Citizen awaiting trial for nine counts of attempted murder in the March 3, 2006 University of North Carolina incident in which he ran over fellow students using a newly purchased SUV, is just one example of many such Lone Wolves in the United States. This has become known as "Sudden Jihad Syndrome." As usual, the University authorities were quick to point out that there was no link to terrorism, but he has quoted several verses from the Quran to justify himself using several of the verses quoted in the opening pages of this chapter.[26]

Support for the notion of Jihad is implicit in many texts that have influenced the Muslim world and that have even spread to become official versions of the

Islamic religion such as Wahhabism. Together, the influence exerted by these ideas is called by the radicals themselves the *Salifiyyah,* after the Arabic denomination of the Righteous Ancestors *al-Salaf al-Salih.* This Salafist inspiration has come from numerous sources and certain ideas which have to be understood to further give meaning to the Global Jihad and to the role that some States may play in this now or in the future, as well as its effect on individual actors, or small cells that are self-igniting and are not commanded by a centralized agency. These ideas and sects or groups upholding extreme versions have vast support in many Islamic communities worldwide.

SEMINAL IDEAS AND MOVEMENTS THAT HAVE LED TO THE GLOBAL JIHAD

Tauhid—The Oneness of God

Islam is a monotheistic religion and one that makes clear that there is only one God. Before the teachings of Muhammed Abd al-Wahhab, who was born in the Eastern part of what is now known as Saudi Arabia in 1703, there had been a rich tapestry of beliefs around Muhammed the Prophet's role, the followers and companions of the prophets, and the rituals and practices surrounding holy places for Sunnah Muslims. It was not clear before Wahhab that *Tauhid* was not an exclusionary idea as it became in Wahhabi thinking.[27] To understand Wahhabi thinking [it was once described by an Aramco (Arabian American Oil Company) analyst as a kind of Arabic Unitarianism; and the work of Abdul Wahhab was described as an "Islamic Reformation," the kind of critical error of interpretation that we are trying to avoid],[28] we need to understand the evolution of this interpretation, its alliance with the House of Saud, and its possible impact on future events.

Al Wahhab's main idea was that Islam had strayed and had allowed *Shirk* or the belief in agents of God or participants/partner to develop, thereby creating polytheists. This would make the believers and practitioners of Islam *Mushrakin*, or idolatrous. Most of the writings that he left are proscriptions (what may not be done). He spent most of his life moving around the Saudia Arabian peninsula inveighing against the demise of Islam and finally found himself expelled from al-Uyayna to Dir'irya, where he was able to conclude an alliance (including marriage between the families) with Muhammad b. Sa'ud, ruler of the city.[29]

In 1802, the predominant style of the new alliance between the future Kings of the Saudi Peninsula and Wahhab's beliefs were seen in raids carried out on the city of Karbala in Southern Iraq. The Saudi Chronicler, writing contemporaneously, described the effect of the results of war against fellow Muslims. "In a morning more than two thousand people were killed, and the Mosque where Imam Husayn had worshipped was sacked. When the booty was given to Sa'ud, he gave one share to every foot solider and a double share to the horseman before returning home." (Abdullah b. Bishr)

Hamid Alger writes "The corollary of identifying Muslims other than the Wahhabis as *Mushrakin* or idolatrous was that warfare against them became not simply permissible but obligatory."[29, p. 34]

Wahhabism is practiced today in Saudi Arabia, where most of the 9/11 hijackers originated from, and it has been the official religious orientation of the House of Saud—the Saudi Monarchy—since this eighteenth-century union.

TAKFIR

The idea of the apostate has had many different implications for the Global Jihad. Bin Laden, as we saw, was clear that anyone not taking part is an apostate—somewhat worse than a Kuffar (from the same root) or nonbeliever. The idea of labeling a Muslim who does not believe in the radical agenda (for example: does not believe in Jihad against innocents) a Kuffar has radicalized parts of the *Ummah or* Muslim community because there is considerable pressure, including excommunication if the Imam or spiritual leader is radical to impose the sanction—one that can mean death.

The idea has been used to characterize Anwar Sadat, who referred to himself and encouraged others to call him, the Pious President. However, having made peace with Israel at Camp David and, almost worse, refusing to enact Islamizing reform in Egypt (almost unbelievably the fact that his wife Jihan danced with Jimmy Carter at the White House was cited as another reason) meant that he was declared Takfir and therefore his death was a matter of time.[30]

JIHAD

The different interpretations for the word Jihad in both Arabic and in Islam are due to the fact that Jihad is much more than a word "struggle" but a concept, rife with the interpretational and evolutionary changes that any concepts are subject to and the additional complication.

In the Koran, the Concept of Jihad is a core concept. It manifests itself at several different levels; and these are very clear, unambiguous, and categorical. There is a personal Jihad that could be translated as "Striving" or an inner struggle to follow the path of God. Almost all religions have a component of this "Inner Struggle." This might be named the Greater Jihad, the idea of an individual's struggle to live a good Muslim life and adhere to the five Pillars of Islam: Shahada (professing the faith), praying regularly, fasting during Ramadan, being charitable, and performing Hajj, the pilgrimage to Mecca, and Jihad, is sometimes known as the sixth Pillar of Islam but does not have this status anywhere officially.[31]

However, this should not confuse the fact that a lesser Jihad is quite clearly denominated in the Koran. The world according to Islam is divided in dar al-Islam (the Land of Islam) and dar-al-harb (the Land of Conflict). The Umma (the Muslim community) must help to expand dar al-Islam so that the rest of the world can benefit from living within the just Islamic order. This is the origin of the Holy war meaning of Jihad.[32]

There are also careful distinctions made about Offensive and Defensive Jihads, not only the inner and the outer Jihad. Here is a passage in the Koran that describes the Defensive Jihad clearly if somewhat aggressively:

> "God does not love the transgressors. Kill them wherever you find them and drive them out [of the place] from which they drove you out and [remember] persecution is worse than carnage. But do not initiate war with them near the Holy Kabah unless they attack you there. But if they attack you, put them to the sword [without any hesitation]." Qurán 2:190–194 (8)

This forms the basis of the Fatwas (religious edicts) issued all over the Arab world when the Soviet Union attacked Afghanistan in 1979. Many people left Arab nations to go and fight the Soviet's in that Muslim country, arguably the most important factor in the modern Jihad of our times—to which we will return later in this chapter.

Then there is an aggressive form of Jihad which has been employed of late by al-Qaeda and many other terrorist organizations in the Muslim world. But long before Sheikh Abdallah Azzam (see page 49) called for Muslims to fight in Afghanistan, and his former student Osama bin Laden took it much further by calling for killing Crusaders and Jews whenever and wherever you find them (see 1998 Fatwa), a group formed in Egypt and made this aggressive Jihad, with the object of imposing Sharia (Koranic Law) and truly Islamic Societies on the World, one of its fundamental tenets. With tentacles around the world, this group has contributed more than any other to the rise of the Global Jihad.

Jahaliyyah

The Muslim Brotherhood and Sayyid Qutb

Jamiat al-Ikhwan al-Muslimun (Arabic for the Society of Muslim Brothers)

> Allah is our objective. The Prophet is our leader. Qur'an is our law. Jihad is our way. Dying in the way of Allah is our highest hope.—Muslim Brotherhood

The Muslim Brotherhood was established in 1928 by Hasan al-Bana, an elementary school teacher in Egypt. In essence the Brotherhood was a society that built on Salafist values about keeping to the older, stricter meanings in Islam and existed to bring Muslims together to achieve this goal. It became, in time, a kind of Islamic Masonic society crossed with very Radical tendencies that we can feel today.

The 1920s saw the end of the Ottoman Empire and a nadir in the fortunes of the Muslim world and marked the complete end of any Muslim empires. This led many in Islam to search for values that might bring the people out of their *Jahiliyya* or corruption and chaos (similar to the state that had existed before Mohammed's preachings). A search for some kind of values that would enable Muslims to rise to grandeur again and escape the Colonial domination that was becoming a reality throughout the Islamic world, and find a way to counter pernicious Western corruption, was seen as the answer. This urge for a purer form of Islam was something that had existed for some time, in particular with the writings of Mohamed ibn Abd-al Wahhab.

In this combustible mixture of a search for a stricter Islam (Salafist) and an austerer form of Islam (Wahhabists) came the influence of Sayyid Qutb of the Muslim Brotherhood. He was to be especially influential on some of the modern al-Qaeda leaders—Ayman al-Zawahiri and Ali Amin Ali al Rashidi, who were Egyptian followers of Qutb, and certainly (by extension) Abdullah Azzam and Osama Bin Laden. Azzam, a Palestinian, was Osama Bin Laden's mentor and heavily influenced by Qutb. His radical writings came in the 1950s and were the beginnings of what we know today as modern Radical Islam.

Understanding his ideas and their influence and how they have spread due to the Brotherhood is a key component of understanding the Global Jihad. Some of his treatises, written while he was in prison, such as "Milestones," sound eerily contemporary. Here are some of the main ideas of Qutb as expressed in "Milestones" written in 1964 and summarizing his thinking[33]:

- The world had reverted to the pre-Islamic Law phase of Jahiliyyah or primordial chaos and therefore even Islamic regimes were not truly Islamic. No regime was fit to live under for a true Muslim. This de-legitimization of Islamic regimes is a key to understanding the use of force. They may be attacked and destroyed and should be.
- The world had reverted to the pre-Islamic Law phase of Jahiliyyah or primordial chaos and therefore even Islamic regimes were not truly Islamic.
- No system of government, and especially not democracy, is suitable for Muslims. Muslims should resist any system where men are in "servitude to other men" as un-Islamic. A truly Islamic polity would not even have theocratic rulers. If a government is necessary, it should be a just dictatorship but certainly not democracy upon which he goes to great lengths to expose as non-Islamic.
- The way to end this primordial chaos was to preach the Quran's meaning everywhere and resort to physical power and Jihad to destroy any remnants of the corrupt order.
- Qutb saw Islam as a way of life, and not a religion. Everything derived from and was in the service of Islam. There was no question that this meant that ideal society would be governed by Sharia or Islamic Holy Law.
- A vanguard would be created that would spread throughout the Islamic homeland and then the entire world. A secret branch of the Brotherhood was created as a result.

- The struggle would not be easy. True believers could look forward to lives of poverty, difficulty, frustration, and sacrifice.

The idea that he would attack Muslims made him a heretic for many schools of Islamic thought, and he was officially declared a deviant after his death. His death at the hands of the Egyptian regime, accused of plotting to kill the Egyptian President, has also given him a martyr's status.

His brother Muhammed Qutb moved to Saudi Arabia and promoted his brother's religious teachings, while one of his students was Ayman al-Zawahiri. Zawahiri has acknowledged and published his views on the importance of Qutb.

The Muslim Brotherhood across the World. Qutb's ideas found forceful expression in the writings of Muhammed Abd al-Salam Faraj, who was the Cairo head of the Tanzim al Jihad (Jihad Organization)—the group that killed Egyptian President Anwar al-Sadat. In Faraj's work the idea of violent jihad replaces the DAWA (evangelizing) and he even goes further and argues that there is no greater and lesser Jihad because this does injustice to the passages about the Sword in the Koran. "Waging jihad against them (those that have illegally seized the leadership of Muslims) is an individual duty.[34]

By the time of the Afghanistan invasion (1979), the ground was ripe for an introduction of a vanguard. The Brotherhood spread to many Muslim and Western countries, thereby spreading these ideas. Most of the leaders emanated from the Egyptian branch of the Brotherhood and already had considerable anti-Government experience.

THE SPREAD OF THE JIHAD ACROSS THE WORLD

The Global Jihad, against any and all targets considered "Crusader or Jewish," began with the Near Enemy or the so-called "apostate" regimes that existed in the Arab world, regimes such as Nasser in Egypt, the Shah in Iran, and even the Hashemite King of Jordan. Indeed, one of Zawahiri's favorite statements was that the "road to Jerusalem goes through Cairo"—or, in other words, that first the leaders of the Arab world that were *takfir* had to be overthrown and the countries have *tajdid* (renewal) and *al-taghullah* (Islamic superiority) imposed on them. These ideas are prominent in Qutb's writing and the Muslim Brotherhood was intent on changing the Islamic regimes by Jihad but made no mention of a war against the West, except for many statements of the disgust with Western corruption. Neither did Faraj, Sadat's assassin, nor other founding members of the *Tanzim al-Jihad* or the *Jamaa al-Islamiyya* in Egypt nor others that advocated local change first.

How did the Jihad become a Global War fought in New York City, London, Madrid, Nairobi, The Philippines, and Northern Africa, just to mention a few afflicted places?

The development of Egyptian Jihadis, along with the dicision to fight against the Far Enemy (the United States), provides a good basis for understanding similar processes that occurred in other countries and areas such as Northern Africa, Somalia, Yemen, and Iraq. The changes in Iran that came about with the Khomeini revolution also played a critical part in taking the Jihad global. In addition, the Saudi Arabian role in exporting Jihad across the world should be taken into careful account.[35]

Case Study of Jihad Development: Egypt

Egypt, under the regime of Gamel Abdul Nasser, was supposed to be the beacon for Pan-Arabic nationalism and secularism and lasted from 1952 to 1970. Upon Nasser's death, Anwar Sadat assumed the Presidency of Egypt. Interestingly, the first conflict between an Egyptian leader and radical Islam occurred in 1954, when the Muslim Brotherhood tried to assassinate Nasser. His real downfall came when Egypt, Jordan, Syria, Iraq, and other Arab States were crushed during the 6-day war with Israel (1967).

One of the first to take up Qutb's call was Shukri Mustafa, who created a group named al-takfir wa'l Hijra (excommunication and exile). In 1978, this early fundamentalist group was prosecuted and Mustafa was killed by the Sadat Regime. Another group in Egypt, led by a Palestinian named Salih Sirriya, illustrates how cross-fertilization of these disparate groups took place. Sirriya was a Palestinian who arrived in Egypt after the failed attempt against King Hussein of Jordan during Black September 1970. In Egypt, he formed a group named the Islamic Liberation Organization and tried to kill Sadat in 1974. He met the same fate as Mustafa.[36]

In these waters swam people like Sheikh Omar Abdul Rahman (the blind Sheikh who was arrested after the First World Trade Center Bombing and planning the multi attacks that were thwarted on New York Cities subways, bridges, and tunnels) and the ever-present Dr. Ayman al-Zawahiri. These two were part of the Jihad Organization (Tanzim al-Jihad).[37]

The spark that created firm contacts between these different but essentially local Egyptian groups and the wider world jihad was provided by the Soviet invasion of Afghanistan in 1979. In a momentous "founding year" for Global Jihad, the Khomeini revolution also took place in Iran and had its effects on other groups such as Amal and eventually Hezbollah in Lebanon.

A Palestinian named Abdullah Azzam, later to be a teacher of Osama bin Laden and his mentor, was instrumental in whipping up the frenzy for participation in the Afghanistan War, seen as a clear case of the invasion of a Godless power into an Islamic country. As Azzam put it: "Before us lie Palestine, Bukhara, Lebanon, Chad, Eritrea, Somalia and the Philippines, Burma, South Yemen, Tashkent and Andalusia "(Southern Spain—once part of the Islamic conquests).

In Egypt, the pivotal event was the assassination of Anwar Sadat in 1981. The resulting crackdown on the Islamist groups resulted in painful, torturous prison sentences for what would become the vanguard of the jihad in later years. Zawahiri, upon his release, went to Afghanistan where he and many others learned their tradecraft—a deadly tradecraft that would be employed against the West during the latter half of the 1990s as the focus from local to Global change, fed by the experience in Afghanistan.

In Afghanistan, associations between future Jihadis were cemented. Mohammed Abu Sittah, better known as Abu Hafs al-Masri (later killed by U.S. Air Strikes in Afghanistan) and Mohammed Mustafa (later to be known as Ali Mohammed, who somehow found his way into the U.S. Army at Fort Bragg), joined forces with other members of the Egyptian Jihad. The connections with bin Laden were also formed in Afghanistan to the extent that in later years a follower would complain that the Egyptians were too favored.

Likewise, on the other side of the Sunni-Shia divide there were other unions being formed. The Khomeini revolution wanted support from the Arab states, but the war with Iraq made that difficult. According to *Asharq Al-Awsat*, the Arabic/English news daily, the Amal (Lebanon) Teheran axis was formed against the background. The supporters of Iran's revolution were Syria, Libya, and the Palestinian Liberation Organization (PLO). Amal, a small Shia Lebanese faction, had just lost their leader, who disappeared on a trip to Libya. This was further complicated by the fact that the PLO had been battling the AMAL or Shia militia in Lebanon, in its attempt to colonize Lebanon for its own purposes. At first, Iranian support for this group was very lukewarm because the Iranians did not want to offend Libya or the PLO.

Later, as the War with Iraq wound down, the support from the Khomeini regime for a little known band that had joined Amal named Hezbollah increased. The Lebanese battlefield had the same effect on Shia factions as Afghanistan had on Sunni Jihadists. It gave them experience. It was Hezbollah's suicide terror missions that were to inspire the entire Global Jihad.

The last factor in the dispersal of the Jihad to the entire world has to be the Saudi Arabian support for (at the best case) Dawa, or evangelizing the Wahhabi sect throughout the world, and the cash and resources that Saudia Arabia has at its disposal to do that. The Saudi regime has, in the best case, been actively involved in supporting groups across the world that are in line with the Wahhabi positions and has, in the worse case, been heavily involved in funding terrorist activity, including bin Laden, and certainly terrorism in Israel. According to British Intelligence, the Kenya and Tanzania bombings that killed more than 240 people and wounded 4500 had two al-Qaeda operatives that drove the suicide vehicle in Nairobi; they were both Saudi as was the planner, Khaled al-Midhar. The rubber dinghy that claimed the lives of 17 American sailors on the USS *Cole* came from Saudi Arabia and U.S. Intelligence apparently concluded that the whole operation had been funded by a Saudi family. Six of the 15 Saudis involved in 9/11 had been through their process of recruiting while they were in Saudi Arabia. Very few, if any, of the 9/11 attackers were veterans of Afghanistan, and much of the preparation took place in Saudi Arabia itself. These are heavy indicators that the prevalent thinking in Wahhabism, detailed in this chapter, plays a role in radicalizing and spreading the Jihad world-wide. Two months before the World Trade Center attack, an email speaking about "a Big Meal that would be impossible for any but the faithful to bear" was circulated.[38]

These same trends, the coalescing of disparate forces in various battles, backed by Arab wealth, can unfortunately also be seen in the Global Jihad in the United States.

The Jihad in the United States

Brooke Goldstein, director of the Legal Project, stated these words at the Middle East Forum in a speech called "Welcome to Lawfare, a New Kind of Jihad":

> The Islamist movement has two wings—one violent and one lawful—which can operate apart but often reinforce each other. While the violent arm attempts to silence speech by burning cars when cartoons of Mohammed are published in Denmark, the lawful arm is skillfully maneuvering within Western legal systems, both here and abroad.

Just before Memorial Day Weekend, in late May of 2008, Solomon Bradman, the CEO of Security Solutions International, found out what "*Lawfare*" means in practice. Security Solutions International has been training U.S. Homeland Security forces since 2004 and trains in everything from terrorist activity on the Internet, to understanding Radical Islam and even tactical training. Bradman said:

> We were not expecting anything out of the ordinary for our training in Seattle just proceeding Memorial Day. Our aim is to train U.S. Homeland Security to protect all Americans, Christians, Jews, Muslims, and any others in our free country, against the possibility of terror attacks—to detect, prevent, and if necessary respond in a professional way with no political agenda.

Bradman claims that the first inkling that there were problems with a training program named "The Threat of Radical Islam to the World" given by a Muslim member of U.S. Law Enforcement was when a local newspaper called him and asked him why his company was associating terrorism with Islam. "I don't believe I can take the credit for linking Radical Islam with terrorism," said Bradman, "the Jihadist did that a long time ago." This was the call to arms for a concerted campaign by a group named CAIR, the Council for American Islamic Relations, to have the SSI training programs banned from the city of Seattle.

Bradman had become the victim of Lawfare. *The formula is simple: Use the laws, freedoms, and loopholes of the most liberal nation on earth to help finance and direct the most violent international terrorism groups in the* world.[39]

According to Goldstein, "there has been a steady increase in Islamist Lawfare and the litany of American researchers, authors, activists, publishers, congressman, newspapers, television stations, think tanks, NGO's, reporters, student journals, and others targeted for censorship is long. ..."

Some of the incidents she mentions are as follows: (a) American Online was sued after permitting an online chat room in which participants discussed Islam). (b) U.S. Congressman Cass Ballenger was sued for describing CAIR as a fund-raising arm of Hezbollah after he had reported the group to the FBI and CIA as such. (c) Andrew Whitehead, an American activist and a blogger was sued for maintaining the website anti-CAIR-net.org; but ironically, after CAIR refused Whitehead's discovery requests, seemingly afraid of what internal documents the legal process (that CAIR had initiated!) would reveal, CAIR withdrew its claims. Goldstein points out more: *Boston Herald*, Fox 25 News, *New York Times*, Police counterterror trainers, and many more have all been victims of legal actions designed to silence their speech on radical Islam. Goldstein points out: "Most of this litigation is predatory, filed without a serious expectation of winning, and undertaken as a means to intimidate, demoralize, and bankrupt defendants."

The Global Jihad in the United States has been characterized by (a) a heavy reliance on fund raising since its earliest days (Abdullah Azzam and Ayman al-Zawahiri conducted fund-raising tours here in the early 1980s), (b) incitement aimed at radicalizing the faithful, and (c) a combination of lawful and illegal activities, usually very well disguised and sometimes on the borders of infiltration but always under the banner of political correctness and U.S. Constitutional rights.

David Yerushalmi, the legal counsel to the Institute for Advanced Strategic & Political Studies and the Center for Security Policy, both Washington, D.C.-based policy think tanks, works hard to uncover "Sharia Compliant" financing, which he views as a Trojan Horse for the establishment of a worldwide Caliphate: "It doesn't matter if you look at Averroes or, much later, Ibn Tamiyaa, and indeed any Islamist text through the centuries, you find a clear-cut relation between Sharia and the obligation to subdue any nonbeliever through the payment of the Jiziya, and through Jihad." Yerushalmi points out that the Mufti Mohammed Taqi Usmani, the son of Pakistan's Chief Mufti, was the legal authority behind the Dow Jones licensed Shaaria Compliant Index. At the same time, he wrote a book entitled *Islam and Modernism*, where Chapter 11 is dedicated to clarifying the obligation of all Muslims to subdue the nonbeliever.

According to Yerushalmi, the Financial Sector in the United States has created what amounts to a "Black Box" through which the Jihadists will have access to capital and the inner workings of Western financial institutions. "They pretend that somehow you can have Shariah finance laws without its laws of Jihad."

If the Courts and Financial sector provide comfortable areas for Jihadists to function, what about the law as it concerns actual acts of terrorism?

Not surprisingly, incidents tend to be broken down into before and after 9/11. Although there have been many incidents of terrorism since 9/11, the classification of incidents seems arbitrary at best. Here are some examples of incidents classified as having no terrorist-related evidence.

1. Mohammed Taheri-azar, despite his claims that he wanted "to punish the Government of the US and to avenge the deaths of Muslims around the

world," is claimed by University Authorities to be deranged when he runs his car into nine students on March 3, 2006 at the University of North Carolina.

2. Naveed Afzal Haq, a Muslim American who said he "hated Israel," forces his way through the door of a Jewish Center while holding a gun at a 13-year-old girl's head and kills one and wounds five. This was declared a hate crime and was dismissed as having nothing to do with terrorism.
3. Suleiman Talovic, an 18-year-old Bosnian Muslim, starts shooting in a Salt Lake City Mall and kills 5. Bruce Tefft, a former CIA counterterrorism official who advises the New York City Police Department, told WND he was "flabbergasted" by the FBI's statement that it saw no possible connection to terrorism (Art Moore ©2008WorldNetDaily.com).
4. In October 2005, Joel Hinrichs, a 21-year-old student, blew himself up outside the University of Oklahoma's football stadium where 84,000 were watching a game. Police insisted it was merely a suicide, but investigators found "Islamic Jihad" material in his apartment, and he reportedly attended a nearby mosque—the same one attended by Zacharias Moussaoui, the only person charged in connection with the September 11, 2001 terrorist attacks.

All four incidents show evidence of radicalization but no definitive (except in the case of Tahari Azar) evidence in the form of statements of responsibility or videotaped testimonials such as is common in cases in Israel and elsewhere.

Success against the Global Jihad in the United States and in the War on the Jihad rests on how the Law deals with the amorphous nature of the Global Jihad here and internationally.

TOWARD A LEGAL DEFINITION OF TERRORISM IN THE UNITED STATES

Since 9/11, it has been difficult to prosecute incidents of terror in the United States and many prosecutors have returned cases sent to them for prosecution.[40] We have seen above how easy it is for quasi-terrorist organizations to function at the same time within our system and actually use it for their ends.

For international terrorism the declination by prosecutors (prosecutors may decline cases as a way of filtering what they view as viable prosecutions) rate has been high, especially in recent years. In fact, data show that in the first eight months of FY 2006 the assistant U.S. Attorneys rejected slightly more than nine out of ten of the referrals. Given the assumption that the investigation of international terrorism must be the single most important target area for the FBI and other agencies, the turn-down rate is hard to understand.

The typical sentences recently imposed on individuals considered to be international terrorists are not impressive. For all those convicted as a result of cases initiated in the two years after 9//11, for example, the median sentence—half got more and half got less—was 28 days. For those referrals that came in more recently—through May 31, 2006—the median sentence was 20 days.[41]

Is there a definition of terrorism, terrorist organizations, and terrorist incidents that will allow our legal system to cope with both the overt and insidious Jihad by organizations that may be thinly veiled fronts for the Global Jihad?

"It is my position that an act of terrorism can be committed by an individual 'lone wolf.' The deciding factor is not how many people it takes to carry out a suicide bombing (one), but who is targeted (civilians) and what appears to be the motivation of the perpetrator (to influence or coerce a government's policies or to instill fear in a population). To point out what an actor's motivational factors are (whether it be religion, jihad, Shaaria law, etc.) is useful in light of the fact that we are facing a version of terrorism steeped in Islamism—the ideology that Islam is a political, legal, as well as religious authority," states Brooke Goldstein, the Director of the Middle East Forum's Legal Project.

Goldstein's definition largely matches the discussion on the International Definition of Terrorism made in the introductory chapter of this book, where 109 different definitions are mentioned.

Goldstein points out that: "There are several definitions of terrorism. My personal definition of an act of terrorism is 'An intentional, unlawful, violent act targeting civilian persons or property and which appears to be for the purpose of (i) intimidating or coercing a population by instilling fear or (ii) influencing the policy of a government.' "

"I do not include the term 'politically motivated' because it is vague and ambiguous and can lead to the omission of acts of terrorism from prosecution as such. For example, if a suicide bomber is motivated by the notion that Allah will reward him with 72 virgins, a prosecutor would then have the challenge to argue this was a politically motivated act and would in the process be required to delve into the psychology of the suicide bomber."

According to Goldstein, "The central focus in determining an act of terrorism should be whether civilians are intentionally targeted, and what appears to be the purpose of the act. What distinguishes an act of terrorism from that of plain murder is whether it has the effect of, or appears to be calculated to, coerce or intimidate a population or government. The motive of the terrorist may be political or religious or both. Using these elements, it is then appropriate to judge an act of terrorism with regard to the greater context in which it was perpetrated; and in the process, one should be able to consider other similar acts of terror by like-minded individuals with like-goals, motivations and religious convictions, etc. This is far more practical in terms of what definition is more likely to result in a just conviction rather than having the determination hinge on post-facto speculation about the political motivations of what may be a very psychologically disturbed individual."

Jerry Goldman is one of the attorneys representing the landmark litigation by the family of John O'Neil and victims of 9/11 in a Class Action suit against the Kingdom of Saudia Arabia. In addition, Goldman has represented Bruce Teft, a counterterrorism advisor to NYPD, now being prosecuted because a voluntary signee to his emails was upset about how these emails characterized Islam.

According to Goldman, "There is nothing wrong with the Law. It's all there but the ability to use the Law, to be able to characterize and tell the story of complicated relationships, and to be able to make the connections is difficult. The Courts are not familiar yet with the nuances; there are challenges in properly conveying the information; there is a learning curve that we have to deal with."

"This is a similar situation to the one we found ourselves after the passage of the RICO act. It took time to be able to come to grips with it and make the cases."

Goldman cannot comment on the Class Action litigation, but he did underline how "very well financed" and how much investment is going into major league defense attorneys for the Kingdom of Saudi Arabia.

We are adapting our laws to deal with amorphous nature of terrorism, and thankfully legal activists are making the case for a sharper understanding of what constitutes a terrorist or a terrorist organization.

FINANCING AND ORGANIZATION

One of the earliest functions of the Global Jihad in the United States was that of organizing a fund raising operation to conduct activities elsewhere. This included everyone from Osama bin Laden to a group that deserves a prime place in the Global Jihad—the Palestinian Islamic Resistance Movement, better known as Hamas, today the elected Government authority of the Palestinian authority in the Gaza strip.

The group was founded by Sheik Ahmed Yassin in 1987. His intentions were always very clear: "Become human bombs, using belts and suitcases aimed at killing every enemy that walks on the earth and polllutes it."[42] Funding for Hamas comes from Saudi Arabia, the Gulf States, Syria, Iraq, and the Emirates; surprisingly, it is very much influenced by Iran—a country that pores money into Hamas. But at one time, the United States was also a prominent donor.

The Holy Land Foundation provides a glimpse into the network of relationships that form the Jihad in the United States. Founded in Texas in 1989, the "charity" collected donations while masquerading as a humanitarian and charitable organization, something it had in common with Sami-al-Arian's WISE (World and Islam Studies Enterprise) and the ICP (Islamic Concern Project) and much else in common besides (al-Arian himself had close ties to Palestinian Islamic Jihad—a different but still deadly group). All these groups or individuals were convicted of crimes related to terrorism.

Apart from both organizations collecting money for "matrydom operations" or suicide terrorists, there were family relationships. Along with Khalid Shaykh Muhammed, al-Arian had been at North Carolina with Mazen al-Najjar—a brother-in-law of al-Arian's, who was deported because of his ties to terrorism.[43] But let's get back to the Hamas side of Palestinian Jihad. Musa Mohammed Abu

Marzook took the example of his friend Abduraman Muhammed Alamoudi—another prominent figure in Washington that had terrorist ties. Marzook, coming to the States at the beginning of the 1980s learned the American way of playing the game. By 1992, the genial Marzook was spreading money around Washington—a lot of money.[44] His association was named the Islamic Association for Palestine and he donated $210,000 to the Holy Land Foundation. Israelis claim that the foundation was linked to al-Qaeda through England and Chicago. Involved in the Holy Land Foundation was one of the founding members of Texas CAIR, Ghassan Elashi, and he was also the chairman of the Holy Land foundation.

When Marzook was detained to look into these links with terrorism, his chief advocate (other than CAIR of course) was none other than Abdurahman Alamoudi. Alamoudi's role is even more entrenched in our system. This man became a goodwill ambassador to several countries and was even invited by the President to join him at a prayer service for the victims of 9/11. Alamoudi turned out to be connected to terrorism through Libya and had ties to al-Qaeda. He also helped appoint the U.S. chaplains to the Muslim population in U.S. prisons, including the holding facility at Guantanamo, Cuba.

This enterprise for raising money is something very much alive and well in the United States today. Whether it is for al-Qaeda, the Palestinian Islamic Jihad, or Hamas, this type of activity goes on in jurisdictions all over the United States. In January 2008, Kenneth Wainstain, the Assistant Attorney General for National Security, said in a statement that the "indictment (of former Congressman Mark D. Siljander) paints a troubling picture of an American charity organization that engaged in transactions for the benefit of terrorists and conspired with a former United States Congressman."

The allegations against Siljander are part of a 42-count indictment handed up by a federal grand jury in Kansas City, which has been conducting an investigation of the charity. It closed in October 2004 when it was added to the designated terrorist list according to the *Washington Post* (WP, January 17, 2008).

The operational arm of the first phase of the Jihad in the United States, (the one that was carefully created during the 1980s when none other than Abdullah Azzam and Zawahiri came to fund raise in the fertile soil of the United States) was best exemplified by the first World Trade Center bombing in 1993. Here decades of activity in the United States, the war in Afghanistan, and a local cell all conspired to bring about not only the incident but planning for a great deal more.

Going back to Egypt, we encounter the name of Ali Muhammed, who is highly educated (two bachelor's degrees and a master's degree), speaks several languages fluently, and was taking part in a foreign officers course at Fort Bragg in 1981 when Egyptian President Sadat was assassinated. Some time after this, as he comes under suspicion in Egypt, he is forced to quit the Egyptian army. He was well-connected to Egyptian Islamic Jihad under Zawahiri and is said to have worked with bin Laden's personal security in Yemen and even earlier in Afghanistan.

He had students in Jihad here too. One of them turned up many years later as the assassin of the problematic Jewish right-wing Rabbi Meir Kahana. El-Sayyid Nosair had some 47 boxes of evidence in his apartment that was not immediately translated but was eventually found to contain plans for the 1993 bombings. Another student of Ali Muhammad's was Mahmud Abouhalima, convicted with Ramzi Yousef, the nephew of Shaikh Khalid Muhammad of the 1993 World Trade Center bombing.

In late 1989, Ali Mohammed left the Army and moved to Santa Clara, California. He shared an apartment with another Egyptian, Khaled Abu al-Dahab. They were extremely busy, and they not only recruited others but also hosted al-Zawahiri in Silicon Valley. Today, Santa Clara has the legacy of these roots.

In 1993, all this came to fruition when a NYC terror cell, which had been led by a blind Egyptian radical cleric named Omar Abdel Rahman, was responsible for the attempt to destroy the WTC. Omar Abdel Rahman was from Egypt and was a member of Zawahiri's Egyptian Islamic Jihad and at one time the presiding cleric named to pass the fatwas needed to kill and maim in the cause of bringing down the Egyptian and other governments.

These were busy years in New York City for the Jihad. Rahman and his cell had planned much more: There was the 1993 attack on five CIA employees; they planned the Day of Terror in New York City—the arrests of Omar Abdul Rachman and his cell saved the United Nations, the Lincoln and Holland Tunnels, and the George Washington Bridge.

But there were other incidents such as in 1994 when an Arab member of the Rahman mosque in Brooklyn opened fire on 15 observant Jews on the Brooklyn Bridge or in 1997 when a Palestinian school teacher opened fire on the observation deck of the Empire State Building.

The FBI had actually photographed Mohammed with his trainees during a 1989 firearms training course.

Look carefully at the origins and the connections here. Not only had Mohammed known the blind cleric in Egypt, but Rahman had been his spiritual mentor and four soldiers from Mohammed's Army unit had assassinated Egyptian President Anwar Sadat. Rahman was charged along with Zawahiri and later acquitted. It is almost a direct line of descent from Qutb to the Islamic Jihad and finally to Brooklyn, New York.

Eventually Ali Muhammed told the courts, when he was charged in the Nairobi and and Tanzania Embassy bombings, that he was an agent of al-Qaeda and Egyptian Islamic Jihad as part of his plea agreement. He admitted to helping plan the embassy bombings. He admitted training al-Qaeda members in the United States and abroad in the arts of terrorism. He admitted coaching al-Qaeda recruits on how to build effective cells. He admitted training bin Laden's personal guards, and he admitted helping Zawahiri visit the United States to raise funds. Despite this, and with much speculation about his cooperation with the CIA, incredibly Ali Mohammed has never been sentenced, nor is it sure that he is in the custody of the U.S. Government despite the Guilty Plea. He has completely disappeared without a trace since 2001.[45]

The strange case of Ali Mohammed underlines the ability of the Jihad to function well within the landscape of U.S. laws and rights and raise money, conduct terror operations, and, in the case of 9/11, get by the best efforts of U.S. intelligence.

The Global Jihad has come to rely on organizations that utilize our own laws protecting freedom of speech, expression, and association; or, as previously quoted, the Jihad is now practicing Lawfare.

As you can see, the ideas, organizations and sometimes the people that have shaped the Jihad Globally are the same that have propelled terrorist actions in the United States both before and after 9/11.

As a result, the biggest growth has been in some of the Muslim organizations dedicated *prima facie* to promoting cultural, economic, and even individual rights issues or warding off hate and discrimination but at heart comprise a very potent arm of the Jihad in the United States because they undermine the very legal foundation of our ability to counter them. More importantly, they are actually active in promoting an agenda that prevents proper training of U.S. forces.

Some, like the Muslim Arab Youth Association (MAYA), have already been declared terrorist associations, but others swim in the waters of freedom with little constraint.

CAIR is said to have a key role in the what has been deemed the Wahhabi lobby—the network of organizations, usually supported by donations from Saudi Arabia, whose aim is to propagate the especially extreme version of Islam practiced in Saudi Arabia, and indeed the Saudi Embassy paid for CAIR's impressive Washington Headquarters.[46]

Unfortunately, they have had many individuals tied up in the network that can best be described as the Global Jihad: those like Ghassan Elashi, a founding board member of CAIR-Texas; Randall (Ismail) Royer, once a communications specialist for the national group; and Bassam Khafagi, the organization's one-time director of community relations. Ties to the previously mentioned Hamas, Holy Land Foundation, and many individuals mentioned in this section are very evident.

This network of organizations makes its way from our courts and law enforcement (such as CAIR does) to the Jails, thereby promoting Dawa but sometimes going much further in incitement such as the Ashland (Southern Oregon) branch of the Dawa organization headed by Pete Seda and Soliman al-But'he. Both men fled the country before they could be tried for terrorist activities.

THE JIHAD: VERSION 2.0

Like the Internet 2.0, we now have Jihad 2.0, which places Terrorism squarely in the twenty-first century.

In 2008, The United States mounted an initiative to reveal how terrorist groups use the Internet. Code named Reynard, this initiative came to light when a recent report by the Office of the Director of National Intelligence was sent to

Congress. In the report, the initiative is described as a "seedling effort to study the emerging phenomenon of social (particularly terrorist) dynamics in virtual worlds and large-scale online games and their implications for the intelligence community."[47]

The global Jihad, however, is beyond seedling efforts when it comes to using the Internet to promote, organize, and execute its mission. Despite the paradox of a fundamentalist Jihad, rooted in the seventh century, efficiently using the tools of a secular twenty-first century, they are doing this with ever greater success daily.

In the 1990s, the Israelis made a national defense effort to go after the head bomb maker of the Hamas, the "engineer" Ihyea Ayash, and finally caught up with him when he answered a call on a cell phone that Israeli intelligence had someone he trusted give him. His head exploded as he answered the call.[48]

Today, taking out the "engineer" or bomb maker would be a fruitless enterprise. There is "Hanbali" and the encylcopedias of "Bajadin"-rich troves of Internet materials and everything from tutorials on bomb making to material acquisitions, as well as analysis of failed tactics and how to remedy this. Whether this is from the legendary Bali bomber that is now thought to be imprisoned in Jordan or an Internet handle, we have a whole online enclyclopedia of bomb-making techniques.

We have mentioned the use of footage taken by incited individuals and uploaded from Dorms in Georgia to gangs in Canada and England. However, there is also a sophisticated organization to the Internet Jihad.

Al-Qaeda has a multi-tiered online media organization.[49] This is something that has evolved from the 1990s when the first sites were opened for incitement. al-Zawahiri knows that we are in a "battlefield of the media." There are regional production centers, As-Sahab media, affiliated with the High Command of al-Qaeda, a Media committee and Sawt-al-jihad—for the Arabian Peninsula. Once content is created by one of the above, it is sent to a "clearing house" such as al-Fajr media center and then on to preapproved sites. The clearing house serves the purpose of authenticity by providing the media logos and banners, and it also serves to confirm the origin of the message. On the constantly shifting forums (see Appendix 1, list of forums) this message will then be transmitted across the Internet. There are thousands of violent Islamist websites, and bombings in Algeria serve to illustrate how virtual becomes real.

Today, terorrists do not even need surveillance video. The proliferation of webcams pointed at many valuable targets, made available by our airport authorities and other helpful institutions, means that an operator anywhere in Pakistan can instruct a comrade in Brooklyn to park his car outside a terminal at JFK and watch from Pakistan if and how authorities react. The tested methodology is just a short step from carrying out an attack using this technique.

The Jihad's Virtual 007

Twenty-two-year-old Irhabi007—a play on the word terrorist and 007—was not trained in Jihadist camps in Afghanistan, nor was he one of the Mujahadeen that learned his trade in the war against the Soviet Union in Afghanistan. But Irahbi007, a son of a Morocccan tourist official and a student of information technology, arguably did more to enable the global Jihad than even al-Qaeda leader Ayman Al Zawahiri.

As a student in Great Britain, he started a frightening and lethal development in the War on Terror: He used the Internet to organize and recruit for the Jihad.

In doing so, he unleashed a dangerous movement that today seems to be replacing al-Qaeda as "The Base" and relegates al-Qaeda to a strictly inspirational role. The new Jihadi networks allow for financing, planning, and executing devastating attacks with relative ease.

Today, the most rudimentary Jihadists can exchange recipes for nitroglycol and thermite, as well as finance, recruit, and learn how to conduct effective Jihad against the West. Tutorials in 20 installments are downloadable and are envied by educational designers in the United States as proper examples of distance learning. Even more disturbing, terrorists recently began distributing flight simulation software—making flight school obsolete. These virtual schools are disseminated with encrypytion software provided on Jihadi forums.

"For example, a school teacher was found on one of the forums asking for help in carrying out a mission," says Gadi Aviran, a former EOD officer in the IDF, a veteran of Israel's Military Intelligence Directorate (AMAN), and today the founder and President of Terrogence, the world's leading actionable intelligence company.

Aviran says the man learned that the U.S. President was visiting his school in the Palestine authority during a recent visit to the Middle East.

"On the forum, the teacher asked what he could do, given that he would be in close proximity to the President," he said. "We take seriously the level of sophistication and what is being done on the Web in the name of the Jihad."

Is the West keeping up with Jihadis?

The Western world, by most accounts, is not keeping up with the Jihadis. Instead, much attention is given to sites like alekhlaas. Though it seems an ominous site due to the extreme nature of member chats, it's not where the serious players post. In addition, poor translations, and even poorer analysis, are giving rise to potentially deadly situations.

For example, the word for "*missile*" and "*rocket*" are the same in Arabic. There are grave consequences, however, of mistaking the intent to use a rocket when a missile is the intended weapon. A mistake such as this was recently made by a U.S. intelligence agency based on a faulty translation provided by a private intelligence company.

As with anything in Internet time, extremism has advanced with lightning speed since 2005.

The antics of "Maximus," a Swedish teenager of Bosnian extraction named Mirsad Bektasevic and inspired by Irhabi007, who along with three others plotted attacks in Europe in 2005, now seem like child's play.

When police recovered 19 kg of explosives, along with video recordings claiming affiliation to al-Queda in Northern Europe, they also discovered

Bektasevic's contact with Irhabi007. The arrests in Sweden led to the arrest of Irhabi007 and a sentence for 10 years after pleading guilty to charges of incitement to murder and conspiracy to commit murder. The gang was also running credit card and other fraudulent businesses to finance other Jihadi sites.

"Today it will be more difficult to track (the terrorists) down," says Aviran. "They are giving each other good guidelines on how to cloak their identity on the Web and how to clean up after a session. It's now a case of drawing and then connecting the dots. Doing this, we can stay ahead of our adversary in this conflict."

Real-Time Threat Example

A real-time example of terrorists' use of the Internet starts with an activist from Algeria who asked online about raw ingredients needed for manufacturing an acetone–peroxide explosive. The activist wanted the resulting explosive to be used as part of a body-worn IED.

In this specific case, the Maghreb activist wanted to know if TATP (triacetone, tri-peroxide explosive) could be easily used in a suicide belt; he also wanted to know specific ingredients, as well as commercial names the raw materials are sold under.

In less than three hours the activist received textual instructions providing him with all necessary information and a link directing him to tutorial videos with "step-by step" visual instructions for TATP "kitchen lab" manufacturing.

The understanding that these support networks may replace the more easily detectable overseas communications with global Jihad organizations, including *al-Qaeda*, is critical for the unveiling of local self-radicalizing cells. It is by tracing this newer type of communication between seemingly anonymous surfers that these may be identified in time.

UNDERSTANDING TECHNO-INTELLIGENCE SIGNATURES

"It is essential that we learn to understand the traces that all activity on the Web leaves," says Aviran. "We need to associate with the entire terrorist life cycle."

Aviran gives an example of a recent incident in Florida: "Two men were fishing in the Everglades to the west of Miami with saltwater rods. A state law enforcement agent, whose job it is to enforce anti-poaching laws in the Everglades, detained the two men. Both were from countries of interest and the officer requested that they show him their car. There were no alligators in the trunk but on the back seat were maps of Miami International Airport, including arrival and departure times of airlines. The Everglades, of course, forms part of the pattern flown by flights into and out of Miami. An RPG could probably bring down an airplane there easily. The officer completely missed the signs, released the men, and wrote up a report. Fortunately, someone in the agency recognized the signs and followed-up."

One of the key factors is to discern the difference between an intelligence threat and a theoretical threat. Aviran's company recently discovered on Jihadi sites a documentary on Paris sewer system. The documentary had been in heavy distribution.

"The film was seemingly benign," he says. "In the hands of a group threatening to create havoc in the French capitol, however, the information on how exactly the location of the sewage system affects the city and where it passes could be devastating if it ended up in the wrong hands. A TV show is a theoretical threat, but when it shows up in forums it becomes an intelligence threat."

Aviran suggests agencies look for the specific signatures that terrorist lifecycle activity leaves. Some signs of that activity include:

- Internet browser history
- Programs found on the suspect's machine
- Membership in forums
- Evidence of videos about potential targets
- Evidence of financial dealings and unusual credit card activity
- Manufacture and access to components and materials for creating IEDs

Aviran also advises agencies to make careful collection lists before approaching the complex problem of getting into closed Jihadi forums. Terrogence specializes in conducting collection projects for clients like the U.S. Department of Defense and European and Israeli intelligence agencies.

Internet Activity and Terrorist Finance—Synergy in Cyberspace

In one of the worst terrorist attacks, more than 200 people died in a 2003 bombing of a nightclub in Bali, Indonesia. They were the victims of Imam Sumadra, a militant Islamist with strong links to al-Qaeda.

As with other terror masterminds who have contributed to the Internet information network, Samudra wrote a jailhouse memoir instructing followers on online credit card fraud to finance terror operations.

The nexus between crime, insurance fraud, credit card fraud, and contrabanding of luxury goods to cigarettes has an inescapable online presence.

What could be done to slow this repository of information on everything from bomb-making to crimes that could finance terror and much more?

Self-regulation needs to be imposed on hosting sites. Many sites, however, are genuinely unaware of the problem. Recently in Florida, an ISP was hosting a forum on which Jihadis could download their own brand of encryption software. Notified by law enforcement, the site closed down the forum. There are also many barriers to controlling the Internet. As a society, we have yet to make unifying decisions about Internet issues, including spam and online criminal activity and fraud protections. A consensus on what and who constitutes a threat to citizens in an Internet democracy may be far away.

THE GLOBAL JIHAD HAS A NEW HOME BASE—THE INTERNET

The expert use of the Internet is evidenced by propaganda issued by groups that espouse the Jihadi ideology. A recent propaganda film (*Lee's Life of Lies*) released by such groups in Iraq uses a real U.S. soldier and gives details about his life (based on stolen documents), to construct a very impressive hour-long, high-value production movie that purports to show U.S. soldiers criticizing the war in Iraq. When we show these documentaries at SSI training sessions, most viewers do not believe them to be fake.

Our challenge is to keep up with the Global Jihad. Failure to do so will have tragic consequences for the War on Terror. We need a concerted national effort—similar to the one made during the Cold War—to train pro-Western Arabic speakers to help decipher terrorist cyber codes. Without that, we are constantly reacting to the advances of the Jihadi's.

Michael Doran, a Near East scholar and terrorism expert at Princeton University, says this: "When we say al-Qaeda is a global ideology, this is where it exists—on the Internet."

CONCLUSIONS

The Global Jihad which began in the hearts and minds of fanatics intent on reestablishing a "golden period" of Islamic hegemony, at the dawn of the Ottoman Caliphate, based on a return to the original values and ideas found in the Quran, has resulted in some of the most violent terror organizations—al-Qaeda, Hamas, Hezbollah, and many others—and may now be spreading to many individuals inspired by the methods and ideologies of these groups.

These groups, aided and sheltered by States that support them, along with a global network of finance and legal organizations within Democratic countries that are sometimes woefully naïve about the Takfiri ideology and how it views other Muslims, let alone the inhabitants of Western Europe or the United States that are not Muslim, are also thriving.

Against this, there are woefully inadequate resources and ever-increasing bureaucratization of the effort. For example, the FBI is woefully short of qualified translators to keep up with the amount of material heading their way from local Law Enforcement and of course, on the Internet.[50] According to agents, the Bureau is "inexcusably understaffed." Meanwhile, the Department of Homeland Security, which might be expected to take on the challenge, has already become ossified and incapable of handling this, giving the task of Intel on Open Source to private contracting companies.

This is despite the fact that experts agree that one of al-Qaeda's main aims is to create havoc either my demonstrating or actually carrying out an attack using weapons of mass destruction.[40] The Director of the U.S. Department of Energy's Office of Intelligence addressed a special forum recently and was very clear about the danger. He points out that al-Qaeda is not the only threat. There are multiple, home-grown possibilities for this type of attack, now that 9/11 has entered the consciousness of terrorists, be they U.S. fringe groups, Jihadi-inspired small independent cells, or even disturbed individuals—which may have been the case with the 2001 Anthrax attacks.

School violence and recent incidents at Virginia Tech have also focused attention on the al-Qaeda perfect-day scenario of multiple attacks in the United States.

In addition to all these threats lies the increasing possibility of the lone wolf we have discussed, which is specifically cited in the Canadian ITAC CIEM bulletin of June 2007 where they point out what we know: The Internet has become an important catalyst for incitement of lone wolves who are not communicating with members of a cell and therefore are almost impossible to prevent.

In order for the United States to defend itself against what may become a lethal combination of lone wolves, Hezbollah sleeper cells, al-Qaeda organization cells, and inspired fringe groups, a consistent national policy similar to that employed throughout the Cold War has to be employed against the Jihad. This means training Law Enforcement by giving them real, hands-on expert training rather than federally mandated, peer-group-reviewed scholarly programs that merely spend federal, state, and local resources without hitting the mark. Cadres of experts in analysis and Arabic, Turkish, Pashtun, Farsi, and other languages need to be trained and employed as they were during the Cold War. Intelligence, in which major organizational moves have been made such as the fusion centers, needs to have the teeth to be able to intercept attacks as they are being planned.

Much progress has been made. But also much still needs to be done. One soldier in Iraq put it this way: The Jetsons are not keeping up with the Flintstones. We are going to have to do so if we want to defeat the Global Jihad.

APPENDIX 1: INTERNET SITES AND THE GLOBAL JIHAD

http://www.m3ark.com/forum/ A partially closed forum supportive of Global Jihad. A vital participant in the distribution of propaganda and weaponry/explosives manufacturing material.

http://hanein.info/vb A forum supportive of Global Jihad and of regional insurgent groups. Prominent distributer of Jihadi films, including self-produced multimedia.

http://www.muslm.net/vb/ A forum supportive of Global Jihad and of regional insurgent groups. This is one of the leading platforms for the distribution of Jihadi multimedia material.

http://www.paldf.net/forum/ A forum supportive of the Palestinian resistance. This site is a prominent distributer of Jihadi multimedia and propaganda produced by Palestinian insurgent groups.

http://www.military.ir/ A general forum in the Farsi language. Members of the forum discuss military and political subjects concerning Iran.

http://www.cecenya.net/ A website in the Turkish language supportive of the Chechen resistance. This site is a distributer of Jihadi multimedia and propaganda concerning Chechnya.

http://www.velfecr.com/ A website in the Turkish language supportive of the Palestinian resistance and of the Lebanese insurgent group Hezbollah. This site publishes news flashes, articles, and Jihadi multimedia.

APPENDIX 2: INCIDENTS IN THE UNITED STATES SINCE 9/11

The common perception is that there have been no incidents of terror since 9/11. In reality, many have been thwarted. This list does not purport to be all-inclusive because many acts are reported as hate crimes and therefore have not been counted in this list.

1. September 18–November 2001: Anthrax attacks. Letters tainted with anthrax kill five across the United States, with politicians and media officials as the apparent targets. The case remains unsolved.
2. 2001: The Center for Urban Horticulture at the University of Washington burned. Replacement building cost $7,000,000. Earth Liberation Front suspected.
3. December 22, 2001: Richard Reid attempted to bomb an American Airlines aircraft inbound from Great Britain. Reid is a professed al-Qaeda member.
4. May 8, 2002: José Padilla accused by John Ashcroft of plotting to attack the United States with a dirty bomb, declared as an enemy combatant. Recently, Padilla was convicted of lesser charges.
5. 2002 Buffalo, New York: Lackawanna Six, The FBI arrested Sahim Alwan, Yahya Goba, Yasein Taher, Faysal Galab, Shafal Mosed, and Mukhtar al-Bakri. Five of the six had been born and raised in Lackawanna, New York. The six American citizens of Yemeni descent were arrested for conspiring with terrorist groups. They had stated that they were going to Pakistan to attend a religious training camp but instead attended an al-Qaeda "Jihadist" camp. All six pled guilty in 2003 to providing support to al-Qaeda.
6. May 2002: Mailbox Pipe Bomber. Lucas John Helder rigged pipe bombs in private mailboxes to explode when the boxes were opened. He injured six people in Nebraska, Colorado, Texas, Illinois, and Iowa. His motivation was to garner media attention so that he could spread a message denouncing government control over daily lives and the illegality of marijuana as well as promoting astral projection.
7. July 4, 2002: Hesham Mohamed Hadayet, a 41-year-old Egyptian national, kills two Israelis and wounds four others at the El Al ticket counter at Los Angeles International Airport. The FBI concluded that

this was terrorism, although they found no evidence linking Hadayet to any terrorist group.

8. October 2002: Beltway Sniper Attacks. During three weeks in October 2002, John Allen Muhammad and Lee Boyd Malvo killed 10 people and critically injured 3 others in Washington, DC, Baltimore, and Virginia. An earlier spree by the pair had resulted in 3 deaths in Louisiana, Alabama, Georgia, California, Arizona, and Texas to bring the total to 16 deaths. No motivation was given at the trail, but evidence presented showed an affinity to the cause of the Islamic Jihad.
9. 2001–2004: The Virginia Jihad. In Alexandria, Virginia, 13 men were arrested for weapons counts and for violating the Neutrality Act, which prohibits U.S. citizens and residents from attacking countries with which the United States is at peace. Of these 13 men, four pled guilty. The other nine members of the group were indicted on additional charges of conspiring to support terrorist organizations. They were found to have connections with al-Qaeda, the Taliban, and Lashkar-e-Taiba, a terrorist organization that targets the Indian government. The FBI stated that the Virginia men had used paintball games as a form of training and preparation for battle. The group had also acquired surveillance and night vision equipment and wireless video cameras. The spiritual leader of the group, Ali al-Timimi, was found guilty of soliciting individuals to assault the United States and was sentenced to life in prison. Ali Asad Chandia received 15 years for supporting Laskar-e-Taiba but maintains his innocence. Randall Royer, Ibrahim al-Hamdi, Yong Ki Kwon, Khwaja Mahmoud Hasan, Muhammed Aatique, Ahmed Abu Ali, and Donald Surratt all pled guilty and were sentenced to prison terms. Masoud Khan, Seifullah Chapman, and Hammad Abdur-Raheem were found guilty at trial. Caliph Basha Abdur-Raheem was acquitted at trial in 2004.
10. April 24, 2003: William Krar is charged for his part in the Tyler poison gas plot, a white-supremacist-related plan. A sodium cyanide bomb was seized with at least 100 other bombs, bomb components, machine guns, and 500,000 rounds of ammunition. He faces up to 10 years in prison.
11. May 1, 2003: Ayman Faris pleas guilty to providing material support to al-Qaeda and plotting to bring down the Brooklyn Bridge by cutting through cables with blowtorches. He had been working as a double for the FBI since March, but in October he was sentenced to 20 years in prison.
12. August 2004: A terrorist cell under the leadership of Dhiren Barot was arrested for plotting to attack the New York Stock Exchange and other financial institutions in New York, Washington, and Newark, New Jersey, and later accused of planning attacks in England. The plots included a "memorable black day of terror" with the employment of a "dirty bomb." A July 2004 police raid on Barot's house in Pakistan discovered a number of incriminating documents in files on a laptop

computer that included instructions for building car bombs. Dhiren Barot pled guilty and was convicted in the United Kingdom for conspiracy to commit mass murder and sentenced to 40 years.

13. 2004, New York: James Elshafay and Shahawar Matin Siraj were arrested for plotting to bomb a subway station near Madison Square Garden in New York City before the Republican National Convention. The New York City Police Department's Intelligence Division helped to conduct an investigation leading to the arrests. An undercover agent infiltrated the group, provided information to authorities, and later testified against Elshafay and Siraj. Elshafay, a U.S. citizen, pled guilty and received a lighter sentence for testifying against his co-conspirator. He received five years. Shawhawar Matin Siraj was sentenced to 30 years in prison.
14. June 2005, Lodi, California: Umer Hayat and Hamid Hayat, a Pakistani immigrant and his American son, were arrested after lying to the FBI about the son's attendance at an Islamic terrorist training camp in Pakistan. The son, Hamid Hayat, was found guilty of supporting terrorism and was sentenced to 24 years. Umer Hayat's trial ended in a mistrial. He later pled guilty to lying to a Customs agent in his attempt to carry $28,000 into Pakistan.
15. October 1, 2005: Joel Hinrichs attempted to enter a University of Oklahoma football game where 84,000 spectators were in attendance. After refusing to let his backpack be searched, Hinrichs walked away and sat on a nearby bench. Hinrichs then detonated the TATP device that was in his backpack, killing only himself. It is unknown if Hinrichs detonated the device intentionally or by accident.
16. December 2005: Michael C. Reynolds was arrested by the FBI and charged with being involved in a plot to blow up a Wyoming natural gas refinery; the Transcontinental Pipeline, a natural-gas pipeline stretching from the Gulf Coast to New York and New Jersey; and a New Jersey Standard Oil refinery. He was arrested when he tried to pick up the $40,000 owed to him for planning the attack. His contact, Shannen Rossmiller, was a Montana judge who was working with the FBI. The FBI found explosives in a locker in Reynolds' home town, Wilkes-Barre, Pennsylvania. Reynolds stated that he was working as a private citizen to find terrorists. Reynolds was convicted of providing material support to terrorists, soliciting a crime of violence, unlawful distribution of explosives, and unlawful possession of a hand grenade and was sentenced to 30 years in prison.
17. August 31, 2005: Kevin James and three others was indicted on charges to wage war against the U.S. government through terrorism in California.
18. February 21, 2006: The Toledo terror plot where three men were accused of conspiring to wage a "holy war" against the United States, supply help to the terrorist in Iraq, and threatening to kill the U.S. president.

19. March 2006: Taheri Azar, an Iranian Psychology student, runs over and injures nine students in North Carolina, claiming he wanted to punish the U.S. Government.
20. June 23, 2006: The Miami bomb plot to attack the Sears Tower where seven men were arrested after an FBI agent infiltrated a group while posing as an al-Qaeda member. No weapons or other materials were found.
21. July 7, 2006: Three suspects arrested in Lebanon for plotting to blow up a Hudson River tunnel and flood the New York financial district. The "plot" was talk, and it was entirely unfeasible because it would require enormous amounts of explosives, and the target was above the water level anyhow.
23. July 2006: Naveed Afzal Haq storms a Jewish center and shoots, killing one and wounding five others.
24. 2006, New Jersey: Conducting online surveillance of chat rooms, the FBI discovered a plot to attack underground transit links with New Jersey. Eight suspects, including Assem Hammoud, an al-Qaeda loyalist living in Lebanon, were arrested for plotting to bomb the New York City train tunnels. Hammoud was a self-proclaimed operative for al-Qaeda and admitted to the plot. He is currently in custody in Lebanon, and his case is pending. Two other suspects are in custody in other locations, and investigators continue to hunt down the other five suspects.
25. April 2006: Syed Haris Ahmed and Ehsanul Islam Sadequee from Atlanta, Georgia, were accused of conspiracy, having discussed terrorist targets with alleged terrorist organizations. They met with Islamic extremists and received training and instruction in how to perform pre-operative surveillance of potential targets in the Washington area. They videotaped places such as the U.S. Capitol and the World Bank headquarters as potential targets and sent the videos to a London extremist group. They were indicted for providing material support to terrorist organizations and have pled not guilty.
26. November 29, 2006: Demetrius Van Crocker, a white supremacist from rural Tennessee, was sentenced to 30 years in prison for attempting to acquire *Sarin* nerve gas and C-4 explosives that he planned to use to destroy government buildings.
27. December 8, 2006: Derrick Shareef, 22, a Muslim convert who talked about his desire to wage Jihad against civilians, was charged in a plot to set off four hand grenades in garbage cans December 22 at the CherryVale Mall in Rockford, Illinois.
28. March 5, 2007: A Riker's Island inmate offered to pay an undercover police officer, posing as a hit man, to behead New York City police commissioner Raymond Kelly and bomb police headquarters in retaliation for the controversial police shooting of Sean Bell. The suspect wanted the bombing to be considered a terrorist act.

29. 2007, Goose Creek, South Carolina: This incident involved the arrest and indictment of two engineering students from the University of South Florida on charges of possessing explosives. Ahmed Adba Sherf Mohamed and Yousef Samir Megahed were stopped for speeding near Goose Creek, South Carolina. Officers found suspicious materials in their car, including what prosecutors allege were pipe bombs, as well as ammunition and remote delivery system. The two were stopped near the Naval Weapons Station Charleston, site of the brig where convicted U.S. terrorist Jose Padilla was detained and interrogated. Mohamed is from Kuwait and Megahed is from Egypt. Both are in the United States legally, Mohamed on student visa and Megahed as a permanent resident alien. According the FBI, this is still an ongoing joint state–federal investigation that is examining the possibility that Mohamed and Megahed's presence near the Naval Weapons Station may be linked to Islamist terrorism.
30. May 1, 2007: Five members of a self-styled Birmingham, Alabama area anti-immigration militia were arrested for planning a machine gun attack on Mexicans.
31. May 7, 2007: Fort Dix attack plot. Six men inspired by Jihadist videos were arrested in a failed homegrown terrorism plot to kill soldiers. Plot unravels when Circuit City clerk becomes suspicious of the DVDs the men had created; he reports it to authorities, who place an informant in the group.
32. June 3, 2007: John F. Kennedy International Airport terror plot. Four men were indicted in plot to blow up jet-fuel supply tanks at JFK Airport and a 40-mile connecting pipeline. One suspect is a U.S. citizen and one, Abdul Kadir, is a former member of parliament in Guyana. "Anytime you hit Kennedy, it is the most hurtful thing to the United States. To hit John F. Kennedy, wow. ... They love JFK—he's like the man." The unraveled when a person from law enforcement was recruited.
33. October 26, 2007: A pair of improvised explosive devices were thrown at the Mexican Consulate in New York City. The fake grenades were filled with black powder and detonated by fuses, causing very minor damage. Police were investigating the connection between this and a similar attack against the British Consulate in New York in 2005.
34. March 3, 2008: Four multimillion-dollar show homes in Woodinville, Washington are torched. The Earth Liberation Front is suspected in the fires.
35. March 6, 2008: A military recruiting station in Times Square was bombed in the early hours, resulting in minor damage to the facility. It is believed that a simply constructed device was used. Surveillance video that recorded the explosion also captured a bicycle-riding man who, prior to the blast, stopped and walked up to the facility. Police are investigating a possible link between this attack and recent attacks on the Mexican and British consulates in New York.

36. 2008, Falls Church Virginia: Abu Ali, 27, grew up in the Washington suburb of *Falls Church* and was valedictorian of a private Islamic high school. He joined al-Qaeda after traveling to Saudi Arabia to attend college in 2002. As a member of a Medina-based al-Qaeda cell, Abu Ali discussed numerous potential terrorist attacks, including a plan to assassinate Bush and a plan to establish a sleeper cell in the United States. He is in prison in the United States, awaiting an appeal on his sentence of 30 years. Appeal judge says the sentence of life in prison was deviated from by a lower court judge.

ENDNOTES

1. United States Senate Committee on Homeland Security and Governmental Affairs, Joseph Liberman Chairman, *Violent Islamist Extremism, The Internet, and the Homegrown Terrorist Threat*. U.S. Senate p. 13, May 8, 2008.
2. Terrorgence, *Homeland Security Bulletin of Open Source Threats* **1–2**, January and May 2008, unpublished, available on a subscription basis only.
3. House of Commons Report, *Report of the Official Account of the Bombings in London on the 7th July 2005*, Annex B, Radicalization in Context, Clause 3, May 11, 2006, p. 31.
4. Steven Emerson, *American Jihad: The Terrorists Living Among Us*. Columbus, OH: Free Press, 2003, p. 110.
5. Harvey Kushner, *Holy War on the Home Front*. New York: Sentinel Penquin, 2004, pp. 1–7.
6. Robert A. Pape, *The Strategic Logic of Suicide Terrorism*. New York: Random House, 2005. Incredibly, Pape has contended that suicide terror attacks have nothing to do with Islamic Fundamentalism but, even in the case of al-Qaeda, are a "land issue." In addition to contradicting everything the Islamists say themselves, this remarkable extrapolation comes from his "database of acts" which must now (2005) be hopelessly out of date. However, it also begs the question of how he can assert that 95% of suicide terror attacks are part of coherent land campaigns, unless he was able to submit a questionnaire to the bombers postmortem. This is a very difficult theory to sustain.
7. Steve Emerson, *American Jihad*. Published by *ICP*, one of al-Arian Journals. Columbus, OH: The Free Press, 2002, p. 114.
8. Internet reference, 2008, http://www.secondworldwar.co.uk/casualty.html.
9. Testimony of Maajid Nawaz, former U.S. member to U.S. Senate Homeland Security and Governmental Affairs Committee, *Congressional Quarterly*, July 2008. In his testimony, Nawaz outlined four main characteristics of Islamism. At the top of the list is that adherents consider Islam to be a political ideology rather than a religion, he said. Additionally, they believe that Sharia, or Islamic religious law, must become state law, that there is a global Islamist community that crosses national borders and has no allegiance except to itself, and that the Islamist community needs a physical bloc, or caliphate. "This state will be expansionist because it must represent that global community," Nawaz said. Furthermore, he said, there are different types of Islamists, ranging from the political, which seek to infiltrate governments through

the ballot box and other institutions, to the revolutionary, which try to infiltrate militaries to stage coup d'états, to militant groups. Nawaz and other experts on Islamism who appeared at the hearing said that a combination of social factors lead to Muslims subscribing to Islamist ideology, a point upon which committee members agreed. www.cqpolitics.com.

10. Testimony of Gary Anthony Ackerman, 2002, United States Senate Committee on Homeland Security and Governmental Affairs hearing on Nuclear Terrorism: Assessing the threat to the Homeland., Quoted by Gary Anthony Ackerman of START, after MEMRI translation, p. 8, http://hsgac.senate.gov/public/_files/040208Ackerman3.pdf.
11. Mark Gabriel, *Islam and Terrorism*. Lake Mary, FL: Charisma House, 2002, p. 30.
12. Raymond Ibrahim, *The Al Qaeda Reader*. New York: Broadway Books, 2007, pp. 18–21.
13. Raymond Ibrahim, *The Al Qaeda Reader*. New York: Broadway Books, Taqiyya, p. 64. "According to this doctrine, Muslims may under certain circumstances openly deceive infidels by feigning friendship or goodwill—even apostasy—provided their hearts remain true to Islam."
14. Hamid Alger, *Wahhabism: A Critical Essay*. Online: Islamic Publications International, 2002, pp. 8–10.
15. Terrorgence, May 2008, Al-Sahab. I am indebted to Terrogence for translating this Internet open meeting on al-Sahab, where al-Zawahiri answers questions on everything from killing innocents, to his own motivations, and even about the al-Qaeda relationship with "Iran of the Magian Rejectionists." In the questions, it is clear that there is frank criticism of al-Qaeda, including asking Zawahiri some tough questions: Did you refuse to pray behind Abdullah Azzam? Why are you killing Muslims? Where does your authority come from? Are you just a show-off? Published on the Internet in April and translated in June 2008, Internet Forum al-Sahab.
16. Raymond Ibrahim, *The Al Qaeda Reader*. New York: Broadway Books, 2007, p. 79.
17. Raymond Ibrahim, *The Al Qaeda Reader*. New York: Broadway Books, p. 123.
18. Raymond Ibrahim, *The Al Qaeda Reader*. New York: Broadway Books, p. 125.
19. Raymond Ibrahim, *The Al Qaeda Reader*. New York: Broadway Books, p. 125, 133.
20. Ayman al Zawahiri, Al Sahab Question and Answer, Internet reference, May 2008.
21. Robin Wright, Inside the Mind of Hezbollah, Internet citing. The entire article reveals only one side of Nasrallah's thinking but is still useful. *Washington Post*, July 2006.
22. *The Economist* online, February 2008. How the Jihad went freelance, http://www.economist.com/books/displaystory.cfm?story_id=10601243
23. Mark Sageman, How the Jihad went Freelance Sageman quoted in *The Economist*, February 2008.
24. Mistrust of America in Europe: Ever Higher Muslim Anger Persists: Survey reports, Pew Research, 2004.
25. Bin Laden More Popular with Nigerian Muslims than Bush, *Internet-Daily Times of Pakistan*, 2003.
26. Sara A. Carter, "As a result, law enforcement should not be too quick to judge their attacks as having no nexus to terrorism," *Washington Times*, January 2008.
27. Hamid Alger, *Wahhabism: A Critical Essay*. Teaneck, NJ: Islamic Publications International, 2002, p. 31.

28. Dore Gold, *Hatred's Kingdom: How Saudi Arabia Supports the New Global Terrorism*. Washington, DC: Regnery, p. 8.
29. Hamid Alger, *Wahhabism: A Critical Essay*. Teaneck, NJ: Islamic Publications International, 2002, pp. 18–19.
30. Fawaz A. Gerges, *The Far Enemy: Why the Jihad Went Global*. New York: Cambridge University Press, 2005, p. 46.
31. Ummat interviews Osama bin Laden, 2001, Interview denying his involvement in 9/11 on September 28, 2001, Jihad is the sixth undeclared pillar of Islam. [The first five being the basic holy words of Islam ("There is no god but God, and Muhammad is the messenger of God"), prayers, fasting (in Ramadan), pilgrimage to Mecca, and giving alms (zakat).] Every anti-Islamic person is afraid of Jihad. Khilafah.com.
32. Encyclopedia of Islam, 2008, Massive *Encyclopedia of Islam Online*. Internet citing, http://www.bible.ca/islam/dictionary/index.html
33. Sayyid Qutb, *Milestones*, 2008, Chapter 7: Islam is the real civilization http://www.youngmuslims.ca.
34. Fawaz Gerges, *The Far Enemy: Why the Jihad Went Global*. New York: Cambridge University Press, 2005, p. 44.
35. Fawaz A. Gerges, *The Far Enemy: Why the Jihad Went Global*. New York: Cambridge University Press, 2005. The reader is strongly recommend to read Gerges' excellent study, including interviews with terrorists and former terrorists that go far beyond rather simplistic accounts by U.S. pundits. In particular, it shows the range of opinions even within hierarchically ruled al-Qaeda of 9/11.
36. Marc Sageman, *Understanding Terror Networks*. Philadelphia: University of Pennsylvania Press, 2005, pp. 25–34.
37. Fawaz A. Gerges, *The Far Enemy: Why Jihad Went Global*. New York: Cambridge University Press, 2005, pp. 103–107.
38. Dore Gold, *Hatred's Kingdom: How Saudi Arabia Supports the New Global Terrrorism*. Washington, DC: Regnery, pp. 182–183. Gold's book traces Saudi Arabian involvement from the beginning of the Royal House of Saud's involvement with Wahhabism to the the role played today and is well worth reading to know what the United States is up against.
39. Steven Emerson, *American Jihad: The Terrorists Living Among Us*. Columbus, OH: Free Press, 2003, p. 111.
40. University of Syracuse, Criminal Terrorism Enforcement in the United States During the Five Years Since the 9/11/01 Attacks, 2006; sentence was much longer, 41 months. See Figure 3, http://trac.syr.edu/tracreports/terrorism/169/
41. Rolf Mowat-Larssen, *The Strategic Threat of Nuclear Terrorism*, Washington Institute for Near East Policy, June 2008, www.washingtoninstitute.org
42. Rachel Ehrenfeld, *Funding Evil*. Santa Monica, CA: Bonus Books, 2003, p. 101.
43. Harvey Kushner, 2004, *Holy War on the Home Front*. New York: Sentinel Penguin, 2004, pp. 1–7.
44. Steven Emerson, American Jihad: *The Terrorists Living Among Us*. Columbus, OH: Free Press, 2003, p. 22
45. Court Records, 2000, Court Reporters Office of the Southern District of NY, http://cryptome.org/qaeda102000.htm

46. Daniel Pipes, New York Post and Online, 2002. Pipes has written many exposés of CAIR's dealings, prosecutions, and networking ties. Daniel Pipes and Sharon Chadha, CAIR: Islamists Fooling the Establishment, *Middle East Quarterly*, Spring 2006, pp. 3–20, http://www.danielpipes.org/article/394
47. BBC News, March 2008, BBC Online Chris Vallance Reported online, http://news.bbc.co.uk/1/hi/technology/7274377.stm
48. Samuel M. Katz, *The Hunt for the Engineer*. Guildford, CT: The Lyons Press, 2002. Worth reading to understand the value of a good improvised explosives expert in the pre-Internet period.
49. United States Senate, Violent Islamist Extremism, the Internet and the Homegrown Terrorist threat, 2008, p. 5.
50. Jason Ryan and Jack Date, May 2008, ABC News, Agent: FBI ill equipped for Terror threats, www.abcnews.go.com

2

ISRAEL'S CONFRONTATION WITH SUICIDE TERRORISM

Amir Kulick

BACKGROUND

Arab–Israeli Conflict

The conflict between Zionism—the Jewish people's movement for self-determination—and the Arab world, particularly the Palestinians, has lasted for more than a century now. At the core of this bloody conflict is control of a territory called the Land of Israel/Palestine and the right to self-determination. The modern Zionist political movement, which sprang up during the last two decades of the nineteenth century, strived for the national renaissance of the Jewish people in its historic homeland after two thousand years of exile and persecution. Its roots lie, to a great degree, in the Jews' experience of failed attempts to assimilate with European society, in the rise and expansion of anti-Semitism and the advent of Nationalism during the Enlightment period.[1] Hence, since the 1880s, thousands of Jews immigrated to the land of Israel with the purpose of establishing an independent, Jewish national home. These waves of Jewish immigration strengthened a consistent Jewish presence that concentrated for thousands of years in the ancient cities at Jerusalem, Jaffe, Tiberius and Safed.

This chapter would not have been written without the knowledge, experience and insights shared with the author by all the distinguished interviewees. The author wishes to express his deep appreciation and sincere gratitude.

Suicide Terror: Understanding and Confronting the Threat. Edited by Falk and Morgenstern

This process of politically aspired immigration was met with Arab opposition that objected to Jews settling in that area. Consequently, the first major waves of Jewish arrivals were already subjected to increasing Arab terrorism and violence. These confrontations were initially localized and mainly characterized by disputes over land ownership or the use of water resources. Starting in the 1920s, more organized waves of anti-Jewish violence took form, targeting the fledgling Jewish towns and villages with pogroms and other violent attacks. Such was the case in the 1921 and 1929 pogroms, when dozens of Jews were murdered by Arab marauders.

Various attempts at a compromise failed again and again due to Arab intransigence; and in 1947, after the Arab leadership in British-controlled Palestine as well as the Arab states rejected the U.N.'s proposal for a two-state solution (Resolution 181), war broke out between the Jewish population and the Arabs of Palestine, supported by the Arab states.

In May 1948 the Jewish leadership in Palestine declared the establishment of an independent state and Israel formally came to be. Thus Israel was actually born in wartime and has since endured violent confrontations without respite. In addition to the 1948 War of Independence, Israel has gone through five other conventional wars with its Arab neighbors and in the Six-Day War of June 1967 it took over the territory of the West Bank (biblical Judea and Samaria) from Jordan and the Gaza Strip from Egypt. These will be herein referred to as "the territories." The line separating "the territories" from the territory of Israel proper is called the "Green Line." Today some three million Palestinian Arabs live in the territories.

In 1987, the Palestinians in the territories started a popular armed confrontation called "the First Intifada"—aimed at ending Israeli presence in, as well as occupation of, the territories. In an effort to arrive at a historic compromise, Israel signed the 1993 Oslo Accords with the Palestine Liberation Organization (PLO), whereby the State of Israel recognized the PLO as the legitimate representative of the Palestinians and the PLO recognized Israel's right to exist. Under this agreement the Israeli military (IDF—Israel Defense Forces) pulled its forces out of the main Palestinian towns and other parts of the territories and in 1994 the "Palestinian Authority" was created instead. This PLO-led entity was to become the nucleus of a future sovereign Palestinian state, to be established once final negotiations between the two sides were completed.

In reality, serious difficulties inflicted this political process from its early stages. The Palestinians were dissatisfied with the pace of negotiations and the extent of the concessions made by Israel. Conversely, several terrorist groups grew in scope within the Palestinian Authority (PA), while its formal security apparatuses, now effectively responsible, to a large degree, for Israel's security, refrained from acting against these terrorist elements. And so within a short time after the establishment of the PA, the first suicide bombing attacks came out of Palestinian towns to target the population centers of Israel. The Israelis viewed the PA's security forces as partners and therefore initially resorted to sharing intelligence information with the Palestinians rather than actively thwarting such attacks inside the territories. This state of affairs remained largely the same until the summer/fall of 2000. In July, the Camp David summit between prime-minister Ehud Barak and PLO chairman Yasser Arafat had failed. As a result, another violent confrontation between Israel and the Palestinians broke out "—the Second Intifada," which will be the focus of this chapter.

Scholars differ as to whether this confrontation was initiated by the Palestinian leadership for political reasons or whether it was the result of growing frustration of the Palestinian "street" with the existing situation. Regardless of the conclusions of this debate, the Palestinian leadership saw this as an opportunity to force certain political concessions on Israel through the use of violence, thus throwing its own security forces into the mix against Israel. Many of the PA's leaders became instigators and financial supporters of terrorism, with many of its security personnel joining the various terrorist organizations that began operating shortly after the Second Intifada broke out. As part of this process, the PA also released the Islamist terror activists that it had arrested since 1994 and thereinafter would not act against them again.

During the Second Intifada, thousands of terror attacks were launched against Israelis and Israeli targets in the territories and inside Israel proper. Within that campaign of violence, suicide bombings became the most lethal and significant tactical weapon employed by the Palestinian terrorist groups. These attacks reached their deadly apex in March 2002, therein known as "Black March."

This chapter concentrates on the Second Intifada (the "Confrontation") and analyzes how and why suicide terrorism became the most popular and lethal mode of Palestinian attacks and how Israel managed to curb the Palestinian campaign of suicide bombings.

Introduction

"BLACK MARCH". Disguised as a women, Abdel–Basset Odeh, a member of the Hamas Iz a Din al-Kassam Brigades, from the West Bank city of Tulkarem, walked into the dining room of the Park Hotel in Netanya and detonated himself with an explosive device in the midst of 250 civilians as they were celebrating the Passover Seder. Thirty people were killed and 143 injured in that March 27, 2002 suicide bombing, commonly referred to as the "Passover Massacre."[2] It marked the peak of an unprecedented wave of Palestinian suicide terrorism. In the wake of the attack, Israeli Prime Minister Ariel Sharon and his cabinet ordered the immediate recruitment of 20,000 reservists in an emergency call-up and the following day launched Operation Defensive Shield.[3]

Fifteen suicide bombings were carried out by Palestinian terrorists on civilian targets in Israel during "Black March"—on average a suicide bombing every

Figure 2.1. Map of Israel.

other day. The attacks resulted in the deaths of 136 Israelis and the injury of an additional 500.[4] In Israel, which has a Jewish population of 5.5 million people, this amounted to an enormous number of casualties If we compare this to the population of the United States (about: 304,500,000 people), we are talking about 7530 deaths per month—in other words, over two 9/11 attacks.

Time after time, the Israeli security apparatuses saw different terrorist organizations preparing suicide bombers, sometimes three or four simultaneously, for their launch to population centers in Israel, and time and again it failed to give an early warning and prevent the terrorist attack. The cost of error was immediate: dozens dead every time. In one incident a simultaneous suicide attack was planned by Na'af Abu-Shrakh from Nablus, at the old central bus station in Tel Aviv; the sounds and echoes of the blasts reached the outskirts of the city. The end was not in sight. Yitzchak Eitan, the commander of the IDF's Central Command (the military unit responsible for security in the West Bank) at the time, describes the times as follows: "There were days we were facing 15 or 18 suicide bombers on their way to Israel; Despite the many successful arrests it was enough for one to reach Tel Aviv in order to erase the effect of our great efforts."[5] One of the brigade commanders in the West Bank at the time defined that time period as "catastrophic."[6]

The effects of the suicide bombings on the civilian population in Israel were devastating. The major cities were the worst hit. Downtown Jerusalem became a virtual ghost town in the winter of 2002. Merchants abandoned their stores and camera crews swooped in on the few who dared to visit the outdoor markets. The restaurant owners, who suffered from a mass abandonment of customers, hired security guards. Many security guards were also positioned on buses.[7] It seemed that the bombers and their accomplices managed to spread collective fear and melancholy over many layers of Israeli society. Israelis avoided going out to the centers of entertainment and shopping malls, while civilian security guards were seen at every street corner. (The number of people employed in security grew by 30% following the bombings and stood at 46,500 people.) Those who could afford it avoided the use of public transportation. Funerals of victims were held daily in different locations of the country.[8]

On the national level, Israel had just begun to digest the financial damages of terrorism. The optimistic prediction of the minister of finance, Silvan Shalom, of a 4% rise in growth in 2002 was refuted (instead, a one percent drop in growth was recorded). The recession in Israel was affected by a lethal combination of terrorism and the world high-tech crisis and the dive of the American NASDAQ index, which forced Israeli technology companies to carry out a large wave of layoffs. Foreign investments in Israel dwindled, as had the flow of tourists (it virtually stopped). The entrances recorded in Ben Gurion International Airport were mostly of Israelis that were residing abroad. International corporations doing business in Israel moved their board meetings to Cyprus and London. The few who still dared to come to Israel were usually American officials. Even the Israeli soccer and basketball teams had to host their home games abroad. At the same time, in the areas controlled by the Palestinian Authority, Fatah and Hamas operatives were flooded by applications from people volunteering to be suicide bombers. "Every day 20 people would call me to volunteer," says Nasser Abu Hamid, the head of a Fatah terror branch in Ramallah. "When I walked the streets of my refugee camp, Al-Amari, children would pull me aside and say 'come on, give me a weapon.' " We didn't have to be brainwashed. Forget it."[9] To paraphrase Churchill: Never in Israel's history, has so much harm been inflicted on so many by so few.[10]

It wouldn't be an exaggeration to say that the Israeli security forces felt at times helpless while facing the wave of suicide bombers. Some of the senior officers of the IDF at the time, including the Chief of Staff, Moshe Yaalon, and commander of the Central Command, Yitzchak Eitan, claimed that the IDF was ready for a Confrontation[11] with the Palestinians.[12] When the Second Intifada (in Arabic: Uprising) erupted on September 29, 2000 the IDF managed to cope with the mass demonstrations and even the sniper shootings from the Palestinian Authority's territories. One of the best indicators of this is the large number of Palestinian causalities in the early stages of the Confrontation. For example, in the first week of the Confrontation (September 29–October 4), 74 Palestinians were killed, as opposed to a small number of Israelis.[13] Nevertheless, nothing prepared the IDF for the wave of suicide bombers that came a month later and reached its peak during "Black March." Furthermore, doubts rose within the army itself at the time regarding its ability to deal with the phenomenon. For example, the IDF's head of the Operational Branch at the time admitted in an interview that he was not certain that "it is possible to deal effectively with such amorphic terrorism [...]."[14] Other statements were even harsher and compared fighting terrorism and suicide bombers to trying to "empty the sea with a spoon." A common cliché at the time was: "There is no military solution for terrorism."[15] Nevertheless, within a few years the Israeli security system managed to reduce the suicide bombings within Israel to close to zero. From a peak of 60 bombings in 2002 to 6 in 2006 and one in 2007.[16]

Is the Israeli Experience Relevant for Americans? It is definitely an impressive success, but is the Israeli experience relevant for Americans? The answer to this key question is "Yes." A number of reasons make the Israeli experience in dealing with suicide terrorism relevant for America:

Suicide Terrorism Is a Worldwide Phenomenon. On the face of it, there is no link between global terrorism, headed by al-Qaeda and organizations of its ilk, and Palestinian terrorism in general and the suicide bombings in particular. Palestinian terrorism is limited and local, while global terrorism knows no borders. Palestinian terrorism, like that of the Chechens, Kashmir residents, and Kurds, has immediate political demands (the end of the occupation). These demands can ostensibly be resolved, in principle,[17] by negotiations between the opposing parties. Global terrorism, on the other hand, is driven by an ideological vision based on a historic long-term perspective and is aimed at reaching its goals in the distant future—Islamic world dominance, without immediate political demands. Indeed, these two forms of terrorism differ in their motives and demands, yet they use the same method to reach their goal: the indistinctive killing of noncombatant civilians.[18] In this context, both in local and global terrorism, the use of suicide bombings stands out. For local or "global" terrorists, suicide bombings are a method, a fighting tactic. In general it can be said that suicide terrorism is used all over the world as a weapon that aims to balance, to a certain degree, the asymmetry that exists in a conflict between a weak side (the terrorist organization) and the strong side (the government or country to which it is opposed).[19]

This is true regarding the Palestinian suicide terrorism, Hezbollah's suicide bombings in Lebanon, the Tamil Tigers of Sri Lanka, the Kurdish resistance in Turkey, numerous groups in Iraq, and the Chechen rebels in Russia. These organizations' goals aren't necessarily related; however, all over the world, the use of human bombs has proven to be a powerful tactic in the hands of terrorists. These seek to use suicide bombings as a means of forcing concessions from superior powers, to fight who they view as occupiers and even expel them, to thwart diplomatic–political processes, to gain international recognition in the media, both for their organization and for their goals, to disrupt daily life, and to spread fear and terror.[20]

The Palestinian suicide bombings in Israel may very well serve as a model for terrorist elements worldwide, and in America in particular. The fact that (as Steven Emerson has shown) an active terrorist infrastructure of fundamentalist organizations, including Hamas and Palestinian Islamic Jihad, already exists in the United States can only support this possibility.[21] Thus, it is no wonder that the fear of "Palestinian-style" terrorist bombings is shared by high-ranked officials in the American administration. For example, as former Vice President Dick Cheney said, there was "a real possibility" that Palestinian-style suicide bombers would carry out attacks within the United States. This assessment was shared by FBI Director, Robert S. Mueller III, who claimed that walk-in suicide bombings like the ones that terrorized Israelis are "inevitable" in the United States.[22]

Suicide Terrorism Poses the Same Challenge for All Security Forces. Suicide bombings pose several key challenges for any country, and these are relevant and true for the Israeli context as well as for the American one. Roughly, they can be divided into three fields:

1. *The Challenge of Deterrence.* Deterrence is the main pillar of both Israeli and U.S. security policy. At the basis of the deterrence concept is a simple principle: "If you hit me, I'll hit you harder". It is true both at the state and individual levels. However, how can you deter a bomber who is set out to die? Moreover, the success of his mission depends on him dying. What harm greater than death can be inflicted on such a person?[23] When the origin of the bombers is unknown or worse, when the bomber is a citizen of the country, who should you harm in an attempt to deter?
2. *The Challenge of Defense.* Suicide bombers set a complex challenge in this field, both on a personal (the guard) and national level. On a personal level, a soldier or other member of the security forces learns that he is allowed to open fire when he recognizes two elements: A means (weapon, bomb, etc.) and intent (an interpretation of reality). However, with suicide bombers the means are always concealed, and therefore are difficult to

recognize, not to mention the intent.[24] Furthermore, a large portion of the suicide bombers' dispatchers' efforts are dedicated to the concealment of both means and intent. On a national level, the most common means of defense is the setting of physical boundaries. These are very effective in dealing with regular armies, but their efficiency is decreased significantly when confronting suicide bombers.[25] Furthermore, in the age of globalization, which is characterized by the lifting of travel and communication limitations, terrorists can cross borders easily, receive instructions and preaching via cell phones and online tapes, move money electronically, smuggle weapons or the technical knowledge to operate them, and all this in an easier and more efficient manner in comparison to pervious years.[26]

3. *The Challenge of Resolution.* Traditionally, a war ends when the adversary surrenders. However, when confronting suicide terrorism the adversary is usually hidden within civilian populations. Furthermore, the enemy is usually a sub-state player. In some cases, this element has no internal hierarchy, but a structure of scattered networks. In this reality, it is difficult to bomb the enemy's headquarters, destroy its weapons of war, and by this to force it to forego its wishes. The United States occupied Afghanistan and Iraq, yet bin Laden and the heads of the Taliban are still free. Therefore, a question presents itself: Can a resolution of a conflict with a suicide organization ever be achieved?[27]

From these three basic challenges stem different dilemmas and problems in different fields: In the political field (How do you explain to the public that the struggle with suicide terrorism will take a long time, and a fast resolution cannot be reached?), the field of intelligence for operations (How do you locate a suicide bomber working in a civilian environment, how do you deal with a target with short life span, etc.?), in the technological field (How do you convert technology designed for the battlefield so it would be relevant for fighting suicide bombers?), the perceptional field (a war taking place on the home front, against civilians), the legal field, and many more.[28]

Israel Is a Democratic Nation that Managed to Confront Suicide Attacks without Losing Its Liberal Identity. Another reason that makes the Israeli experience relevant for Americans is the liberal-democratic character of both countries. When democratic regimes face suicide terrorism and terrorism in general, similar dilemmas of values, law, and government arise. At the core of these dilemmas stands the state's wish to effectively fight terrorism without losing its liberal-democratic character and hurt its democratic core values: human rights, honoring the rights of the minority, avoiding hurting innocent people, and so on.[29] Israel has been portrayed by the various international media channels as a brutal state on a number of occasions. Israel is often depicted as using unrestrained and nonproportional force against the Palestinian population. In such a country, one can argue, the security forces have no limits in their war against suicide terrorism, and therefore a democratic country like the United States, where human rights are revered above all else, has nothing to learn from Israeli methods.

Nothing can be farther from the truth, reality, of course, is completely different. Israel is a liberal-democratic country, with an open society including: an influential public opinion, channels that the public can use to influence the decision-making processes, limitations on penalties, limits on the activities of the security forces, competitive and independent media, a democratically elected parliament, and one of the most respected judicial systems on earth that serves as checks and balances. One of the topics most exposed to public and legal criticism in Israel is the Israel Security Agency's (ISA) interrogation methods (the ISA, also known as the General Security Service, the Shabak or Shin Beit, is the internal security mechanism in Israel responsible for confronting homeland terrorism). In the late 1980s a national committee of inquiry (the Landoy committee) that defined strict limitations on the use of force in the course of interrogations was established. The report defined what means were acceptable and on whom they can be applied ("only on a person with information that can lead to saving human lives").[30] Following the report, the legal advisor position at the ISA was designated as a head of a department (equivalent in status to the rank of general in the military) and a member of the ISA's headquarters. As the former legal adviser of the ISA testified: "I was a partner to most of the major processes the ISA went through in the last 12 years, while following many issues as they were happening and not in retrospect."[31] These limitations were often criticized internally in the ISA,[32] but remained standing and even got a seal of approval in a decision given by the Supreme Court in September 1999.[33] In the verdict, six "physical" interrogation techniques that the ISA used were banned from use, including the "shaking" of a detainee, sleep deprivation, and forcing a prisoner to remain in a frog squat position.[34]

The judiciary criticism was aimed not only at the interrogation methods, but also at different counterterrorism methods—specifically, targeted killing. In a verdict on this issue, the Supreme Court defined the area in which using targeted killing was legitimate and set guidelines of using them.[35] In the military, similarly to the ISA, the Military Advocate General became a member of the highest decision-making forum (the General Staff forum). These days, one can find legal advisors even in the division headquarters. As one of the military analysts of one of the influential newspapers in Israel noted: "During the years, in the IDF, an attitude developed [...] sometimes to the extreme, of lawyers meddling in operational issues."[36] The IDF's Military Advocate General confirms this trend and announced that today the army consults with its lawyers much more, "including on issues that in the past no one would ever let a lawyer have access to."[37]

Alongside the judiciary system, the Israeli public opinion is also used as a means to put pressure not only on politicians, but also on senior military officers. For example, former Chief of Staff, Moshe Yaalon, notes that the army is accountable to its citizens on its methods of operation: "I have to take into account a soldier's parents, who are civilians, and I must have their understanding that our actions are correct, that the How of our actions is correct, so there are many moral criteria I have to adhere to." Yaalon emphasized that due to public pressure he decided to put a stop to the demolition of houses on the Israeli-Egyptian border (an action meant to stop Hamas' weapons smuggling into the Gaza strip). "If there is no public support I won't do it" said Yaalon, "I'll win the battle and lose the war."[38] Nongovernmental organizations operating in Israel, such as "The Public Committee against Torture," "Betselem," "Law," and others also add a significant dimension that limits the actions of the security forces and exposes them to public scrutiny. And yet, Israel has managed to confront suicide terrorism and mitigated the level of such attacks. All these lead to a conclusion that "Israel proved that it is possible to cope effectively with terrorism while at the same time preserving the liberal democratic nature of the country."[39]

Israel Is a Model of Success. Finally, there are two things that are hard to argue with: numbers and success. Israel has been confronting terrorism in its different forms for more than 60 years—from the moment of its establishment and even before. Israel has been confronting suicide attacks for over a decade—since 1993. In that time, Israeli security forces managed, relatively fast, to lower the number of Palestinian suicide attacks from a record 60 in 2002 to one suicide attack in 2007 (Figure 2.2). In accordance, the number of Israeli fatalities caused by suicide bombers dropped from 188 in 2002 to 3 in 2007. As the former head of the Military Intelligence, Farkash notes "in three years we've had an incredible success."[40] It is true that some aspects of the terrorist threat that the Unites States faces are different from the ones faced by Israel, but there are still many similarities, especially at the operational level that American law enforcement officers, investigators, and clerks can definitely learn from. Israel did not come to the war on terrorism with a clear strategy or organized plan of action. As will be described later, Israel's capability to mitigate the threat was built brick by brick and was largely a result of trial and error rather than deep organized thinking processes.

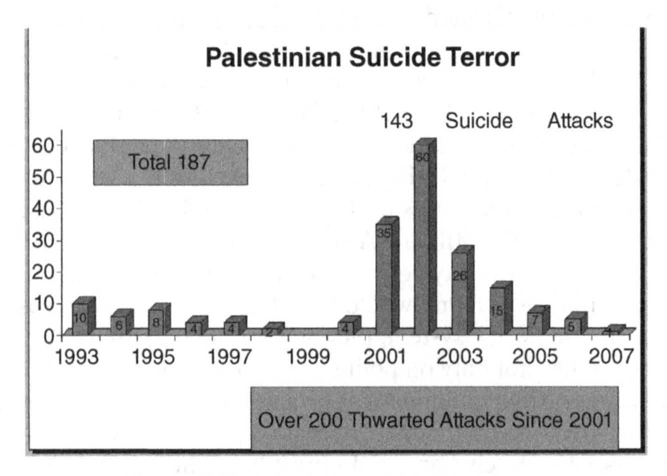

Figure 2.2. Palestinian suicide terrorism.

As a result of mistakes made along the way, many Israelis, as well as innocent Palestinians, lost their lives. Yet, at the end the day, the Israeli security system found effective methods of confronting suicide terrorism. The motivation behind Palestinian terrorism has not diminished, but its suicide bombing capabilities have been substantially curbed for the time being. The results are there to be seen, and they are mostly reflected in the ability of the citizens of Israel to continue to maintain a relatively normal and democratic way of life. What are these methods, how did they evolve, and what mistakes has Israel made that America can learn from? All these topics will be discussed in the next parts of this chapter. However, it seems that first we must understand the nature of the threat Israelis are dealing with, and explain how suicide terrorism became an integral part of the Israeli–Palestinian conflict in recent years.

Who Are the People Behind the Suicide Attacks in the Palestinian Arena and What Are Their Motives?

The Evolution of Terrorist Attacks in the Palestinian Arena—A Short Historic Background. The crude yet effective tactic of modern suicide terrorism was initiated by Hezbollah, the Shiite fundamentalist organization that Iran established in Lebanon in the early 1980s. The tactic was first put in use in December 1981, when a suicide bomber, probably a Shiite, detonated a car bomb near the Iraqi embassy in Beirut; 61 civilians were killed. In April 1983 another Shiite suicide bomber detonated a truck loaded with explosives, this time at the entrance to the U.S. embassy in Beirut. Some 60 people were killed. In October of the same year, Hezbollah detonated a truck bomb at the entrance to the American Marines base in Beirut; 241 American soldiers and 56 French soldiers were killed from the blast. One month later, 33 Israeli soldiers died from a car bomb that exploded at the entrance to the IDF headquarters in Tyre. The driver was a 15-year-old Shiite boy. The Palestinians didn't rush to imitate this form of terrorism. Only a decade later, in April 1993, did a Hamas explosives expert Yihia Ayash (a.k.a. "the engineer") send the first Palestinian suicide bomber on his way. The bomber drove a van laden with 150 of homemade explosives and blew himself at the entrance to a truck stop. The attack killed a local Palestinian and injured seven Israeli soldiers. By the end of that year, eight more suicide attacks and attempts were carried out, mostly in the Gaza Strip.[41]

An escalation in the Palestinian use of suicide bombings began in February 1994 after an incident of mass shooting against Muslim worshipers by a Jewish settler at the Tomb of the Patriarchs in Hebron; 29 Muslims were killed. The new wave of attacks demonstrated the power of suicide bombings. In April 1994, eight Israelis were murdered in a suicide attack in the city of Afula. More attacks followed, creating shock waves in the Israeli public and the International community. On the other hand, in the Palestinian public, suicide attacks were viewed by many as an effective tool and a valuable means for reaching their goals. The wave of terrorism that began in 1994 created a violent cycle of action–reaction. Thus, in January 1996, when Yihia Ayash was killed, his supporters avenged his death with a series of suicide attacks, including two attacks on buses in Jerusalem (February–March 1996), an attack in Askelon, and an attack at a shopping center in Tel Aviv by the Islamic Jihad.[42] Following this wave of attacks, Israel by various means, including forcing the Palestinian Authority to arrest Hamas activists, managed to damage the Hamas infrastructure in the territories, and until 2000 there was a relative calm in suicide attacks. This, however, was short-lived. The Palestinians learned the effectiveness of suicide terrorism, and would use it again on a wider scale once the conditions became ripe again.[43]

In total, during the 1990s, 35 suicide bombings and bombing attempts were carried out, killing 163 Israelis.[44] However, these numbers would pale compared to what took place in the violent Confrontation that began in September 2000. The failure of diplomatic negotiations between Israel and the Palestinians renewed the fighting between the two sides. In the setting of this was also the precedent established by suicide attacks, as an effective method to attack Israel's military superiority. An intensive propaganda campaign laid the foundations for the return of the suicide bombings. The Islamic Jihad was the first to renew these attacks, shortly followed by Hamas.[45]

That being said, the adoption of suicide attacks as a central Palestinian fighting tactic would occur gradually. In the beginning of the Intifada, the Palestinians preferred guerilla warfare tactics: standoff firing, drive-by shooting, laying explosive charges, and attacks on IDF positions. This was based on the thought that fighting similarly to Hezbollah would put public pressure on the Israeli government to conceade land. Soon Arafat delegated responsibility to the various Palestinian factions. The Islamic groups—Hamas and Islamic Jihad—were given a freedom of action to execute suicide attacks against civilian targets in Israel. However, from a Palestinian point of view, the guerilla warfare's gains in the first year of Confrontation were few, especially if the length of the Confrontation and the number of attacks were taken into consideration. The Palestinians carried out more than 1500 shooting attacks between 2000 and mid-2002, on Israeli vehicles in the territories, killing "only" 75 Israelis. They attacked IDF posts over 6000 times, but killed "only" 20 soldiers. They launched over 300 antitank grenades, killing no Israelis.[46] On the other hand, suicide attacks were proving themselves as the most lethal and effective weapon and an answer to the Israeli superiority. In suicide attacks carried out by Palestinian fundamentalist groups—Hamas and Islamic Jihad—in the first year of the Second Intifada, 85 Israelis were killed. In the second year the number grew to 220 deaths. Furthermore, despite the small number of suicide attacks, they caused most of the Israeli casualties. For example, in the year 2004 alone, suicide attacks accounted for 0.4% of all attacks, yet they caused 47% of the fatalities.[47]

The Islamic organizations' operational success created tension among the Arafat-led Fatah organization—the largest secular organization in the territories, which led the Palestinian national struggle since the 1960s. The Fatah's field operatives watched nervously as the Hamas and Islamic Jihad grew in popularity throughout the Palestinian street. This reality finally pushed the Fatah to carry out attacks against Israel. The beginning of the process was a modest local initiative: the laying of an explosives charge in a soldiers' hitchhiking stop at the end of May 2001. The attack was carried out by Albed Al-Karim Awiss, the head of Fatah's terror organization in Jenin. According to him "the officials wanted us to focus on a popular struggle and attacks in the territories. We decided to disobey them simply because we had no choice. We knew that if we didn't start carrying out attacks beyond the Green line [the line separating Israel and the West Bank], we would lose the street to Hamas and Islamic Jihad. Barghuti, Al-Sheikh, and even Arafat [heads of the Fatah movement] simply didn't interest me." Nevertheless, the efforts of constraint by Palestinian security forces and the Fatah political ranks still prevented further attacks in Israel by the organization. However, the future direction of operation was already set: Awiss and his companions, the local operatives, gave more weight to the opinions of the residents of the neighborhoods and refugee camps in which they lived than to the warnings of the heads of Fatah and the Palestinian security forces.[48] Sure enough, six months later, in November 2001, Awiss sent, in collaboration with Islamic Jihad, the first secular suicide bomber, accompanied by a religious bomber. The attack killed one man and one woman and again, according to Awiss, was carried despite Fatah leader's orders. The immediate motive for the attack was in retaliation for Israel's killing of one of his squad members.[49] This attack was another step in transforming suicide attacks from the sole property of Palestinian Islamic organizations to a central Fatah tactic. Unlike Hamas and Islamic Jihad, Fatah networks were spread in every city and village in the territories. When the Fatah initially unleashed its suicide bombers, Israel failed to face the diversified wave of "human bombs" that swept its streets. From a Palestinian perspective, the catalyst to this process was given by Israel itself, in its eagerness to kill Ra'ed Carmi, one of the high ranked operatives of Fatah's military wing in the territories.

Ra'ed Carmi, the head of Fatah's organization in Tulkarm, represented an imminent threat to Israeli security. During 2001, his men murdered 12 Israelis in various terrorist attacks, some of them very sophisticated and daring. In that sense, Carmi was an unusual figure in the terror landscape: charismatic, dangerous, and mainly reckless. The risks he took in his private life and in sending his people to confront the IDF seemed at times on the verge of madness. Among other things, he had affairs with women married to operatives in the gangs he led. In the city itself, Ra'ed Carmi gained status and prestige and soon became the lone ruler of Tulkarm. The high-ranked officials of the Palestinian Authority in town feared him. The heads of the official Palestinian security forces respected and honored him; the youths of the refugee camps admired him.[50] The Israeli security system tried to put its hands on Carmi for a long time. Eventually the ISA managed to form an operational plan that would put an end to his activities. The ISA monitored his moves and found his vulnerabilities. Carmi would regularly visit one of his lovers, the wife of a local Fatah operative, on a daily basis before noon. He never varied the route he took to visit her. Routine killed him. Palestinians reported that a demolition block was laid inside a cemetery wall, on the route from his safe house in the eastern neighborhood of Tulkarm to his mistress' home. The charge detonated on Monday, January 14, 2002 at approximately 11 A.M. Carmi, who walked along the wall regularly because he feared helicopters, died instantly. Thousands took part in his funeral, which was held on that very day, as cries for revenge were hailed. A few hours later an IDF soldier was killed by Palestinian gunfire west of Nablus. This was but a first sign of what was to come in the following days. Carmi's killing not only abruptly cut the slow decrease in violence, but also pushed Fatah to pursue a path, which until then it seemed hesitant to follow: suicide attacks against Israelis.[51]

Three days after Carmi's killing, Fatah's terror organization in Tulkarm avenged the death of its leader. On January 17, a terrorist who left the city walked into the "Armon David" functions hall in Hadera, which was hosting a family celebration. The terrorist opened fire, killing six civilians and injuring about 30, before guests stormed him with their bare hands and knocked him down. Policemen called to the scene shot and killed him. In the following weeks Fatah members launched terrorists to two sacrifice attacks in Jerusalem and also sent the first female suicide bomber, Wafa Idris, to carry out a suicide attack in the downtown area.[52] For the sake of clarification, it should be emphasized that the adoption of the suicide terrorism mode by the Fatach would have occurred regardless of the Carmi targeting, but that incident did in fact serve as a catalyst.

At the time, January 2002 seemed to be a peak of the confrontation. In that month, 17 Israelis were killed in three suicide attacks. But the next couple of months would be even worse. One after the other, Palestinian suicide bombers infiltrated Israel. In February, 22 Israelis were killed. In horrible March, the worst month of the conflict, 17 suicide bombers blew themselves up against Israeli civilians; 136 Israelis were killed. Soon, a diabolical competition was started between Fatah and the Islamic organizations, to see who would send the most suicide bombers to Israel. At the height of this competition, the attacks seemed to strike everywhere: Not only in Jerusalem and Netanya, two cities that became favorite targets for the Palestinians, but also in Haifa, Tel Aviv, Wadi Ara, the Judean desert, Gush Katif, and even in the North road, an area that was usually calm (Hezbollah sent two Islamic Jihad operatives from Lebanon to shoot at cars near Kibbutz Matzuba. Five civilians and an IDF officer were killed). Particularly lethal attacks were carried out in the Beit Israel neighborhood in Jerusalem where a suicide bomber disguised as a religious Jew detonated himself outside a synagogue (11 killed, including 5 children, most of them in the same family); in the "Moment" coffee shop in Jerusalem, a stone's-throw away from the prime minister's residence, 11 Israelis were killed (Hamas from Ramallah claimed responsibility); in a bus in Wadi Ara an Islamic Jihad suicide bomber exploded using an explosive belt, murdering 7 Israelis; and in the "Matza" restaurant in Haifa, a bomber sent by Hamas detonated himself, killing 15 civilians.[53]

As previously noted, until the second Intifada broke out in 2000, the use of suicide bombings was limited. Now this tactic became widespread and suicide itself became a mass phenomenon. Thus, alongside 60 suicide bombers who exploded in Israel in 2002, 115 more were killed or caught on their way to a suicide attack.[54] This means that at least 175 Palestinians were sent to carry out suicide attacks. This phenomenon was not short-lived. According to the security system's data, 160 potential suicide bombers were arrested in 2005.[55] If we take into consideration that not everyone wishing to carry out an attack fulfilled their wishes, we are faced with an unprecedented social phenomenon. The phenomenon had some notable features, the most salient was the volunteering of Palestinian youngsters to carry out suicide attacks. In the 1990s, the process of locating the bomber and preparing him would take a long time and mostly consisted of a mental–religious preparation, using radical religious clerics. Today, according to Avi Dichter, former Israeli Minister of Internal Security, launching suicide bombers had become a "production line with a waiting list."[56] A similar mood is reflected in the words of Amjad Ubeidi, one of the noted dispachers of suicide bombers for the Islamic Jihad in the Jenin area: "The volunteers always came to me. The operatives in the area knew that they had to refer potential suicide bombers to me. There were enough people who wanted to carry out attacks. I checked the volunteer's willingness to sacrifice themselves, asked about his religious beliefs and his family's condition. I tried to find out if I was dealing with a man of good mental health. There were candidates I dismissed because their answers didn't satisfy me."[57] A., a Palestinian suicide bomber who failed to carry out a suicide bombing attempt, confirms Ubeidi's claims regarding young people's volunteering for suicide attacks. According to him, a month before he set out for his attack he sought out the Islamic Jihad operative in charge of his area in order to volunteer for a mission and kept being rejected. This was until "I once went to him straight from work and he asked me if I was willing to carry out an assignment on that very same day. I said yes, showered, prepared myself and returned to him [...] He (the Jihad operative) showed me several ways to detonate the charge. There was a side switch and a main switch and a fuse you can light. ..." (The bomber was caught on his way to Israel, based on intelligence information.[58])

How Did Suicide Terrorism Become a Widespread Phenomenon in Palestinian Society? How did suicide bombings turn from being the territory of a handful of Islamic fundamentalists to a prevalent social phenomenon? Different studies showed that for a suicide bombing to take place, not to mention for it to become a widespread phenomenon as it is in the Palestinian society, three important elements are required: motivated individuals, accessibility to the organizations who's goals are to carry out suicide attacks, and a community that provides the suicide bombers with a hero's status and views their actions as valiant acts of resistance.[59]

The First Circle: The Palestinian Suicide Bomber and His Motives

The Profile of the Palestinian Suicide Bomber. A recently published report by two counterterrorism experts from the NYPD determined that "There is no useful profile ... to predict who will follow this trajectory of radicalization" because those who end up being radicalized begin as "unmarkable" individuals "from various walks of life." ... "The subtle and noncriminal nature of the behaviors involved in the process of radicalization makes it difficult to identify or even monitor from a law enforcement standpoint."[60] Similarly, Israel has also encountered difficulties in profiling the suicide bomber. Ilan Paz, formerly the commander of the Ramallah brigade and head of civil administration in the West Bank, notes that in the beginning of the Confrontation "the IDF thought it could profile the potential suicide bomber: young, bearded (or newly shaved), childless, religious, all of these are signs that should alert suspicion [...] Several months later, the characters changed and women, children, and married men joined the cycle of violence."[61] The security system needed the bomber profile in order to try and set criteria that the IDF and the ISA could use to prevent Palestinians, whose personal information reflected a security threat, from entering Israel. However, as a high-ranking ISA official noted in frustration: "Every time we tried to redefine it, draw the limits in a different criteria, reality came and bit us. When we set 30 as the maximal age, we encountered a 31-year-old suicide bomber; when we allowed for married men in the profile, fathers of toddlers carried out attacks. Later we saw women suicide bombers and attempts to dispatch children. The peak was in a failed attack in Jerusalem: The suicide bomber was a 48-year-old man from the village of Artas near Bethlehem; he had seven children. It got to the point that it seemed easier to profile a bicycle rider."[62]

Profiling the suicide bomber in order to identify him became even more difficult when the suicide bombers and their dispatchers started to try and "fool" the "average bomber profile" with a variety of means: shaving their beards, dying their hair blond, wearing Israeli army uniforms, and even disguising themselves as orthodox Jews.[63] For example, suicide bombers dressed up as ultra-orthodox Jews exploded at the entrance to a Jerusalem synagogue (March 2, 2002, 10 killed), on bus number 6 in the French Hill in Jerusalem (May 18, 2003, 17 killed), and on the number 14 bus on Jaffa street in Jerusalem (June 11, 2003, 17 killed). The height of the art of disguising suicide bombers was reached in the deadly attack in the "Park" hotel in Netanya (March 27, 2002, 30 killed). The bomber, sent by Hamas, was disguised as a woman in order to pass the security check at the entrance to the hotel. To do this, he wore make-up, a pair of women's blue jeans, a brown leather coat with a leopard print collar, matching women's shoes, and a wig. Once he entered the hotel's dining room, he exploded, killing 30 Israeli civilians and injuring 143 others.[64]

Nevertheless, when all of the suicide bombers are examined, and these unfortunately were abundant during the long years of confrontation between Israel and the Palestinians, one can generally characterize an average suicide bomber's profile. This profile is not a "winning model" or "magic cure" for identifying any suicide bomber, yet it can indicate the bomber's social identity and general features. According to the Israeli ISA's data, based on the analysis of the suicide bombers' characteristics in the 4 years of the conflict (between 2000 and 2004), approximately 85% of the bombers were single and approximately 80% were between the ages of 17 to 24. More than one-third of the bombers had a high school education; about one-fourth had a college education.[65] The profile also matched similar analyses done on Palestinian suicide bombers, such as by the Intelligence and Terrorism Information Center.[66] Regarding the suicide bombers' religious beliefs, it can be said that as time went by this component became gradually less dominant. During the 1990s and in the early days of the Intifada, Palestinian suicide bombers were not only older and more experienced, but mostly more devout and had a religious education.[67] However, out of dozens of interviews held by Berko in prison with failed suicide bombers, it appears that "both the dispatchers and the bombers are not fanatically religious. For some, the strengthening of their religious beliefs happens inside the jail, where religious studies are held by one of the prisoners." However, for all of them, religion plays some part in the preparation process (organizational or personal) for suicide.[68]

The Motives of the Palestinian Suicide Bomber. If religion is not as dominant as one would assume, what are the main motivates of these Palestinian bombers to kill themselves? There probably won't ever be one clear and overriding answer. People that choose to kill themselves in suicide attacks are motivated by different reasons, some complementary or intertwined with one another. Furthermore, with different suicide bombers, different motives can be identified as being more dominant. Identifying and classifying the different motives is not the topic of this chapter; however, in the reference and general literature, three types of personal (as opposed to the organizational) motives can be identified.[69] The first is a psychological motive: Trauma, personal identity problems, and feelings of humiliation and victimization create a strong will for revenge.[70] Wishing to restore dignity in view of the personal and social humiliation of the Israeli occupation, along with different social processes in which the individual loses his personal identity to that of a group plays into the terrorist organizations.[71]

These two motives—the personal and the social—blend into each other, as pointed out by the deputy head of the ISA, who said in an interview that the lack of expectation and hope that many Palestinians feel increased the suicide bomber phenomenon and gave terrorist leaders an opportunity they could exploit for their own purposes.[72] That being said, most suicide bombers have no mental handicap causing them to confuse reality with fantasy, although some have been diagnosed with personality disorders, sometimes suicidal thoughts, and even previous attempts of hurting themselves.[73] Another possible psychological characteristic of suicide terrorists arise in personality tests (Rorschach test) that were carried out by IDF psychologists on a small group of would be suicide bombers arrested on their way to their mission. The results of these tests reveal that these would-be suicide bombers had difficulty in seeing a wide perspective on matters, or to grasp the complexity of issues. Those would-be suicide bombers viewed reality in terms of black or white. Although this study cannot serve as a scientifically representative sample, it may point to something that is part of sucide bombers' personality.[74]

Another type of motive is the religious motive. At the essence of this lies the perception of suicide as a form of self-sacrifice for Allah. According to interpretation of Islamic law by extremist clerics, the suicide bomber is promised various benefits upon his arrival at heaven: pardon for sins, redemption of the pains of death, protection from entering hell, a crown of glory coated with gems "worth more than the world and all it contains," marriage to 72 virgins, and the ability to bestow these privileges upon 70 relatives. In that sense, it seems that heaven grants privileges to the suicide bomber that he could only dream of in this world.[75] A third type of motive is political and national motives. The suicide bomber and his dispatchers, according to this approach, seek to achieve a concrete political goal. For example, former commander of the Intelligence Branch, Major General Ze'evi-Farkash, claims that a suicide bomber's personal feeling that he can change things and advance the establishment of a Palestinian state is the first factor that motivates Palestinian suicide bombers.[76] This type of motive is also reflected in the words of Jamal Abu Al-Hija, a senior Hamas leader in Jenin, who explained that the choice of suicide bombings came from a combination of feelings of revenge with the wish "to change the Israelis' perception, who thought they could go on forever with the occupation. The diplomatic negotiations didn't change anything. On the other hand, the bombings caused Israelis to feel the pain we have been feeling. We wanted them to pressure their government to stop its operations [...]."[77]

Of course, these motives all blend in the mind of the potential suicide bomber. This phenomenon is reflected in the words of N., a Palestinian suicide bomber who was asked about his motives: "... First of all ... the first goal is a martyr's death and the precious reward from the almighty ... the second thing is the occupation and the killing and the friends who died in the Intifada." N. tells that following a meeting with a religious person he started praying and worshiping God. That person "is the one who makes me want to meet the Lord in any way there is. ... When I reached that level, I wished to die a martyr's death. All this added to the occupation and intimidation and killing and curfews, until no life remained at all."[78] In other cases, the suicide bomber himself finds it hard to define the reasons that motivated him. For example, M. was a resident of one of the villages in the West Bank and was a farmer until he decided to carry out a suicide attack. When asked why he set out to kill himself, he answered: "That's what's screwed up. Everyone says there's a reason here, but no. I don't know what I was thinking. The truth is that beforehand I saw images of killed and injured children on the TV screens. ... But to tell you I had one reason or several ... no. ... My cousin came to me one day and said: 'I want something from you.' I asked 'what?' He said: 'What to do you think about becoming a martyr?' I told him 'I wish.' I was sure he was joking. But the next day we went into town and sat in a restaurant, eating hummus and ful with a guy, and later I went along with him and tried on the explosive belt and he told me this would be in Fatah's name."[79]

In sum, there are those that believe that lack of hope and socioeconomic hardship foster suicide terrorism. Or in the words of a former high-ranked ISA official who stated: "Once you come to the conclusion that there's nothing to live for, you immediately find that there is something to die for."[80] That may be a facilitating factor at times; however, in light of the above analysis and other studies of past case studies, it is clear that there are many motivating factors that lead one to commit a suicide attack. There are various motivations and many different profiles of attackers. Limiting counterterrorism to finding a single factor or profile of suicide terrorism motivation is looking for the easy yet insufficient manner of coping with a complex phenomenon.

The Second Circle: The Organizational Wrapping

These motives, strong as they may be, are not sufficient for carrying out a suicide attack. A motivated individual needs an organizational wrapping in order to be a successful suicide bomber. This support usually strengthens the individual's motivation to carry out an attack, but more importantly supplies him with the means and resources: weapons (explosive belt and charge), a plan of attack (destinations and ways to get to them), and assistance in executing it (a support system and someone to transport him). Finally, this system is committed to help the bomber's family (generous financial support, building a house for it in the event that the previous one is demolished by the IDF) and to make the bomber a legend in his former community (publishing obituaries, proclamations glorifying his name, a mass funeral/march to commemorate him, etc.). The apparatus that supports the Palestinian suicide bomber is led by several terrorist organizations:

Hamas. The literal translation for Hamas is "enthusiasm" or "zeal" and is also the acronym of "the Islamic Resistance Movement"—Kharakat Al-Muqawama Al-Islamiya. Hamas is a religious fundamentalist organization that was established in December 1987 on the basis of the Muslim Brotherhood movement in the Gaza Strip.[81] Like other fundamentalist groups around the world, Hamas aspires to establish an Islamic state that would run based on Islamic laws. This state is meant to be established in all of Palestine and would bring forth the destruction of the state of Israel. Accordingly, Hamas opposes any durable peaceful solution to the Israeli–Palestinian conflict. In order to meet those goals, the organization works in two reciprocal paths. The first is the path of Da'wa (educational and social activity). In an attempt to spread its doctrine in the Palestinian territories, Hamas has established over the years a large social and financial infrastructure. This consists of a line of charity associations (Jama ya Hiriya) and committees (Lejan Zakath). These bodies, generally referred to as "Da'wa," supply medical services, education, religious services, welfare (money and food for the needy), youth after-school activities, and other services for the community. These services are given free of charge or in exchange for a symbolic fee. This social activity is a means of spreading Islamic fundamentalist ideology in the Palestinian society and to recruit more supporters.

The Path of Terrorism. Alongside its civilian activities, Hamas operates a terrorist body referred by it as the "military wing" named "The Izz Al-Din

Al-Kasam Brigades."[82] The "brigades" were founded in the late 1980s and have been continuously active until this very day. Their operatives are the ones who carried out the very first suicide attack in Israel in 1993 and are responsible for most of the suicide attacks in Israel in the 1990s. This body is built in a highly organized hierarchy and includes all of the movement's terror operatives. In the Gaza Strip, the operatives are organized based on a geographical base and hierarchal structure. In charge of each area is a senior military operative.[83] Alongside the terrorist attack mechanism, Hamas operates other operational bodies, such as: a policing mechanism, a weapon manufacturing body, an acquisitioning body, a smuggling body, and more. However, in the West Bank because of Israel's preventative efforts, Hamas' military wing is more spread out, less hierarchal, and structured in networks. In practice, most West Bank cities have terrorist cells loosely connected to each other. There is a tight link between the military and supposedly civilian systems: Money intended for charity activities is transferred to fund terror attacks, the men of Izz Al-Din Al-Kasam are recruited out of the pool of Hamas' social activists, and often the suicide bombers originate from the same place.[84] The civilian bodies and the terrorist wing of the movement all answer, at least in theory, to a governing body, called "the Political Bureau." The bureau located in Damascus has been headed by Khaled Mash'al since 1996. There is also an organized local leadership body in the territories. According to a high-ranked ISA official, there is a constant dialogue between the movement's leadership in the territories and the one that sits in Damascus; organizational decisions are made with the consent of both sides and are enforced down to the very last operative.[85] Additionally, the movement has representatives in various Arab countries, in Iran, and in other countries around the world. The movement's funding comes from several sources, the main ones being various Islamic charities working in Europe, the United States, Saudi Arabia, and the Persian Gulf.[86]

When the second Intifada began in the end of September 2000, most of the violent incidents that took place were between Fatah and the IDF. However, Hamas gradually restored its military wing's abilities, which were damaged mainly as a result of Israel preventative actions during the 1990s. The release of the movement's operatives who were detained by the Palestinian Authority gave a significant boost to this process. Within a number of months, Hamas succeeded in establishing lethal terrorist networks in most of the West Bank cities. These networks upheld a continuous relationship with the movement's leadership within the territories and abroad and also cooperated with each other. This cooperation manifested itself in the preparation of explosives, reciprocal transferring of knowledge, recruiting suicide bombers, and dispatching them to attacks.[87] Therefore, it is not surprising that after a few failures to hit the Israeli home front, it was Hamas that managed to execute the first lethal suicide attack in the Confrontation. A suicide bomber sent by Hamas infrastructure in Nablus detonated himself in a bus station in the city of Netanya, killing three Israelis and injuring 55.[88] As time went by and the Hamas gained more field experience, the efficiency of its terrorist networks improved. This is especially noteworthy when the movement's performance is compared to those of other terror organizations. For example, in the first quarter of 2002, 55 Israelis were killed in three Hamas suicide attacks, as opposed to 24 Israelis killed in 11 Fatah attacks. This ratio repeated itself in the second quarter of the same year.[89] Up to 2006, Hamas operatives carried our 58 suicide attacks (approximately 40% of all attacks).[90]

Fatah. The Fatah organization (reverse acronym for "Palestinian National Liberation Movement") was founded in Kuwait in the late 1950s by a number of Palestinian youngsters and headed by Yasser Arafat. In the 1960s the Organization took over the Palestinian Liberation Organization (PLO), the framework organization of the Palestinian national movement. Led by Arafat, the organization carried out widespread terrorist operations against Israel, from the mid-1960s and until the 1990s. Following a political and ideological process that the Fatah went through, Arafat agreed to declare that he acknowledges Israel's right to exist in peace and security and to end the Israeli–Palestinian conflict by means of compromise. Following this development, a peace agreement was signed between Israel and the PLO in 1993. This brought to Fatah's terrorist units being dismantled and assimilated into the security forces of Palestinian Authority, which was established in 1994. When the diplomatic process failed, and the second Intifada broke out in September 2000, the Palestinian Authority led the mass demonstrations and violent protests that characterized the first weeks of the Confrontation.[91] However, Fatah found itself without an organized terror branch.

Gradually, and thanks to the Fatah's younger generation's initiatives, with the Palestinian Authority's funding, the organization established a terrorist networks in most of the Palestinian cities. These were named "the Al-Aqsa Martyrs Brigades." In practice, the "Brigades" lacked any hierarchy and were in fact a bunch of scattered terror cells. The members of these cells did perceive Fatah leaders and especially Arafat as their main leadership, but some also received directions and money from Hezbollah in Lebanon. As the Confrontation with Israel grew longer, the Palestinian official leadership institutions disintegrated along with Fatah's organizational structures. In accordance, its leadership had less influence over the terrorist networks. A horde of conflicting personal interests, personal rivalries, and gang wars over control characterized the Al-Aqsa Martyrs Brigades' activities. In Bethlehem, for instance, there was a Fatah terrorist network that was descended from a Bedouin tribe named "Ta'amra," whose members reside to the east of the city. During the Confrontation, another network was established that belonged to Fatah and consisted of people from both the refugee camps and the city itself. Each gang ran an independent policy. Hassin Al-Sheikh, a senior member of Fatah's West Bank leadership, admits that the organization lost control over its men and blames this on the Israeli target killing policy.[92]

At the early stages of the Confrontation Al-Aqsa Martyrs Brigades avoided carrying out suicide attacks or attacks in Israel. In these initial stages, most operatives obeyed commands of their leadership, and the latter decided to avoid the hitting the Israeli home front for political reasons. This situation was, in retrospect, only temporary, and a year and a half after the second Intifada erupted, Fatah joined Hamas and Islamic Jihad in carrying out suicide attacks in Israel. Furthermore, due to its widespread layout, Fatah in the West Bank cities reached record numbers in sending out bombers. In March 2002, Fatah managed to send more suicide bombers to Israel than Hamas and Islamic Jihad combined. It's hard to overemphasize the severity of this development and its implications. For the first time, a secular movement in the Middle East led a campaign of suicide bombers, by the dozens.[93] What caused this radical change? How could an organization, which was secular at its core, make suicide bombings such an important factor in its struggle? There are a number of reasons that are related to each other. The first one is the Palestinian street and the different groups' competition to gain its support. Fatah members watched anxiously as the Islamic movements grew in popularity by launching suicide bombers to Israel's centers of population, striking Israel's underbelly. This reality was the reason why Fatah operatives came out with sporadic local initiatives to carry out attacks in Israel, in the first year of the Confrontation, despite of the directions of the political ranks of the movement.[94]

Another cause for Fatah's radicalization was the Palestinian Authority's campaign of incitment. From the very beginning of the conflict, the PA ran an intense propaganda campaign against the state of Israel. Arafat's slogan "Le'Al-Quds ra'i'hin shuhada bil-malaiin" (millions of martyrs are marching to Jerusalem) became known to all and set the stage for Fatah's joining to the production line of Palestinian suicide bombers.[95] A third cause of Fatah's radicalization was Israel's forces' entries into the Palestinian Authority's controlled territories ("Area A"). The IDF's repetitive incursions into Area A, along with the violent operations carried out in them, added to the desire to attack Israel. Fatah leaders headed by Marwan Barghuti encouraged this mindset. A fourth element that pushed Fatah terror cells to carry out suicide attacks was Israel's target killing policy. This added a dimension of personal retribution to their actions and to the Fatah leaders set of considerations. The case of Ra'ed Carmi, the head of Al-Aqsa Martyrs' Brigades in Tulkarm, stands out in particular. As noted, he was killed by an explosives charge by Israel in January 2002. Following the hit, Fatah began launching suicide bombers to Israeli cities systematically and on a daily basis. Nasser Abu Hamid, for instance, head of a terrorist network operating from Ramallah, claims that the response to the assassination wasn't necessarily related to the national struggle. "It became personal. You killed our friend, and we sent 30 suicide bombers as revenge." Nasser Awiss, head of the Fatah terror organization in Nablus claims that "When they killed Carmi, we went mad. He was a friend, but also a symbol. We realized that if Carmi was killed, we are all targets."[96]

Finally, Fatah's adoption of the suicide bombing tactic was a result and a symptom of the disintegration of the organization and the Palestinian Authority's leadership. The latter, after a year of violent conflict with Israel, had in practice lost its direct control over the street and the daily operations of the Al-Aqsa martyrs' brigades' terrorists. For example, Ahmad Mughrabi, who headed one of Fatah's terrorist cells in Bethlehem and was responsible for sending four suicide bombers, admitted that "we had no strategy." Ahmad Barghuti, the assistant of

Fatah's West Bank leader, Marwan Barghuti, was even blunter, explaining: "Our [suicide bombing] missions were an expression of an organizational anarchy. Our organization was purely tactical—it linked a suicide bomber with a dispatcher and a Kalashnikov or explosives charge."[97]

Interviews given by Palestinian terrorists need to be taken with a grain of salt. There are other views concerning the PA's ability and willingness to control their operatives. Moshe Yaalon, the former IDF Chief of Staff, for example, said on a number of occasions, including in an interview for this book, that Arafat controlled the level of the flames, and Benjamin Netanyahu, the former Prime Minister of Israel, emphasized that Arafat could stop almost all the terrorist activity immediately, and he gives an example from September 1996, when after warning that if the attacks don't stop within 30 minutes, his regime would be overthrown the violence stopped immediately.[98]

Islamic Jihad. The Islamic Jihad was another organization that launched suicide bombers against Israel. Islamic Jihad is a Palestinian fundamentalist organization that is pro-Iranian. The Jihad was founded in the Gaza strip in the early 1980s by Palestinians who were educated in Egypt, where they were exposed to radical Islamic ideologies. At the same time, they were inspired by the success of the Islamic revolution in Iran in 1979. They were especially impressed by Khomeini's approach that aimed to unite the Islamic world and that set the Palestinian problem at the top of its priorities.[99] Accordingly, the organizations ideology is both (a) Pan-Islamic—aiming to establish an Islamic empire (Caliphate) in the Middle East—and (b) Palestinian—aiming to destroy Israel as a step in establishing this empire. During the 1980s and 1990s, several Palestinian factions operated under the name "Islamic Jihad," some even supported by Fatah. However, the main faction that survived was the one established by Fathi Shkaki. Shkaki himself was killed, probably by the Israeli Mossad, in Malta in 1995. Today the organization is headed by Ramadan Shalah, who sits in Damascus. The organization enjoys the sponsorship of the Syrian regime and generous Iranian support. During the second Intifada, its operatives carried out dozens of suicide bombing in Israel. Up to 2006 these attacks represented 27% of all suicide attacks.[100]

The Popular Front for the Liberation of Palestine (PFLP). A small number of suicide attacks were also carried out by the PFLP. This organization supported communist ideas and during the 1970s and 1980s carried out dozens of terrorist attacks, including the hijacking of airplanes. However, following the Soviet Union's collapse, the organization began to fade. During the second Intifada, it also joined the violent struggle and even succeeded in assassinating Israel's Minister of Tourism, and former IDF General, Rehavam Ze'evi, in October 2001. As part of the Palestinian suicide terrorism trend, the popular Front also joined the wave of suicide bombings and launched a number of suicide bombers to Israel.

Iran and Hezbollah. Another factor that facilitated the execution of Palestinian suicide bombings was Iran and its Hezbollah proxy. Iran viewed the second

Intifada as a strategic opportunity to weaken Israel, and facilitated the activities of Palestinian terrorist organization, from the outset of the confrontation. Iran transferred substantial sums to the Islamic Jihad and Hamas, and it trafficked weapons to the territories. A key case in that context was the "Karin A" vessel that was on its way to transfer the Palestinian Authority tones of munitions before being intercepted by an IDF commando unit on January 3, 2002.

Almost immediately after being elected Palestinian Prime Minister, Hamas leader Ishmael Hanyia paid a visit to Teheran and declared that Iran serves as the Palestinian's "strategic home front."[101]

In addition to Iran's direct hastening of Palestinian terrorism, it also used its Hezbollah proxy to facilitate Palestinian terrorist actions, especially those carried out by Fathah. As early as the Confrontation's initial stages, Hezbollah established a special body that dealt with supporting Palestinian terrorism in the territories.[102] This body contacted Palestinian terror operatives, mostly from the Fatah. Through them it began to transfer money, bomb-building knowledge, and instructions to carry out attacks. The organization's efforts bared fruit; and as early as the winter of 2001, Hezbollah was behind the first Palestinian squad that successfully launched mortar bombs on the Jewish Settlements in the Gaza Strip.[103] Later the organization managed to motivate and finance the activities of several Fatah squads that carried out many suicide attacks in Israel. Notedly among these is the double suicide attack at the central bus station in Tel Aviv, on January 5, 2003, which caused the deaths of 23 civilians and the injury of 106 others. The attack itself was carried out by Fatah's network from Nablus headed by Na'ef Abu Sharakh.[104] One year later, several other Fatah squads were arrested in the West Bank, who were operating under Hezbollah's sponsorship and who either took part or were going to execute suicide attacks in Israel.[105] This process reached its peak in the summer of 2006, during the war in Lebanon, Hezbollah put enormous pressure on the Palestinians to carry out "sympathy attacks" within the Green line. At least eight suicide bombers were caught on their way to attack at the time.[106] Furthermore, Hezbollah was involved at times in coordinating squads that were operating in different areas. This was especially prominent when the terrorists' movement became difficult because of the IDF's widespread checkpoints in the territories. For example, at one opportunity the organization's men hooked up an explosive belt manufacturer from Nablus with a suicide bomber from Tulkarm and a transporter from Qalqiliya. In the Gaza Strip, Hezbollah initiated joint attacks by Hamas, Fatah, and Islamic Jihad and flooded the territories with "education" materials for the improvement of the weapons the Palestinians had, relaying knowledge and techniques consolidated during its war in Lebanon.[107]

The Third Circle: The Social Wrapping

Personal motives and an organizational system that allows them to be realized by carrying out suicide attacks can only evolve in a compatible social environment that turns suicide bombers into heroes and views their actions as a model of national or religious sacrifice. This social climate developed in the territories during the second Intifada.[108] Prior to September 2000, suicide attacks lacked a widespread support from the Palestinian street and all its organizations; however, as the Confrontation drew longer, suicide bombings were adopted as a social ethos and a powerful symbol of Palestinian resistance.[109] This was illustrated, among other things, in public opinion polls taken in the territories, according to which between 60% and 70% of the Palestinian public support suicide attacks and see them as an effective tool in realizing their national goals.[110] A number of factors united to form this reality. First, the Palestinian Authority's support of this process. In the beginning of the Intifada, this support was expressed in the release of Hamas and Islamic Jihad detainees who were jailed in its facilities, and they were given a green light to carry out their actions.[111] At the same time, the Palestinian security bodies refrained from taking action against those who carried out attacks. Furthermore, funding was transferred to the heads of the Fatah's terror organizations by the commanders of the security forces and by Arafat himself.[112]

Gradually and systematically, the Palestinian Authority and the terror organizations created the right mindset infrastructure that supported suicide terrorism. Central to this was an ensemble of steps and ceremonies that could be dubbed "worship of the suicide bomber." The pictures of the suicide bombers were hung on the walls of houses, on bulletin boards, and in schools.[113] In the Friday sermons in mosques, religious preachers praised the act of martyrdom and also praised the suicide bombers and their actions. Every suicide bomber received an impressive funeral, attended by masses, in his hometown. Palestinian children wore dog tags on their necks with pictures of martyrs on them. The PA's official media ran loops of video clips praising martyrdom—the Shuhada. The suicide bombers' video-recorded wills also got extensive airtime in the local and Arab media. This and more, occasionally the PA published official obituaries in its newspapers for suicide bombers who blew themselves up in Israeli cities.[114] All these gave the suicide bomber an upgraded social status. In a culture where respect and personal prestige are important, the suicide bomber became the personification of the highest level of these, a motivator in its own right. It's not hard to imagine deprived kids in Gaza, hoping to one day get the glory that a suicide bomber receives. Given the harsh conditions that exist in the territories, suicide might be their only way to reach that status.[115].

Alongside the glorification of the suicide bomber, the terror organizations and Palestinian Authority carried out a wild hate campaign against Israel and against Jews in general, which dehumanized Israelis in the Palestinian society. For example, Berko, who interviewed would-be suicide bombers and their dispatchers in Israeli jails, notes that she was told many times by her interviewees that "until they were in prison, they never saw Jews as human at all." Jews were described by many phrases such as "the killers of the prophets," "the bloodsuckers," and "the enemies of God." The demonization of Israel and the West in general is what allows, according to Berko, to recruit more and more living bombs.[116] Another important factor in creating a social climate that supports suicide attacks was the improved social and financial status that the families of the suicide bombers gained. The different terror organizations transferred generous financial aid on a monthly basis to the suicide bombers' families. Other parties, such as Saddam Hussein, also donated generous grants to these families. For example, during the Confrontation, the Iraqi dictator transferred via the "Arab Liberation Front"—a Palestinian pro-Iraqi organization—25 thousand dollars to the family of every suicide bomber. Though no link between financial distress and suicide bombings has ever been proven,[117] the commander the IDF's Operations Branch notes the following: "In a place where the gross national product per capita stands at less than one thousand dollars a year, this was significant assistance."[118] The family of the suicide bomber or the bomber's dispatcher not only gained financial benefits, but also became more "popular and respected," as noted by Mahmud, a dispatcher of suicide bombers interviewed in prison.[119]

The three circles discussed thus far—the personal circle, the organizational circle, and the social circle—created a climate that encouraged the Palestinian suicide terrorism phenomenon, made the bombers accessible, and gave them the tools and resources to carry out their death wish. How were the suicide attacks carried out in practice—or, in the words of former IDF Chief of Staff, Moshe Yaalon, what was "the attack's food chain"?[120] The understanding of this process is important for two reasons: The first is so that we can understand the Israeli method of confronting suicide attacks. In that context it is important to understand which links in the chain are the weakest or the most significant, and which links should be the focus of the thwarting operations in order to get the best results. The second reason to analyze the "chain of attack" is so that it can be used to learn about how suicide attacks in other places in the world were carried out. It is reasonable to assume that similar attacks would include similar elements at some level or another. And as many people in the Israeli security system have learned for themselves, understanding the problem is half the solution.

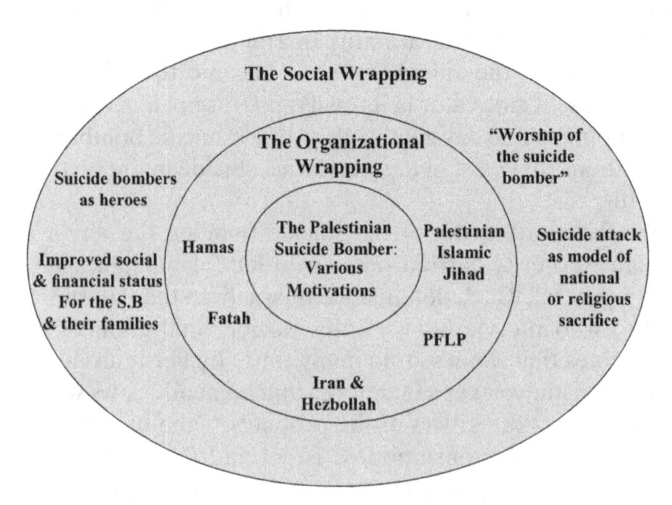

Figure 2.3. How did suicide terrorism become a widespread phenomenon in Palestinian society?

How a Suicide Attack Is Carried Out—The Palestinian Model. Mordechai and Tzira Schijveschuurder wanted to break a routine of security tensions with a day visit in Jerusalem. The couple, who have been residing in the settlement of Nerriya for the last seven years, had been subject for many months to a reality of near war: Only the day before yesterday a resident of Nerriya was shot at near the settlement of Nili; a week and a half ago, three members of a family from a neighboring community, Dolev, were injured. Mordechai and Tzira thought that a day out in downtown Jerusalem with five of their children would restore some of their energy for the risky road ahead, living in the small community halfway between Modi'in and Ramallah. At approximately 1 P.M. the family—father, mother, and five of the eight children—reached the Bell building in the heart of Jerusalem. The kids still managed to hear the giant black bell, set in the wall of the building that faces King George Street, play the tune of "Jerusalem of Gold." Later they headed to the adjacent Sbarro restaurant. At 2 o'clock, a large blast was heard; a terrorist blew himself up inside the restaurant. Five of the seven family members were killed: the father, Mordechai, age 43, his wife, Tzira, age 41, and three of their children—namely, 14-year-old Ra'aya, a student in a girls' college, who only the day before came back from the "Ariel" youth movement's summer camp; 4-year-old Avraham Yitzhak, and 2-year-old Hemdat. Nine-year-old Haya was badly injured and 11-year-old Leah suffered minor injuries. At a nearby table sat 8-year-old Tamara and her mother Laly. They didn't survive the blast either, along with eight other Israelis (altogether 15 Israelis died in the attack and 140 were injured).[121]

As mentioned, this suicide attack took place in the Sbarro restaurant in Jerusalem on August 9, 2001. The attack's target, outcome, and the preparations that preceded it are typical of many other attacks were carried out against Israel during the second Intifada. Such is the following case. The suicide bomber, Izz Al-Din Al-Masri, was an activist in Hamas' political organization in Jenin (a Palestinian city in northern Samaria). A few weeks earlier, he was approached by Keis Aduwan, the commander of the movement's military wing in Jenin, who suggested that he should volunteer for a suicide attack. Getting from Jenin to Nablus and from Nablus to Ramallah didn't present him with real difficulties. The IDF forces had been besieging the West Bank cities at the time, but the Hamas' terrorists knew ways around them. On August 8, Al-Masri met two of the organization's operatives, Bilal Othman and Muhammad Daghlas, in a Ramallah mosque, in a meeting that was set in advance. Othman recruited Daghlas, who served in the Authority's "Force 17" and had been taking media classes at Birzeit University, to the Hamas' military wing. In the weeks before the mosque meeting, Daghlas was given relatively simple assignments: to rent a flat in town and buy a guitar. Later he was asked to find a youngster that would help the organization's activities. Daghlas chose his girlfriend, Ahlam Tamimi, a 21-year-old Palestinian student who studied with him and with whom he was having an affair. Tamimi was sent on July 30 from Ramallah to Jerusalem with a beer can booby trapped with a small charge in her possession. She placed the booby trapped can in a mini-market on Jaffa Street in Jerusalem and walked away. The blast didn't result in casualties, but her handlers were pleased. It was proven that, when necessary, it was possible to get around the IDF's checkpoints and enter Jerusalem with a charge. The facilitator Tamimi and intended suicide bomber, Al-Masri, were never introduced with the organization's real leaders: Sheikh Ibrahin Hamed, the head of the military wing in Ramallah, a mystery man who believed in extreme compartmentalization, and the explosives expert, Abdullah Barghuti. Barghuti would leave the charges he prepared in stashed spots, from which they were collected by the network's facilitators. While Daghlas met the bomber in the mosque and handed the explosives charge hidden inside a guitar over to him, Tamimi set off on another test ride to Jerusalem. She returned to Ramallah and reported to the pair that there were IDF checkpoints on the way, but they didn't completely prevent access to the city.[122]

Al-Masri spent the night before the attack at a safe house rented by Daghlas. The next day, in the early hours of the morning, the suicide bomber and his transporter, Tamimi, left Ramallah in a taxi toward Jerusalem. The two wanted to avoid taking any risks. Tamimi got out of the cab with the explosives charge at the Kalandia checkpoint, went around it by foot, and got back into the cab on the other side of checkpoint. She did the same at the A-Ram checkpoint. The two left the driver at Damascus Gate and walked up Jaffa Street, in a route that was familiar to Tamimi from her previous visits in town. Passers-by noticed a couple with a Western appearance conversing in English. The man was carrying a guitar case. Just before they reached the Sbarro restaurant, Tamimi said goodbye to the bomber, not before she pointed out the selected target. At exactly 2 P.M. Al-Masri entered the restaurant, stood at its center, and detonated himself.[123]

Generally, it can be said that the preparation of such a suicide attack takes place in two parallel routes: (1) "the suicide bombers route"—spotting him, mentally preparing the attacker, arming him/her, and transporting him/her to the destination; and (2) "the logistical–operational route"—preparing the explosive belt/charge and locating a transporter and a target for the attack. The two routes merge in the final stretch before the attack. The bomber is hooked up with the explosive belt and transporter, and these two are sent to one of Israel's centers of population.

THE SUICIDE BOMBER ROUTE. Spotting the suicide bomber is not a set process with clear rules. There are three main ways that terror networks use to spot a potential bomber: In the first method, the head of the terrorist cell himself approaches a person who seems suitable. For example, Fuwaz Badran, a Hamas operative from Tulkarm, who was behind the first suicide attack in the Confrontation, picked the suicide bomber from his own students. Badran was a teacher in the religious school in the city and chose Ahmad Alian, age 23, a resident of Nur Al-Shams refugee camp, who studied under him and was deemed suitable by him because of his religious zeal. Alian did indeed blow himself up in a crowded bus station in Netanya, killing three Israelis and injuring 53. Similarly, Salah Buhari, the head of Islamic Jihad's terror network in Nablus, recruited Shadi Bahlul, a 19-year-old suicide bomber, who was arrested on his way to carry out a suicide attack in Israel.[124] Another way that is used to recruit suicide bombers is through one of the network's members or a person associated with it. Often the recruitment is done by someone in the suicide bomber's close environment, including friends, who are personally connected to both the dispatcher and the potential suicide bomber. In the case of one suicide bomber that was interviewed in the Israeli prison, the connection to the terror network was done by her best friend.[125] In a similar case, a student named Manal Sab'ana recruited several suicide bombers from her school friends for the Islamic Jihad network in Jenin. In Nablus, Nasser Awartani, a 16-year-old resident of the Kasbah, convinced four of his friends to set out to commit suicide bombings for the Fatah's terror network (one of them, his best friend, died and injured a soldier in the blast; the others were arrested).[126].

A third way that a suicide bomber is recruited is by his own initiative. As the Confrontation between Israel and the Palestinians escalated, this pattern of recruitment became more common. Nasser Abu Hamid, a leader of one of Fatah's terror networks in Ramallah, stated: "We were flooded with potential suicide bombers. Every demonstration a serious group of people approached us, who wanted to carry out suicide attacks. For example, after the Ra'ed Carmi killing [head of the Fatah's Tulkarm terror network, who was eliminated by Israel] many volunteered to become martyrs. From the moment a young person approaches us and asks to become a suicide bomber, he is automatically considered as a potential person to send out."[127] Amjad Ubeidi, a suicide bomber's dispatcher and the head of Islamic Jihad's military wing in Jenin, says similar things: "Volunteers always came to me. The operatives in the area knew that they had to refer potential suicide bombers to me. There were enough people who wanted to carry out attacks."[128] In many cases the number of volunteers exceeded the terror organizations' abilities to send them out for attacks, and often it was the potential bombers who encouraged the launchers to find ways for them to kill themselves.[129] In that respect, the ISA notes, the recruitment of the bomber became the easiest part of the process: At no point of the Confrontation was there a shortage of suicide bombers.[130]

As for the process of the mental preparation of the bomber, in the past, mainly in the 1990s, from the moment the suicide bomber was recruited, the organization started to put him through a long preparation process. This included indoctrination, religious lessons, and a process of mental and spiritual purification that included fasts, nightly prayers, payment of all financial debts, and requests of forgiveness from all he hurt or offended. In the final stage, in which the suicide bomber vanishes from his home environment, he goes through several days of operational training and prepares a recorded or videotaped will.[131] This process grew constantly shorter as suicide became more accepted and popular in the Palestinian society. In this reality, the processes of choosing and preparing the bomber became faster and more technical. Checks were made to verify that the bomber is serious in his intent.[132] An example of this kind of preparation or lack of it is provided by the testimony of A., a resident of a refugee camp who was caught on his way to an attack. According to him, a month before he set out for his attack he sought out the Islamic Jihad operative in charge of his area in order to volunteer for a mission and kept being rejected. This was until "I once went to him straight from work and he asked me if I was willing to carry out an assignment on that very same day. I said yes, showered, prepared myself and returned to him [...] He (the Jihad operative) showed me several ways to detonate the charge. There was a side switch and a main switch and a fuse you can light...."[133] M., another suicide bomber, tells a similar story. In this case the potential bomber was approached by a family member who offered him to carry out a suicide attack. After M. agreed, "I asked him when I could carry out the attack and he told me—tomorrow we'll arrange things. In town I prepared a videotape, said that the attack was on Fatah's behalf, and was told not to confess if I get caught."[134] That summed up the complete process of his mental and operational preparation.

The Logistic–Operational Route

Putting Together the Explosive Belts. The belts or charges are prepared by "explosives experts" operating in the territories. These operatives either were trained in training camps outside of the territories, mainly in Syria, Lebanon, and Sudan, or received a professional education from other operatives with demolitions knowledge that were working in the territories. Abdullah Barghuti, for instance, who made the explosive charge that was detonated in the Sbarro restaurant, was a chemical engineer, who was educated in Jordan. Until his arrest in March 2003, Barghuti managed to prepare charges for a large number of "successful" attacks, in which 66 Israelis died and approximately 500 were injured.[135] Barghuti's case—an "expert charges engineer" with a wide knowledge base—is mostly a feature of Hamas' explosives engineers. These were usually more skilled and got their training either abroad or from professional operatives that were smuggled into the territories for that purpose. In this context, the Hamas headquarters in Damascus had an active role in both coordinating the training abroad and smuggling of the operatives back into the West Bank and Gaza Strip. For example, Ibrahim Beni Udda, a "charges engineer" from the village of Tamun near Nablus, underwent his initial training in Islamic Radicals' camps in Afghanistan and was smuggled into the territories by the Hamas leadership in Jordan in the late 1990s.[136] Fuwaz Badran, Hamas' charges manufacturer in Samaria, went through a crash course in explosives in Jordan, taught by a Chechen operative. Badran built a number of explosive belts, including the ones used in the suicide attacks in Netanya in March 2001 (3 Israelis killed and 53 injured) and at the entrance to Hasharon shopping mall, also in Netanya (5 Israelis killed, 85 injured), in May 2001.[137] These were not unique cases; in late 2001, Israeli security forces arrested three Palestinian explosives experts who underwent training in Jordan and Syria. The three, students of computer sciences and electronics, were smuggled into the territories in order to carry out attacks and to train more explosives experts.[138] As opposed to Hamas, the other organizations—Islamic Jihad and Fatah's Al-Aqsa Martyrs Brigades—preferred to settle for local knowledge that was transferred by word of mouth between terror operatives. The source of the knowledge was both from personal experience and from Hezbollah and Iran. These, according to sources in the security forces, have put great efforts in recent years into improving the terror organizations' demolition abilities. An abundance of technical knowledge was transferred through the Internet or on portable memory devices.[139] The explosives themselves were mostly made from chemicals purchased in pharmacies; and later on, when Israel banned the import of medical supplies that could be used in the manufacturing of explosives into the territories, the "explosives experts" started using agricultural fertilizers.[140]

The Bombers' Transporters. The main criteria for choosing the bomber's transporters were first and foremost their ability to access Israel and to find their way around it. That is, their ability to pass the bomber through the Israeli checkpoints and get him to his final destination. This accessibility was mostly achieved by means of Israeli IDs or smugglers of illegal workers. For example, Sa'id Al-Hutri, who carried out the suicide attack at a dance club on Tel Aviv's promenade on June 2001, in which 21 teenagers were killed, reached his destination with the help of Mahmud Nadi, who was a member of a family of Palestinian collaborators and therefore had an Israeli ID and a car with Israeli license plates.[141] Similarly, Hamas' network in Tulkarm used a transporter named Nihad Abu Kishk. Kishk had an Israeli ID since his mother was born in Israel, and therefore he could enter Israel legally. Because of this ability, he was also recruited to the Hamas organization in Nablus, and he transported a suicide bomber on its behalf to a suicide bombing in the central bus station in Kfar Saba.[142] In a different case, a Media studies student from Nablus named Bassam Handakaji used the journalist card issued to him to transport a suicide bomber to an attack in a market in Tel Aviv. Handakaji was assisted in this by an Arab taxi driver who used to smuggle Palestinians to work in Israel.[143]

Choosing the Target. Choosing the suicide attack target does not always entail a long and detailed process of gathering intelligence, and it usually requires a limited amount of information gathering on the target. The fact that the bomber's goal is simply to kill as many people as possible allows for flexibility in the target choice. The leading criterion for the choice of target is therefore it being "crowded with people," as Ahlam Tamimi, the Sbarrro suicide bomber's transporter, notes.[144] That being said, in some cases Hamas (the more professional and organized) can be distinguished from the other organizations. Fatah and Islamic Jihad usually settled for giving the bomber general instructions. Amjad Ubeidi, a dispatcher of suicide bombers for Islamic Jihad in the Jenin area, said that he used to settle for giving general directions in regard to the target of the attack and left the choice of the specific location to the bomber's judgment.[145] On the other hand, Hamas chose the targets in advance, usually using the bomber's transporters, who used their access to Israel to serve this purpose.

Yet even in this case, there were no set rules. For example, the Park Hotel, in which 30 people were killed in a suicide attack, was chosen after the transporter and bomber drove around for 5 hours looking for a crowded location. They reached the hotel itself at the suggestion of the bomber himself, who previously worked in the city. Similarly to the Park Hotel bombing, the double suicide attack in Be'er Sheva in August 2004, in which 16 Israelis died, was carried out based on previous knowledge of the area by one of the bombers.[146] Only in rare cases did the attack come after a detailed gathering of intelligence. In that context, the Hamas suicide attack at the central bus station in Kfar Saba in April 2001 stands out. It was carried out after an extended period of information gathering, carried out by the terror network's bombers' transporter. The intelligence was gathered over a period of 5 days straight, at which point the transporter came to a decision about the best day to carry out the attack.[147] However, this case is, as previously noted, an exception to the rule.

It is often held that detailed planning and information gathering is required prior to attacks. Past case studies imply that al-Qaeda often carry out detailed information gathering. For example, al-Qaeda planned their first attack in Tanzinia against the American Embassy for 5 years. Nevertheless, as seen in the Palestinian case, in cases where local networks or "homegrown" activists facilitate in the attack, the information gathering process can be significantly shorter.

The power of the Palestinian suicide attack model lies in its simplicity. The suicide bombers are available: volunteers for the mission or recruits from the cadre of people associated with the organization. In both cases the bomber's preparation takes up little time or resources. The explosive belt is the main component that takes time and money and is a bottleneck in the attack's execution. Once the belt is made, all that is left is to spot a suitable transporter. It is not a very complex mission considering the vast connections between the Israeli Arabs and the Arabs of the territories, or the large numbers of residents of the territories that carry Israeli ID's. The freedom of action given to the terror networks also helps make carrying out the suicide attack easier. Apparently, because of an organizational decision in Hamas' case or a lack of control in Fatah's and Islamic Jihad's case, the leaderships of the organizations didn't interfere with the operational details of every single attack. Mostly, they settled with giving general directions for either escalation or de-escalation and financing the activity. The rest of the work was left to the field operatives. These were given a freedom of action and judgment. This led to the decentralization of the terrorist activity and more flexibility in planning and carrying out an attack. As one of the heads of the Islamic Jihad's terror network notes: "I made the decisions myself, according to my own judgment, and if I needed assistance, I had a connection with the organization, they knew they could depend on me. I prepared everything—the explosive belts, sending out people to recruit the Istishhadis for me, and I would check them before they went out and decide who would go and who wouldn't. It's not as hard as it seems."[148]

Decentralization, simplicity, and freedom of action were the keys to the terror networks' success in sending hundreds of suicide bombers in a relatively short period of time. And so, by 2007, 171 suicide bombers were successfully sent

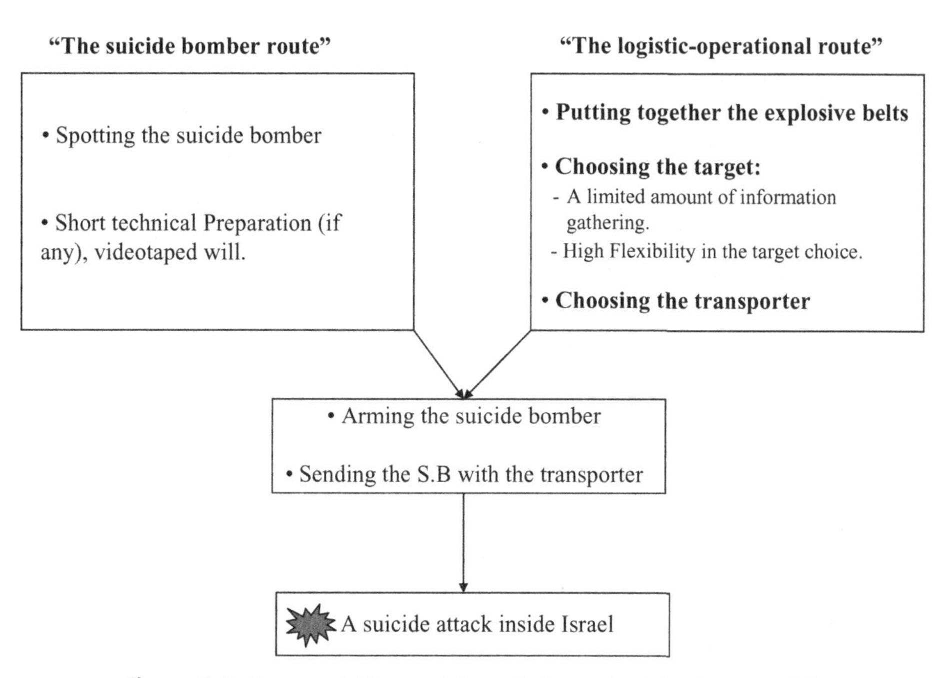

Figure 2.4. How a suicide attack is carried out: the Palestinian model.

to Israeli cities. These not only took the lives of 526 Israeli citizens and injured over 3000,[149] but left a great psychological effect on Israeli society. Nevertheless, in a relatively short period of time, the Israeli security system managed to confront the challenge and decrease the number of suicide attacks from a record 60 attacks in 2002 to eight in 2005 to one in 2007.[150] How did it accomplish this mission, which methods proved themselves effective, and which were wrong? And what can other democracies learn from all of this? See Figure 2.4.

How Has Israel Confronted Suicide Terrorism?

As detailed above, Palestinian suicide terrorism is a complex and multidimensional phenomenon, consisting of the individual dimension (the bombers motives and mental state), organizational dimensions (resources and abilities) and social dimensions (a social environment that encourages the phenomenon). Faced with all of these, which of them should first be confronted? Do we confront with the personal and social "motivators" or the "capabilities" first, meaning the terror organizations and suicide bombers themselves? Furthermore, if you choose to focus on the terror organization, then the moment of the bomber's explosion is only the tip of the iceberg of an organizational process that includes money transfers, directions from the leadership, the smuggling and manufacturing of weapons, bomber recruitment, facilitators and transporters, target selection, and so on. On what part of this chain should the security effort focus on first? What link should most of the thwarting efforts be placed upon? What are the most effective operational methods that achieve the desired goal in the shortest amount of time and minimal risk to the combatant forces?

The Israeli security system took a long time to formulate answers to these questions. In fact, many experts are of the view that during most phases of the Confrontation, Israel lacked an organized strategy of confronting suicide attacks, and its citizens paid the price until a successful strategy was formed. As Cooperwasser, the former head of the military intelligence's research department, notes: "Along the way many courses of action failed or at the very least didn't present the proper answer; for example, targeted killing began early in the Intifada, supervision was tightened on the access routes, limited operations were carried out in A areas [areas controlled by the Palestinian Authority], the military presence in the travel routes was increased, and a large attempt was made to stop the suicide bombers by operating against the Palestinian Authority itself and its facilities. The bottom line was that none of these stopped the phenomenon."[151] Furthermore, even today, after the number of suicide bombings in Israel has significantly decreased, it's not completely clear that Israel has a set strategy. That being said, there is definitely a concept of how to confront Palestinian suicide terrorism and more effective ways to apply force against suicide terrorism than in the past. One thing is certain: In the debate between confronting the bombers' motivators and the society in which they grow or confronting their direct capabilities, Israel decided in favor of the latter. As Farkash former head of the military intelligence branch says: "We decided in advance that we wanted to live, and since we couldn't deal with the motivation—it takes a long process and needs to concentrate on education and persuasion of the public—we chose to focus on the capabilities, and our experience proves that this is possible."[152] In this sense, the term "terror capabilities" received a broader meaning in the Israeli context. In other words, facing the terror capabilities meant not just stopping the suicide bomber but also striking the various components of the operational process—direct (members of the terror network) and indirect (the organization's leadership, sources of funding, etc.)—that allow him to carry out his mission.[153]

The Israeli Perception of Dealing with Suicide Terrorism[154]**.** Israel's method of confronting suicide terrorism is based on three primary spheres, referred to as security circles: Prevention, Delay, and Consequence mitigation. Each layer of activity has its own characteristics and complements the other components. The main differences stem from a few basic variables:

The Dimension of Time. When to operate, how much time the force has to prepare itself, and how much time is needed to carry out the operation.

The Geographic Space. In the suicide bomber's and network members' homes, in the area surrounding his home, or in the area targeted for his attack.

The Type of Forces Operated. In every security circle there are different forces that are operating: regular military forces, special units, police forces, the border patrol, civilian security officers, and so on.

The Kind of Intelligence Required. Every security circle requires a different kind of intelligence to launch an operation—specific information, general information, threat assessments.

Finally, the circles of coping with terrorism differ in the required outcome, meaning the definition of "success"—the elimination/arrest of the bomber or network member, prevention of passage, foiling the transfer of weapons or funds, preventing the bomber from reaching a crowd concentrated location, and so on.

The Prevention Circle

This is the most important security circle and where most of the effort is invested. Its goal is to eliminate or arrest the suicide bomber "in his bed," as former Chief of Staff of the IDF, Moshe Yaalon, puts it.[155] This means that the required result in this circle of coping is to thwart the attack before the suicide bomber is unleashed. Zohar Dvir, former commander of the Yamam, the Israeli police elite unit, stated the following: "The security forces' mission is to prevent the bomber and the civilian from meeting, and the sooner it's done, the better."[156] Beyond the efforts focused on the suicide bomber himself, this circle also includes operations against the members of his terror network, in an effort to hinder its capabilities. Priority in this case is given to striking centers of operational knowledge: the head of the network ("the planner and executor"), his deputies, or the explosive belts' maker. Geographically, this circle includes the suicide bomber's or network members' close living environment. Most of the suicide attacks are thwarted at this circle, and it is here that the forces' operations are the most focused and effective. Since the environment is usually in a hostile urban setting, the forces operating within this circle are mostly special units (intelligence and operational units) and regular military units with special capabilities. The duration of the actual attack operation itself is short and often may take less than an hour (including covert approach to the target, execution, and exit). The time at hand for the operation's preparation varies. Sometimes the force is required to act immediately—for example, if information is received that a terrorist is making his final preparations before setting out. In other cases, like an operation against a senior wanted terrorist, the preparations can take a long time and could include gathering tactical and other information for the operation, operating agents, concealed patrols in the target's area, and more. In the circle of prevention, the Israeli security system takes even further steps that are designed not to physically stop the bomber or terror network, but to inflict wider damage to the organization operating them. To achieve this, Israel—in accordance with the assessment of the situation and considerations of cost—targets senior leaders of terrorist organizations, in order to influence the organizational decision-making process, and it also works to prevent the smuggling of arms and the transfer of money to finance terrorism.

The Delay Circle

The delay circle takes place in the area between the bomber's extended living environment (the village/city's outskirts) and the entrance to Israel's city centers. The nature of the activity in this circle is more defensive and is based on set checkpoints, surprise checkpoints, a security fence, and police activity (patrols, checkpoints, investigating suspicious persons). This circle involves the routine military operations, border patrol, and police forces. Sometimes, when a community or city needs to be closed off, the activity also involves local authority officials such as regional or city security officers. The activity itself takes place on a regular basis regardless of specific intelligence information. However, the activity can intensify when general information is received regarding a terror network that is in its final preparations to carry out a suicide attack, or information regarding a dispatched suicide bomber. The activity within this circle is designed to delay the bomber—that is, to lengthen the amount of time it takes him to get from his home to his target. This delay is meant to fill several purposes:

To Get Specific Information. In other words, "to buy time" for the intelligence units. Checkpoints and heavy traffic can increase the "intelligence signature" of the suicide bomber and transporter and allow the intelligence units to produce more specific information regarding the identity and exact location of the suicide bomber making his way toward the target. On several occasions, says Zohar Dvir, former commander of the police's elite unit (Yamam), Israeli forces managed to reach the suicide bomber's car thanks to accurate information that was received just minutes before and that was produced while the bomber was delayed at a checkpoint or at a traffic jam in the entrance to the city.[157] For example, in one case that took place during the suicide boming's peak period, the Israeli police received general information regarding a dispatched suicide bomber leaving for Israel. Based on this information, the police deployed large forces to the west of Jerusalem, a police helicopter was sent out, and eventually the suicide bomber was arrested in the car that was supposed to transport him to a suicide attack in the town of Beit Shemesh.[158]

Increasing the Suicide Bomber's Room for Error. The delaying efforts make the logistic and operational preparation more cumbersome. The need to find alternative routes, to plan longer travel stretches on the way to the target, and to find a terrorist's transporter who can pass through checkpoints create psychological and practical difficulties that increase the probability of an operational or intelligence error being committed by any one of the chain's links, which would lead to the thwarting of the attack. This principle is nicely demonstrated in the story of the capture of Shadi Bahlul, a suicide bomber for the Islamic Jihad. On February 6, 2003, Bahlul left Nablus on his way to carry out an attack in the city of Netanya. The bomber was joined by two Jihad operatives that were supposed to deliver him to his destination. The three took back roads from Nablus to Tulkarm (which lies on the Green Line) without encountering any checkpoints or military patrols on their way. At the time the ISA had an alert that Bahlul was on his way to the center of Israel to carry out a suicide attack. Following the alert, a siege on Tulkarm was tightened, which was noticed by the three as they approached the city. They turned to a small town called Faruun, to the southeast of the city, and from there walked into Taibe (an Arab city in Israel that is near the West Bank). In the meanwhile, Taibe was also encircled by security forces and the three weren't able to find a taxi driver willing to drive them to Netanya—the target of the attack. Eventually the three decided to return to Nablus. On their way back they were arrested at a surprise checkpoint near the city.[159]

Channeling the Bomber to an Electronic Device or the Open Field. From past experience we know that the reservoir of potential suicide bombers is large. Without any specific information regarding a certain suicide bomber, the Israeli security system tries to channel the movement of the hostile population toward an electronic device, meaning a regular border crossing or checkpoint. In these places, the means of detection make the identification of the bomber possible. Alternatively, as it often happens, out of fear that the suicide bomber would be arrested in a checkpoint/border crossing, the terror networks prefer to send out the bomber through the open field or back roads. The IDF carries out regular random security operations in these areas, both overt and covert, which deter or surprise attackers and their transporters.

The Consequence Mitigation Circle

When a suicide bomber isn't stopped by preventative or delaying activities, the "consequence mitigation" circle is the last line of defense. This circle includes police work, along with the work of governmental and private security units. The activity is carried out on a regular basis, based on a general threat assessment and not only on specific intelligence information. The activity is mainly focused on the protection of government and public facilities, institutions, and the public transportation system. To this end, substantial financial resources were invested in Israel, by both the public and private sectors. Private security guards protect the entrances to businesses, restaurants, shopping malls, cinemas, theaters, and almost every other public accessed areas. A special security unit, though small in size, operates in the public transportation system, and guards hired by the local authorities are posted at the entrances to schools. Community police units patrol between schools and kindergartens.[160] This security activity has two main purposes:

Consequence Mitigation. This involves preventing the suicide bomber from detonating himself in a congested area. That kind of an explosion significantly increases the lethality and potential number of victims. In a significant number of cases, suicide bombers chose to detonate themselves on the street or at the entrance to a public institution rather then risk being detected by the security guard. For example, in the attack at a dance club in Tel Aviv on June 2001, the bomber chose to blow himself up at the entrance to the club. The attack's outcome was horrific: 21 teenagers were murdered. Yet, according to Uri Bar Lev, the Tel Aviv district police commander at the time, had the terrorist managed to go past the security guard and detonate himself inside the club, the number of victims would have been significantly higher, probably resulting in hundreds of casualties.[161] In other cases, security guards managed to prevent the entry of suicide bombers. In this circle the ISA gives professional guidance to the security guards protecting governmental facilities, and the Israeli police force is in charge of regulating the professional circle of the private security guards (though not responsible for training them).

Granting a Sense of Security. An important factor in the war against suicide terrorism is "public resilience" or how much its citizens are able to endure. One of suicide attack's main objectives is to make the threat personal, to undermine the citizens' sense of personal safety. Therefore, the purpose of the security measures is to demonstrate to the citizen that its personal safety has not been discarded; and even if a "successful" attack takes place, it is "only a deviation and not an inevitable evil," as Lior Lotan, former head of the IDF's negotiations unit, phrases it.[162] In Israel it is believed that the strengthening of the sense of personal security, significantly helps strengthen public resilience.

The circle of "consequence mitigation" also includes handling the attack's aftermath, crisis management, and tracking the transporter of the suicide bomber. This activity includes the cooperation of the different emergency agencies (rescue, the fire department, and medical crews) as well as police work.

THE INTELLIGENCE LEVEL

The Intelligence Challenge in the War on Suicide Terrorism[163]

The Israeli experience shows that in all three circles of dealing with suicide terrorism (prevention, delay, and consequence mitigation), the key factor is intelligence. Indeed, intelligence is an important factor in any war or conflict, yet in facing suicide terrorism, its importance can't be overstated. As the former head of the Yamam says: "At the end, without intelligence we are like a blindfolded giant."[164] Confronting suicide terrorism presents the intelligence services with several challenges that either differ significantly or don't exist in the conventional battlefield:

Detecting the Enemy. The first question the intelligence officer has to answer is, Where is the enemy? In the conventional battlefield, in a confrontation between two armies, this question is relatively simple. The enemy is deployed in a certain geographical space, in a certain layout (defensive or offensive). In confronting suicide terrorism, the assignment is much more difficult. The terror network always operates from within the civilian community, using a high degree of compartmentalization and secrecy. Furthermore, the suicide bomber himself is often an "average Joe" who seems to lead a normal lifestyle. Often the potential bomber goes on with his daily routine almost until the very moment he leaves for his attack. He doesn't wear a uniform and certainly doesn't introduce and identify himself as the next "human bomb."

Avoid Harming Noncombatants. Because terror networks operate within civilian populations are often harbored and sheltered in it, any violent action against the bomber or the network's members might result in compensating noncombatants. Beyond the moral issue, there is a great importance to avoid harming the other side's civilians in order to preserve the legitimacy to act both "from home" (the domestic arena) and "from the outside" (the international arena). In a conventional war between two symmetric forces, civilian casualties are perceived as a "necessary evil." In the war against suicide terrorism, because of the asymmetrical nature of the campaign (state vs. terror organizations), civilian casualties are often viewed as a result of disproportional or even brutal use of force.

Information about People, not Buildings. In a conventional battlefield, the information required from the intelligence agencies has to do with the enemy's headquarters, buildings, infrastructures, and other inanimate objects. This type of information is much less relevant in the confrontation with suicide bombers. In suicide terrorism, the operatives or bombers are the key and striking them is what thwarts the attack or hinders the terror network's capabilities. Therefore, houses or other buildings are used only as reference points for locating the operative or bomber.

Fast Processing of Information and Sending it to the Relevant Recipients. In confronting suicide terrorism, one of the basic requirements of an intelligence array is that it will provide early warning that would enable us to either prevent an attack or set up defensive preparations against it. That means that the intelligence services have to process and send out information on time and to the relevant recipients. Meeting this challenge is even more important when the system wants to strike at or arrest a suicide bomber or terrorist cell that is sending out suicide bombers. Every target has a "target lifetime," a window of opportunity in which an offensive operation can be carried out (destruction, elimination, or arrest). Targets in a conventional battlefield, a company of soldiers, a tank brigade, or a headquarters—have a relatively long "lifetime." For example, a tank company will stay at a certain location for several hours, or if it's on the move it will be found in a radius of a few kilometers from where it was initially discovered. This distance can be easily calculated. On the other hand, the "target lifetime" of a potential suicide bomber or any terrorist target can be very brief. Because the bomber is located within a civilian environment, and because of the low "intelligence signature" he leaves, the window of opportunity for an attack or arrest can at times be limited to a few minutes (for example, the amount of time at hand to hit a moving vehicle before it disappears in a maze of alleyways or merges into heavy traffic) to several hours (for example, the amount of time a suicide bomber stays in a safe house the night before he's dispatched). For this reason, the intelligence system needs to be able to process a lot of information in a short time period, turn it into intelligence relevant for an operation, and to distribute it to the relevant recipients on time.

Israeli Intelligence

The Israeli intelligence community is composed of three main intelligence agencies: The first of these is the military intelligence (the Intelligence Branch or Aman), which is in charge of the national intelligence assessment and, to this end, gathers information both in the territories and in Arab countries. This agency is subordinate to the Minister of Defense, although the head of the military intelligence also functions as the government's advisor on matters of intelligence and takes an active part in government meetings. Alongside Aman, which acts as both a collection agency and an information assessment agency, two other agencies operate: The Mossad (the Institute for Intelligence and Special Operations), which is in charge of collecting information and carrying out special operations abroad, and the ISA (the Israel Security Agency), which is in charge of collecting information in the domestic arena, counter-espionage, and the prevention of terrorist attacks. The "Arab wing" of this agency is in charge of thwarting Arab terrorism and insurgency. In addition its security department is in charge of the security of key facilities in Israel, government offices, and important administration figureheads.[165]

As noted by Tucker, unlike the often referred to rivalry between the FBI and the CIA, Israel's intelligence agencies usually cooperate in intelligence collection and share information that is relevant for confronting terrorism. Thus, as part of the daily work, vast amounts of information flow from one intelligence agency to another, in accordance with the defined areas of responsibility. The three agencies (Aman, ISA, Mossad) share a continuous cooperation, which is, for the most part, operational as well. Representatives from one intelligence organization occasionally stay in other organizations for specific projects. For example, the ISA does background checks on Aman and the Mossad's personnel, as well as for other IDF units. The three main intelligence agencies are represented in the Heads of Services Committee (Varash), which convenes once every few weeks. In this forum, issues regarding the working relations and interfaces between the three agencies are debated. These meetings are attended by the head of the Mossad, the head of the ISA, the head of Aman, and the prime minister's military secretary.[166]

Despite all this, the Israeli intelligence community was not ready to deal with suicide terrorism on an intelligence or operational level. During the first year and a half of the recent confrontation wave with the Palestinians, from September 2000 to April–May 2002, the Israeli security system managed to thwart only a small percentage of all suicide attacks. For example, in the first quarter of 2002—the height of the wave of the suicide attacks—the percentage of attacks thwarted stood at 11%.[167] Ilan Paz, the former commander of the Ramallah brigade, best describes the intelligence situation at the time: "Up until operation Defensive Shield [the reoccupation of the West Bank cities in April 2002] the times were catastrophic ... we lived with nonspecific warnings almost every day. The intelligence was very shallow and the ability to pair the intelligence with a special unit that would come and thwart the attack was close to zero."[168]

The source of this situation is in a number of developments that took place in the decade that preceded the second Intifada, in September 2000. In 1993, a peace agreement was signed between Israel and the Palestinians (the Oslo Accords, 1993). Following those Accords, the IDF withdrew from Palestinian populated cities throughout the territories and the Palestinian Authority was established. The Palestinian Authority was responsible for governing the local Arab population in the territories. Following these developments, one of Israel's main sources of information on the Palestinians was dwindling: Human Intelligence ("Hum-Int"), intelligence produced from the recruitment and operation of agents from within the adversary's population. The reasons for this lie in the blows handed to two key elements in source recruitment: the personal interaction of the ISA intelligence officer with the local population and the local government's interaction with him. The interaction between the intelligence officer and the target group allows the former to personally know the target population and mainly to identify new opportunities to recruit agents. Personal acquaintance is the key factor in the process, as Nissim Levi, a former high-ranked ISA official, puts it: "an intelligence officer in charge of a refugee camp of 18 thousand people knows everybody. He's the prince of the area. An intelligence officer needs to be a bit of a 'gossip.' He has to know everybody, know what they look like, what they're wearing, what their nickname is, how many kids they have, and what their origin is."[169] Another key element that is equally important is "a deep understanding of the field and culture" says Yisrael Hason, former deputy head of the ISA, who started his career as a intelligence officer in the Jerusalem area.[170] Before the Oslo Accords, ISA facilities were located in the heart of the Palestinian population and the intelligence officer could patrol the area freely and know it and its residents intimately. As former head of the ISA, Yaakov Perry, expressed:

> I served in Nablus for almost two years. I learnt to know every street and ally, every roof that connected to another roof in the Kasbah, every hole that led to another escape hole, every neighborhood and sheikh's grave, every stream, every grove and abandoned field. I could walk around the area in total darkness and find my way easily. I sat in coffee shops a lot and talked to the locals. I saw the hate in their eyes, but they always carried out the commandment of good hospitality to its fullest, and never tried to harm me when I was a guest at a private residence or when I sat in a coffee shop in Nablus.[171]

All of this came to a near halt with the exit of the Israeli security forces from the Palestinian population centers in the second half of the 1990s. All of the IDF camps were evicted and rebuilt outside of the cities. The ISA facilities were moved into Israeli territory. Now, the entry of every ISA intelligence officer for a tour of the field was accompanied by heavy security.[172] Worse, as far as the recruitment process was concerned, the Palestinian authority took the place of the Israeli government. This resulted in a decrease in the Palestinian citizen's dependency on Israel. In the past, the ISA was able to exploit his ability to grant a long line of benefits other than money for recruitment: Israeli work permits, jobs, a pay raise, a decrease in prison time for criminals, and so on. Now the Palestinian Authority's institutes arranged for these, or alternatively were an official mediator between Israel and the Palestinian citizen. As a result, the ISA officers' ability to identify opportunities for recruiting new agents and operating old ones also decreased. Following the new situation, the amount of intelligence relevant for counterterrorism shrank. Early signs of this could already be seen in the suicide bombings of the 1990s, but this weakness was fully exposed by the waves of suicide bombers that started early after the onset of the second Intifada.[173]

Beyond the dwindling of human intelligence sources, the intelligence services were required to change their collection emphases. In the years before the September 2000 Confrontation broke out, the Israeli intelligence focused mainly on thwarting attacks (primarily suicide attacks) carried out by the Islamic terror organizations. After the signing of the Oslo accords, the Fatach was no longer perceived as an enemy and was not high in the ISA's list of priorities. However, a year after the violent conflict broke out, it was Fatah terrorists—the Al-Aqsa Martyrs Brigades—who were responsible for a significant number of suicide attacks. In March 2002, they even launched more suicide bombers than all the Islamic organizations (Hamas and Islamic Jihad) combined. Furthermore, the ISA grew accustomed to work in cooperation with the Palestinian security services and pass on to them missions to arrest suspects in the Palestinian Authority's territories. Now these ties weakened, trust was broken, and still it took a relatively long time until Israel recounted that it could only trust itself. Initially it was lacking in sources: There was so much violence that carrying out meetings in the field with Palestinians who reported to the ISA became difficult. Some took the opportunity to switch sides and retake a part in the national struggle. Several months passed before the ISA adjusted itself to the new nature of operations and took it upon itself to lead the confrontation against terrorism. Just as importantly, it took the Israeli intelligence time to understand how the new terrorist cells, which kept sprouting during the Intifada, worked. This understanding is key in understanding (a) how to allocate the intelligence resources, (b) where an ISA intelligence officer should focus his collection efforts, and (c) how to decide on a priorities list for operational care—in other words, who gets eliminated/arrested and when.[174]

The Israeli answer to this situation was given in several different levels; the first one is a formal organizational level that settles the division of labor between the intelligence agencies. This issue is often a weakness in the operations of intelligence services around the world. In Israel, given the new situation created by the Oslo Accords, the Israeli intelligence services decided to create think tanks to examine the new reality. This decision was especially relevant to the ISA, but it was also discussed in the Heads of Services Committee (Varash)—a governing body composed of the head of the Mossad, the head of the ISA, and the head of Aman, which is in charge of coordinating the different agencies of the intelligence community. The think tanks' conclusions caused major changes, of structural and substantial nature, that were known as "Magna Carta 1 and 2." In the structural area a clearer division of labor between Aman, the ISA, and the Mossad in the Palestinian Authority's territories was established. Its purpose was to increase the cooperation between the agencies, to establish geographical and functional borders between the different collection agencies, and to avoid overlapping and unwanted frictions. According to the new understandings, it was agreed that the overall responsibility for intelligence gathering in the Authority's territories would remain with the ISA. However, some tasks were passed on to Aman's responsibility, among them the political analysis of that area. It was also decided to allow Aman's agents unit, 504, to operate in the Authority's lands. This activity was exposed following the June 14th murder of Lieutenant-colonel Yehuda Edri, the deputy commander of 504's Jerusalem section, while meeting with a Palestinian agent near Bethlehem.[175]

The clear division of jurisdiction between the IDF and the ISA regarding the territories—alongside the growing frequency of suicide bombings and, with it, a sense of national urgency—tightened the cooperation between the two bodies and allowed for a better flow of intelligence in the system. As a result of several decisions made by the IDF chief of Staff, Shaul Mofaz and the head of the ISA, Avi Dichter, a large amount of the barriers in passing of intelligence, between the two bodies and between them and the police and other intelligence and operational bodies, were lifted. This allowed not only a better understanding of reality and a fast response at the headquarters level, but also a faster flow of information to the field. Thus, today as opposed to the past, detailed information of warnings reaches the commanders of forces in the field and isn't stopped in the ISA or Aman. Brigade commanders and regiment commanders have direct lines of communication to the ISA intelligence officer in their area. The effective links and sharing of information, especially in the West Bank, have affected the speed of response at the face of immediate warnings.[176]

The link between ISA officer, Aman officer, air force operator and police officer is greatly improved and as efficient as ever. The information first flows horizontally, to those who need it for immediate use, and only later does it go upwards, to the executives.[177] In fact, as former IDF Chief of Staff, Moshe Yaalon, put it: "the breaking of barriers between organizational layouts is our crowning achievement."[178] The Israeli intelligence system managed to integrate or even fuse information by forming working channels between the layouts and different organizations both in the headquarters level and in the working ranks and field levels. The head of Aman, Ze'evi-Farkash, calls this organizational reality "an integrated intelligence bathtub" into which information flows from all sources: "Hum-int," "Vis-int" (visual information—information obtained from observations and different means of collection), "Sig-int" (signal information—information produced from listening in on different means of communications), and information from detainee interrogations.[179] This situation not only allows for a better flow of information, but also provides the ability to concentrate an intelligence effort on a daily basis against significant terrorist targets—for example, the heads of terror networks or suicide bombers that hadn't been launched. It's true that flow of intelligence and sharing of information are principles that are relevant for any kind of combat, but in the conventional battle field they merely give an advantage, whereas in confronting terrorism it is a critical component. Only by using these principles can a terrorist or suicide bomber be distinguished from his civilian environment and can be arrested/eliminated in real time.[180]

This integration of intelligence is not only a result of elevated organizational and work level barriers, but also of advanced development of technological abilities—especially when it came to C4I systems (command, control, communication, computers, and information), which could transport raw material (such as aerial photographs or integrative intelligence products) and to distribute them to all of the intelligence community in real time.[181] This area's importance is shown by the fact that one of the main functions in the ISA's headquarters is the information systems department. More than 10% of the organization's purchasing budget is directed at preservation and development of information infrastructures, as opposed to single percentages in civilian organizations.[182]

The cooperation within the intelligence community was joined by another factor during the second Intifada, which was just as important: a fast connection between the intelligence and the executing element—the shooter (helicopter, sniper), or unit making an arrest.[183] A fast flow of information from the intelligence body that produces or processes it to the executing element is critical when dealing with terrorist targets. A terrorist or suicide bomber is, as previously mentioned, a "low signature" target, meaning that they blend into their civilian environment without leaving a sign. Furthermore, the target's "lifetime"—the window of opportunities to hit or arrest a terrorist target—is very brief (minutes to hours). Therefore the immediate translation of intelligence information into an operation is critical for the operation's success or the thwarting of a suicide attack. As Moshe Yaalon, the former IDF Chief of Staff, explains: "The trick is to be able to take everything you've collected intelligence wise… and turn it into real time knowledge of 'Muhammad the terrorist is at that place right now' and in real time, not 'was there yesterday,' but is there right now, and 'should be there at a certain hour' is even better. You should have the ability to get forces to that place quickly."[184].

As the Confrontation went on and the quality of the intelligence improved, the ability of the Israeli system to respond also improved, as the commander of the IDF's Ramallah brigade during the Confrontation says: "The ability to hook up a special unit to a terrorist's arrest increased."[185] How this principle works de facto can be learned from the words of Zohar Dvir, the commander of the Yamam, Israel's police special counter terrorism unit: "As the Yamam's commander I had far-reaching freedom of action. … I could get a phone call from the ISA and to call in the unit with no questions asked, and to update the Police

Chiefs only after or during the act. It's legitimate, part of the rules of the game, or else you couldn't 'close circles' [an expression meaning to turn intelligence into an operation] so quickly."[186].

Alongside the improvement in intelligence cooperation and the flow of information to operational elements, the Israeli intelligence also vastly improved its technological abilities. This process began in the late 1990s, and it truly bore fruit in the second Intifada. Following the Oslo Accords, a former head of an ISA department stresses that "because of the circumstances [the withdrawal from the Palestinian cities and dwindling of human sources], a perceptional change took place; the organization, which was mainly based on 'Hum-int,' improved its technological ability and made it more efficient."[187] The cooperation with the military intelligence in this field eventually resulted in, as the head of Aman's research department defines it, "the development of an intelligence ability of the highest order."[188] The technological supremacy turned into a clear intelligence dominance as a result of the regular control of the field that was achieved following the reoccupation of the West Bank cities in operation "Defensive Shield" (April 2002, see below). The control of the field enabled the translation of the intelligence information into a massive arrest of operatives. The operatives' interrogations, in turn, provided valuable information and enabled the Israeli security forces to know of most of the terror plots before the suicide bombers left their houses on route to Israel.[189]

Alongside the systemic issues, a few more topics can be mentioned that improved the Israeli response to suicide terrorism. Among others, we should mention the improvement in agent recruitment during the confrontation. This ability that, as previously mentioned, was compromised during the 1990s, was significantly restored as the Intifada went on and the financial situation in territories worsened. The severe economic distress that the population in the territories found themselves in made them more willing to consider collaborating with the ISA.[190]

Another important step was understanding of the manner in which suicide attacks were carried out and how the terrorist organizations were organized. The new terrorist cells that formed after the Confrontation broke out in September 2000 were different from those that operated in the 1990s in many ways: Primarily, the human composition became more diverse. Prior to the Intifada, it was mostly "professional" terror cells operating in the territories, which were composed of Hamas and Islamic Jihad members—religious fundamentalists. These were mostly people known, to some degree, to the Israeli security forces. In the first months of the Confrontation they were joined by a different kind of operatives—youngsters, members of Palestinian security forces, Fatah field operatives, and various criminal elements. Into all this fit in handlers on Hezbullah and Iran's behalf as well as Fatah leadership elements, who were not known for their connections to terrorism. All of these cooperated with each other. For example, one could find terrorist cells that included activists from the Palestinian Authority's security forces cooperating with an Islamic Jihad terrorist cell. Such was the case of Abd Al-Karim Awiss, an officer in the

Authority's General Security Forces who headed a terrorist squad in Jenin and sent a suicide bomber on a double suicide attack with members of the Islamic Jihad.[191]

Not only did the human composition change, but the modus operandi of these terror networks was different and new. For example, the Fatah's Tulkarm terrorist organization included Palestinian Authority's security officers and Fatah activists alongside criminals who used to steal cars in Israel. The latter used the knowledge they gained to infiltrate suicide bombers or explosive charges into Israel in the car theft routes. In one case, this terror network sent a car thief named Muhammad Abu Jamus, who was known as the greatest car thief in the territories, on a mission in the Israeli home front. Abu Jamus broke into the home of a naval officer, Colonel Natan Barak, in Ra'anana and placed a large explosives charge in it (the information about the house came from Palestinian workers who were doing renovations work in the neighborhood). Barak woke up in the morning, heard a peculiar ticking sound, and took the charge out of his bedroom to his backyard.[192] The Israeli intelligence officers learns to identify a routine, and a deviation from that routine provides them with signs—the enemies' activities, movements, and expressions, all of which could signal his intent.[193] Now the Israeli intelligence officers needed to relearn how a terror network operates in the Intifada. As Cooperwasser, the head of Aman's research department admits, it took the Israeli intelligence a long time to figure out how a terrorist attack was carried out (money, transporter, weapons, etc.). Ilan Paz, the commander of the Ramallah brigade in the Intifada, notes the following: "Only after two years of conflict was the chain of attack preparation starting to be understood."[194] These insights had ramifications on critical decisions in the system, such as the prioritization of the intelligence effort, the level of alert, and the ability to investigate attacks in retrospect.

In sum, Intelligence is probably the most important factor in counterterrorism. When referring to the issue, former IDF Chief of Staff said that if he were responsible for allocating resources, he would invest the lion's share of the security budget on intelligence.[195]

THE OPERATIONAL LEVEL

The Evolution of Israel's Responses to Suicide Terrorism—A Brief Historic Review

Based on its intelligence capability, the Israeli security system came up with a line of steps and operational methods for confronting the suicide bombers phenomenon. However, the evolution of the operational answer for this form of terrorism was inconsistent and was more a process of trial and error. A number of factors influenced this process. The first was the understanding that the political reality had changed. It took the Israeli security forces a long time to understand that the Palestinian partner—the Palestinian Authority and Fatah leadership—had turned into enemies. In the years after the signing of the Oslo Accords, a close cooperation was established between the Israeli and the Palestinian security bodies. This relationship included the regular passing of information regarding terrorists and terrorism plots. In many cases the Palestinian security forces worked on Israel's behalf and arrested terrorists or potential suicide bombers based on information produced by the Israeli intelligence. Even when the violent Confrontation broke out, Israel focused its attention on putting pressure on the Palestinian Authority, in an attempt to demonstrate to the latter the cost of its support of terrorism and to make it act against the terror organizations. At any rate, the Israeli activity was mostly responsive in nature and was carried out only after big attacks in population centers or after severe attacks in the territories. The response included limited strikes from the air or the ground against Palestinian facilities or checkpoints. At that time, says Ilan Paz, former head of the Ramallah brigade, "we developed the attitude of 'the attack came out of Jenin; and Ramallah, as a governmental city, will pay.' Then we would attack Palestinian strongholds and checkpoints. If we could prove an attack came from a specific checkpoint, we would attack it."[196] Starting in September 2001, measured nightly raids were carried out on the Authority's headquarters in the cities, with the forces leaving at dawn.[197] "We'd march on checkpoints with tanks and armored vehicles, or just pass through Bituniya's main street as a demonstration of power," as Paz describes the nature of these operations.[198]

Additionally, in a futile attempt to prevent the suicide bombers passage into Israel, several steps were carried out to limit the movement in Judea and Samaria: Roadblocks and checkpoints were placed, the supervision on exit routes from the territories was tightened, and the military presence on the roads was increased.[199] All these measures did not provide an answer to the suicide bomber's phenomenon, because as Cooperwasser, head of Aman's research department, notes, "the idea was that we continued to entrust some of our security in the Authority's hands."[200]

Indeed, in a press conference, about a year after the Confrontation broke out, the head of Aman, Amos Malka, still believed that "a combined system of pressures—by Israel and the international community—might change Arafat's policy."[201] A hard debate, which in historic perspective seems meaningless, was taking place in the Israeli intelligence community around the question, "Is the Palestinian Authority's chairman, Yasser Arafat, unable or only unwilling to operate against the terror organizations?" Only at this point, approximately a year and three months after the Confrontation broke out, after dozens of suicide attacks and hundreds dead, did the Israeli security services come to the conclusion that the Authority's thwarting efforts are "one big act."[202] In simple words, it took too long for the Israeli security establishment to realize that no one else would do its work for it. From this point on, aggressive operations though on a limited scale were carried out to confront the terror infrastructures that blossomed undisturbed in Palestinian cities. These operations reached their high point after the suicide attack on Passover in Netanya (March 27, 2002). A suicide bomber dressed as a woman, who was sent by Hamas, blew himself up in the

Park hotel's dinning room during the festive meal. The explosion killed 30 people and injured over 143. Following the attack, on March 29, 2002 Israel set out on operation "Defensive Shield," in which the West Bank cities were re-occupied. From this point on, Israel took its fate into its own hands and started to fight the suicide bombings' terror network itself, in a systematic and intensive manner.

Another factor that affected the evolution of the Israeli response to suicide bombings was the issue of internal and international legitimacy. From a Palestinian standpoint, suicide bombings were meant to force Israel to make further land concessions and to create a debate in Israeli society regarding the continuation of their presence in the West Bank and Gaza Strip. However, as the number of Israeli casualties in suicide attacks grew, the opposite opinion grew stronger throughout the Israeli public. Most Israelis felt that the severe attacks reflected a murderous Palestinian passion and an attempt to take over the lands within Israel itself (the "1967 borders"). This mindset allowed the government and Israeli security forces to act more aggressively against terrorism.[203]

At the same time, as the international TV screens were filled with pictures of Israeli terror victims, the international legitimacy for stronger action against the Palestinian Authority grew. Gradually, as it became apparent that "there's no one to talk to" in the Palestinian Authority concerning the fight against terror, Israel's need for "freedom of action" was accepted by the United States. That being said, the Americans still frequently condemned, sometimes harshly, Israel's target killing operations. Ron Preshauer, who served as Israel's diplomatic advisor in the Israeli embassy in Washington, was occasionally summoned for reprimand meetings. The American administration even pressured Israel to stop using American weapons for the target killing. A hinted threat of stopping the supply of spare parts for "Apache" helicopters (which were sometimes filmed launching missiles in the territories) was made, and there was even a demand to avoid using American-made sniper rifles for the assassinations. Those demands vanished after the September 11th attacks. The United States understood that terrorism, in all its manifestations, must be confronted with all the legitimate means at our disposal. This also affected Israel's freedom of action and allowed the Israeli security forces to run a freer, more aggressive campaign against the terror infrastructures in the territories.[204].

Finally, it seemed that the evolution of the Israeli response to suicide attacks was also affected by different psychological or perceptional blocks in the Israeli security forces themselves. Primarily, the Israeli armed forces were hesitant to enter the Palestinian Authority's territories (referred to in the Oslo Accords as Area A). The Israeli intelligence warned of a large number of casualties when entering Palestinian territory; and the army, which still harbored the "Lebanon trauma," was extremely cautious. It seems that the IDF's negative experience in Lebanon was indeed a significant factor in shaping the Israeli response. Until May 2000, the IDF controlled a narrow strip of land in southern Lebanon based on the assumption that this would help in preventing attacks in Israel. During that time the IDF suffered casualties in a daily war of attrition with Hezbollah. Eventually, mainly because of the pressure of the Israeli public opinion and in view of the many losses suffered by the army, Israel withdrew to the international border. Hesitant of another entanglement similar to the one in Lebanon, the army avoided a massive invasion of the Authority's territories. Operations in Palestinian areas were carried out with a great degree of caution and mainly in the area's outskirts.[205] Even when this activity was gradually expanded and the IDF carried out nightly raids in the Palestinian cities, the refugee camps—the terrorists main strongholds—remained off limits.[206] In this context, it seems that the IDF officers displaced thought patterns that they used in the fighting in Lebanon, and they copied them to the fighting in the Palestinian territories. As Hagai Peleg, the Yamam's commander in the beginning of the second Intifada, notes: "It was a natural but wrong process." The concepts of time and space in the operational activity in Lebanon were completely different. "There it took six hours for a force to walk into an ambush in the Wadi. Here, you can get from Kibbutz Eyal [in Israel] to Qalqiliya in five minutes." Compared with Lebanon, the rescue forces' work in the territories was also much easier. For all these reasons, the IDF's activity in the first months was slow and cautious.[207]

Gradually, following an increasing amount of suicide attacks in Israel, the IDF was pushed to change the nature of it actions. The public demanded results in the fight on terror, politicians exerted pressure, and in February 2002, a year and five months after the Confrontation broke out, the IDF set out on two large-scale operations in two refugee camps that formed the spearhead of suicide terrorism: the Jenin refugee camp and the Balata refugee camp in Nablus. Aviv Kochavi, commander of the paratroopers brigade at the time, notes that in the IDF's perception these places were still thought of as "the dark forest that you never enter." However, both operations turned out to be even more successful than planned. The forces entered the heart of the camps and suffered relatively few losses: In Jenin a soldier from Golani's anti-tank company was killed; in Balata a soldier from the paratroopers' reconnaissance company was killed. The Palestinians suffered dozens of fatalities, mostly of armed militants. The resistance, especially in Jenin, was greater than anything the IDF had ever encountered in the West Bank: Dozens of shootings and charges were set off; however, the commanders left the camps with the feeling that the challenge could be met and the Palestinians witnessed the IDF's superiority even at their home court in a crowded environment. It was, said one of the IDF officers, a watershed. "We realized what we were capable of."[208]

This realization was finally translated into widespread military activity at the end of March 2002, but not only because of the army's initiative, but mainly because of the pressure from the Israeli public. On March 27, 2002, a Hamas suicide bomber exploded in the Park hotel in the midst of the Passover festive meal. The attack killed 30 people and injured 143. The Passover night massacre was the "straw that broke the camel's back." It was the peak of a wave of unprecedented suicide attacks that took the lives of hundreds of Israelis. Targeting dozens of Jews observing a religious ceremony created a wide public consensus for a large-scale military operation in the Palestinian Authority's controlled territory, which had become out of control. It was high time, it was said, to show the

Palestinians the IDF's true power—and remove the limitations put on it that hampered its activities in the previous months. "We are at war" said Prime Minister Sharon in a special speech to the nation, "the war on our home." We did everything possible to reach a ceasefire, claimed Sharon, "and in return all we got was terror, terror and more terror."[209] Following the government meeting on March 28th, a military move, unprecedented in its scale, had begun: "operation Defensive Shield," during which the IDF took over most of the West Bank cities and all of its refugee camps within two weeks.[210] The objective of the operation was defined as the dismantling of the terrorist infrastructures, but in practice it was much more meaningful: In principle, for the first time since the Oslo Accords, Israel took back the security responsibility over all of the West Bank—and stopped recognizing the Palestinian Authority and its bodies as partners in this responsibility. From the operational perspective, Israel moved from defense to offense and took the fight to the opponent's court.[211]

The operation's outcome was dramatic: Approximately 6000 Palestinian militants were arrested, 54 potential suicide bombers were killed, and 58 suicide bombings were thwarted. The effect on the terrorist activity was felt immediately. In the first quarter of 2002 (the time period just before "Defensive Shield") 40 suicide attacks were recorded within the green line; in the second quarter, 23; in the third, 17; and in the fourth, 12. In the first quarter of 2003, only one suicide attack took place. It is true that in the following months some of the terrorist networks managed to rebuild themselves and renew the suicide attacks; however, the free hand given to IDF following the operation allowed the gradual dismantling of these networks as well. The physical ability to reach every site in the Palestinian cities enhanced the Israeli security system's success in preventing suicide attacks. In fact, the ratio between suicide attacks that succeeded and those thwarted was reversed, as the head of the Operations branch at the time, Dan Harel, notes. If in the year before operation "Defensive Shield" the ratio stood at two to three bombers who "succeeded" for every one arrested, the trend reversed after the operation. "Out of 300 suicide bombers, we caught 90% of them after 'Defensive Shield,'" says Harel.[212] The great freedom of action that the IDF enjoyed since the operation allowed the Israeli security system to continue with systematic preventative operations. Ever since, the terrorist organizations in the West Bank never really managed to rebuild their capabilities and were forced to invest more resources into survival and self defense. Accordingly, the number of Israeli casualties in terror attacks decreased and stood at 5 in 2006, as opposed to a record 453 in 2002.[213] All of this is fittingly summarized by Moshe Yaalon, the former IDF's chief of staff: "It has to be clear that it's not a one time thing. We started entering the territories before 'Defensive Shield,' bought control since 'Defensive Shield,' and the war is going on ever since with arrests and continuous preventative operations."[214]

The Israeli Methods of Action

As previously mentioned, the Israeli methods of action gradually evolved as the conflict escalated and as the number of Israeli casualties from suicide attacks rose. These methods can be basically categorized as defensive-passive and offensive-active. Within the Israeli offensive actions, one can mention the targeted killing that reached an impressive degree of accuracy during the Confrontation; raids in the Palestinian population centers; operating special units in various operations in the Palestinian urban landscape; and arrest operations. Alongside these, Israel developed, or rather perfected, several more defensive methods of action, such as checkpoints and a security fence. This category also includes different measures taken in the Israeli cities themselves, including securing facilities and public areas. These measures, offensive and defensive, were mainly meant to deal with suicide bombers and their dispatchers. On the other hand, another set of measures was developed for dealing with suicide attacks on other levels: (a) the financing of the attacks and (b) deterring the bomber with punitive measures against his family (expulsion and house demolition—all in accordance to international law). The degree to which these methods were effective varied from time to time. However, with the perspective of several years, it seems that the key factor in the success in the fight on suicide terrorism is preemptive and offensive action against the terrorist networks' capabilities. The offensive action has to be focused, meaning that it has to be based on specific information and include mainly arrests; and when these can't be carried out for various reasons and the threat is imminent, the bomber, dispatcher, or explosives expert must be stopped by other means before they can carry out their murderous plots. It seems that the offensive action must be accompanied by a series of defensive steps that are based on the setting of a physical barrier (checkpoints, security fence) and a human barrier (security personnel) between the suicide bomber and his target. All these will be elaborated upon in the next part of the chapter.

Targeted Killing

A Brief Historic Review. It was late at night. The Israeli ISA intelligence officer of the city of Bethlehem finished another meeting with an agent and got into his car. The cell phone rings, "Abu Ali," as the officer is known by the Palestinians, answers. On the other side of the line is one of his junior agents, the report is short and laconic: "Hussein is sleeping at home tonight." This is the report the ISA have been waiting for. For the last several weeks they have been trying to track Hussein Abayat, the head of the Fatah's military wing in Bethlehem, who was responsible for the shooting at Jerusalem's southern neighborhood and for the murder of several Israelis and who, based on intelligence reports, was plotting an imminent attack. Now there is a lead. Abu Ali thanks the agent and immediately calls one of the research personnel in headquarters. A short check in the databases reveals that Hussein's house is known and so is the car he drives——a silver jeep. Equipped with this information, Abu Ali places another call, this time to Mustafa—one of his more trustworthy field agents. Mustafa is sent to patrol Al-Saf Street and check if any jeeps are parked on it. After half an hour, a report comes in: "A silver colored jeep is parked near one of the Abayat family's houses." "Just one more thing," says Abu Ali, "tomorrow at 6 A.M. tell me who gets in the car." At 1 A.M. the phone rings in the house of Avi Dichter, the head of the ISA. The head of the Bethlehem district is calling (district—a geographic unit in the ISA that usually includes a major city or several villages[215]): "An intelligence circle has been closed on Hussein Abayat." Dichter, an experienced intelligence officer, didn't need anything else, "O.K. good luck," he approves. Meanwhile the IDF gets the report; an operation order is issued from the air force headquarters to an Apache helicopters squadron. At the same time, final authorization is given by the prime minister. 4 A.M.—two pairs of pilots in the briefing room. Final details are decided. An ISA command post is opened to receive the most up-to-date intelligence reports. A check is made with one of the IDF's collection units-Hussein spoke on the phone in his house during the night. A senior air force officer arrives at the air force's command and control center. 5 A.M.—a MRPV (mini remote-piloted vehicle) hovers over Bethlehem. 6:15 A.M.—Abu Ali answers the phone. Mustafa reports: "Hussein Abayat entered the jeep and drove towards the city center." The ISA command post checks with the air force; the MPRV identifies that the silver jeep is heading south. There is one man in the vehicle. 6:18 A.M.—two pairs of Apache helicopters rise above the eastern ridge of hills overlooking Bethlehem. The leader identifies the target. No other vehicles in the vicinity. Requests the control center's permission to execute. 6:20 A.M.—two Hellfire missiles hit the silver jeep. Hussein Abayat, head of the Fatah's military wing in Bethlehem, is killed. The date: November 9, 2000—the age of targeted killing has begun.[216]

Many Israeli methods of confronting terrorism are considered controversial. Of these, it seems that the most controversial is the targeted killing. This method isn't, as many believe, an outcome of the Palestinian Intifadas or suicide attacks. The targeting of Arab terrorists were practiced since the beginning of the Israeli–Palestinian conflict. However, as an organized policy, the beginning of targeted killing can be seen as December 27, 1947. A month after the UN decided on the establishment of two states, Jewish and Arab, and before the bloody confrontations between the two sides escalated into war—Israel's war of independence—the Hagana (an organization that would later serve as the main foundation for the IDF) issued a operation order called "Zarzir." This order can be considered the first comprehensive operation plan of what would be referred to decades later as "targeted killing." For the first time a nationwide operation of "targeted killing" was planned, with clear operational rules, limits, and criteria. Twenty-three Arab local political leaders and officers were included on the list.[217]

The use of this tactic continued in the 1950s. The effort was focused against intelligence officers and Egyptian attachés who organized Arab gangs in order to kill Israelis. Thus, in July 1956, Colonel Mustafa Hafez, Egypt's military intelligence commander in the Gaza Strip, who was responsible for sending out the Fedayun ("infiltrators") from the Gaza Strip to Israel, was eliminated. Hafez was killed by an explosive package hidden in a book and handed to him by an Egyptian double agent. The use of explosive envelopes became a major tool in the

1960s, mainly in the killing of German scientists, some of which served the Nazi regime and who were involved in the development of advanced weapons in Egypt. For example, Dr. Heinz Krug, who was involved in an Egyptian missile building project, was killed in this manner. In the 1970s the use of eliminations was expanded. After the murder of the 11 Israeli athletes in the 1972 Munich Olympics, Prime Minister Golda Meir ordered the Israeli Mossad to go on a mission of revenge in which all the people involved with the murder would be killed. The intelligence prepared a "bank of targets," a special ministers committee authorized it, and by the late 1980s, most of the relevant terrorists were killed.[218] In the 1990s the amount of targeted killings decreased, although the quality of the targets improved. That is how Sheikh Abbas Musawi, the secretary general of Hezbollah (killed in Lebanon along with his wife and son by the air force's helicopter fire), Islamic Jihad's secretary general Fathi Shkaki (killed in Malta by gun shots from an operative on a motorcycle, an operation attributed to the Mossad), and "the father of Palestinian suicide attacks," Yihiya Ayash (killed in Gaza by an exploding cell phone handed to him by an Israeli agent), were all eliminated.[219]

So what's changed? The essence of the change is the turning of targeted killing into a whole doctrine, part of the ISA, IDF, and police's fighting doctrine. In fact, it's about developing an organized operating perception that starts with an "indictment" of sorts that specifies all of the operative's crimes, continues with getting authorizations, legal follow up, and inter-organizational work procedures, and ends with specific intelligence and closing the circle of fire on the terrorist himself. Targeted killings are complicated operations that are coordinated by a designated op-center in the ISA. From the moment the decision is made, this kind of project can take days, weeks, and even months.[220] In this context, it seems that the watershed in the evolution of targeted killings took place in May 2001 in a small nightly meeting that the prime minister at the time, Ariel Sharon, held with the IDF's chief of staff, the head of the ISA, and his deputy. Sharon ordered the transfer of all necessary budgets needed for target killings to the ISA. "We need to strike the Palestinians everywhere at the same time. They need to wake up every morning and find out that they suffered 12 fatalities, in various operations—without understanding how it happened," he told the security officials.[221] And indeed in the coming months the pace of the targeted killings grew faster. In July alone, seven targeted killings were recorded. After 4 years of confrontations the targeted killings numbered at 95.[222] The tactic itself had become, as Jones says, "a standard operational procedure."[223]

The Criteria for Executing a Targeted Killing and the Selection of Targets. Even when targeted killing had become a more common tactic, they were still carried out based on several clear guidelines. These had already been established in the early stages of the Confrontation and later received a legal seal of approval in a Supreme Court verdict dealing with targeted killing*. The first principle in this context determines that targeted killings can only be carried

*The Public Commitee against Torture in Israel V5, The Government of Israel (HCJ 769/02)

out based on specific information. The method of cross-referencing information that the Israeli intelligence system had developed "had achieved amazing results," says Giora Eiland, formerly the head of the National Security Council. "Of hundreds of targeted killing attempts during the second Intifada, there was never even one intelligence error—meaning a mistake in the identification of the wanted man or vehicle."[224] A second principle for targeted killing says that such an operation would be carried out only in a case of "a ticking bomb," meaning as a way of preventing an attack rather than as a punitive measure for crimes of the past. A third principle is that a targeted killing would only take place if there is no reasonable possibility for an arrest, meaning that an arrest would involve a severe risk of soldiers lives. Finally, it was determined that a targeted killing would be carried out only when the collateral damage is "reasonable" when weighed against the benefit that would come from the terrorist's elimination ("the principle of proportionality").[225]

Targeted killings are meant to be used against three kinds of terrorists, primarily against suicide bombers before or on route to the target of their attack ("a ticking bomb"). The second kind included in the targeted killings involves the centers of knowledge whose elimination would severely hinder the terrorist network's abilities ("a ticking infrastructure"), such as the head of the network, the explosives expert, or the suicide bombers' recruiter. A third kind of target is the terrorist organization's leaders. The first targeted killings in the Intifada were carried out because of an operational constraint—an inability to stop bombers on their way to an attack, as former IDF chief of staff Moshe Yaalon notes: "We had come to need of targeted killings because of a lack of accessibility and lack of control of the field."[226] Since the Palestinian Authority was founded in 1994, the IDF avoided entering its territory. Information regarding bombers or terror operatives was passed on from Israel to the Palestinian security forces, and these had dealt with the suspects. When the Intifada broke out in September 2000 the security cooperation ended, and some members of the Palestinian security forces even joined the fighting themselves. Faced with this, the IDF avoided entering the Palestinian cities because of political and operational limitations. This turned them into cities of refuge for terrorists and suicide bombers. Therefore it was only natural that Israel would adopt the weapon of targeted killing first and foremost to strike "ticking bombs." As Yaalon puts it: "Once you can't surprise the terrorist in his own bed, you are forced to get to him in another way before he attacks innocent civilians."[227]

That being said, as the fighting went on, the group of potential targets was expanded. This included not only "ticking bombs" but also what the ISA called "ticking infrastructures." This included the explosives experts, for example, out of the assumption that targeting them can prevent or stop attacks. As the fight went on, says a former deputy head of the ISA, it became apparent that more and more terrorist organization leaders had become involved with giving directions for terrorist attacks and it was decided to add them to the pool of targets.[228] A prime example of this kind of assassination was the operation against the secretary general of the "Popular Front," Abu Ali Mustafa. In the 1970s the

Front led the airplane hijackings and was considered to be second in power only to the Fatah. However, after the Soviet Union collapsed and the peace agreement with the Palestinians was signed, it had become a minor player. After the Confrontation broke out, the organization's operatives laid a few booby-trapped cars within the green line, but could not compete with the Hamas. According to some elements in the territories, Abu Ali Mustafa was a political figure. However, intelligence relayed by the ISA to Prime Minister Ariel Sharon in late 2001 indicated that the Front was planning an ambitious attack at schools in Jerusalem on September 1st, the first day of school. The ISA claimed that Mustafa himself leaned over the maps that the targets of the attack were marked on. Mustafa entered his office on August 27th, unaware that the ISA has been following him for several days. At 9:30 A.M. the air force's Apache helicopters fired several anti-tank missiles into the office, through the window. Mustafa, who was sitting at his desk, died instantly. One of the missiles had decapitated him. Following the killing, on October 16th, a Popular Front hit squad assassinated Israel's Minister of Tourism, Rehavam Ze'evi, who was staying at the Hayat hotel in Jerusalem. This evoked a public debate in Israel around the question whether Israel crossed a red line by killing a figure that was perceived in the territories as political.[229]

The public debate was eventually resolved within the security system. In October 2002, after a series of hard debates with the legal advisor to the government, it was agreed that heads of terrorist organizations that preach for murder or encourage it would also be considered legitimate targets for targeted killing. The separation that was drawn until then between military operatives and political operatives, says Giora Eiland, formerly the head of the National Security Council, was relevant for "countries which have accountability. In a terrorist organization, it can't be that Shiekh Yassin [Hamas leader—A.K.] would give the orders, but not be held accountable."[230] Moshe Yaalon, the IDF Chief of Staff at the time, notes that the decision should have been made a lot sooner since it was the leadership which decided what the attack policy would be. Therefore "you shouldn't make a distinction between a political spiritual leadership and operational ranks."[231] A similar opinion is held by the former head of the Mossad, Danny Yatom.[232] Following this decision, most of the Hamas' leadership in the territories was eliminated within a few months.[233] In any case, the target killing of senior operatives is not carried out automatically but by an operational priority: suicide bombers, their dispatchers, explosive belts' makers, senior military operatives, senior political operatives. This of course is paired with different cost–benefit considerations, such as: What is the organization's current attack policy? Will the assassination have repercussions on a regional level, the diplomatic level, and so forth. According to Yaalon, the conclusion from all of these is conclusive when there is an operational opportunity and when the cost-benefit considerations lean towards benefit—we should strike.[234]

At times, operational temptation seems to override political wisdom. The preparation of a targeted killing operation often takes a long duration (weeks to months), especially when targeting senior terror operatives. Furthermore, the "window of opportunity" for the operation's execution could be very brief, sometimes mere minutes or even seconds. In such a case, when an operational opportunity to eliminate the target arises, the temptation of the operations successful outcome may override other important considerations. Untimely targeting may have far-reaching implications. This is what may have happened with the killing of Ra'ed Carmi, head of the Fatah's Tulkarm terror network, in January 2002. Carmi and his followers were responsible for many acts of murder during the Confrontation. The Israeli security forces had been trying to get their hands on him for a long time. Eventually, after his vulnerability was detected in his daily routine (a regular route he took to his mistress), the ISA put together an operational plan for his elimination. A demolition block was placed in a wall, and Carmi, who feared helicopters, walked close to it as usual and he was killed on the spot. The operation's outcomes were far-reaching. First, the operation led to an end of a fragile cease fire that Arafat declared several months previously, which manifested itself in a sharp decrease in the number of the Palestinian attacks. Even worse, the killing gave the Fatah substantial excuse to carry out suicide attacks in Israel, which it had avoided doing up until then. Hundreds of Israelis paid with their lives. Up to this day, the heads of the security system are undecided about the operation. Its approvers, Prime Minister Sharon, IDF Chief of Staff Shaul Mofaz, and the head of the ISA, Avi Dichter, believe the killing was necessary; and had an arrest been possible, it would be preferred. Others, such as Giora Eiland, who was the head of the IDF's Planning Branch at the time, Yitzchak Eitan, commander of the Central Command at the time, and others, believe that the timing was miserable and had created, at least in the short term, severe problems.[235]

Methods of Execution. How is a targeted killing actually carried out? First we must distinguish between two kinds of targeting that are often confused: (a) the "classic" targeted killing, which requires a long preparation and a complex set of authorizations, and (b) "hunting" methods, or its widely used term "target hunting." The latter is part of the routine security measures and its purpose is the destruction of opportune targets, mainly rocket firing squads, but also, as the need arises, suicide bombers on route to an attack. The intelligence process that comes before the "hunt" is also simpler and can include a report about a suicide bomber and his car's details, for example, which would be followed by an immediate deployment of a collection means and a helicopter or ground interception team.[236]

As opposed to "hunting," targeted killing is composed of several stages: the collection of intelligence, a process of authorizations (up to the level of Prime Minister), planning, and execution. In the intelligence process, the ISA and the IDF collect continuous information regarding the terror networks and organizations. From this database, the system suggests potential targets. These are presented in a limited decision-making forum, primarily with the Minister of Defense. At this point the "operative" is either approved or not; if approved, the mode of operation and details are ironed out (indoors, outside, open field, urban environment, etc.). Additionally, a lower-ranking officer is given the authority to make the decision to execute or abort. The next level of authorizations is with the prime minister. In special or sensitive cases the target is also brought for approval to the cabinet. Additional information is gathered about the target, and the best method of attack is decided upon. From this point on, the target is inserted into what is referred to as "the targets bank."[237]

A target is drawn out of the "bank" when an operational opportunity arises, for example, an indication is received that the target is in his home or will be in the home of a known suicide bomber's dispatcher in the next few hours. From this point on, that operation starts moving along in two parallel channels: The first, a quick authorization channel, receives a "green light" to execute from the prime minister and minister of defense, usually via the telephone; the other is involved in "closing the circle of fire." This is the most significant stage in the process and is the essence of the Israeli success: passing information and the ability to create an intelligence and operational situation report to the field level rank of decision making.[238]

The method of execution may vary between different targets. In the past, most targeting (mainly aimed at the organizations' leaders or senior operatives), both in the territories and abroad, were based on what could be referred to as "hit men." According to foreign publications, these were mostly members of the IDF's special units, Yamam (the police's elite unit) members, and operational members of the Mossad. Today, it can be concluded from media reports that the "targeted killing" policy is implemented mainly by the IDF and the YAMAM, in a variety of different means—sniper shots, launching missiles from helicopters and fighter planes.[239] All of this is in accordance with the character of the target, the environment it operates in, and the need to both guarantee the operation's success and avoid, as much as possible, hurting civilians. For example, one method that terrorists use to try to protect themselves from an Israel strike is to surround themselves with children. To overcome this method, one must often use creative operational thinking. This was the case of one of the suicide bombers' dispatchers for Hamas in Tulkarm. The operative, who was responsible for sending the suicide bomber that attacked a dance club in Tel Aviv in June 2001 and killed 21 teenagers, used to occasionally go up to the roof of his house, but always made sure that several kids would accompany him. Apache helicopters were sent out several times, but the firing of missiles was not approved because of the concern about hurting children. Eventually, the ISA recommended an alternative solution, which was cheaper and less dangerous to his surroundings. Snipers shot the man from a distance of almost 600 meters and killed him.[240]

Whatever the method of execution may be, the phase of "closing the circle of fire" is where the essence of the Israeli success lies—the ability to break organizational barriers and to fuse information in the lowest levels. This is especially necessary when dealing with an opportunity or fast targeted killing. In the decision-making rank, sometimes a brigade or division commander, all the relevant intelligence and operational information is concentrated. This information is received rapidly, since it first passed horizontally and downwards and only later is it passed to higher ranks. The reason for this lies in the fact that every targeted killing can develop into dozens of scenarios that require a quick decision; and when dealing with terrorism, as Avi Dichter, former head of the ISA, notes, "time costs lives."[241] Thus, in order to provide the director of the operation all of the tools and the relevant information, the representatives of the military intelligence, ISA, air force, and other units involved sit alongside him. This allows the creation, in the most relevant place to make a decision, of both an accurate status report and an ability to control all of the units in the field. Giora Eiland, former head of the National Security Council, stated that "The overview and integration of the data that was for the Prime Minister only is now accesible at the field level."[242]

Alongside the information fusing ability, another key to success lies in the Hum-int information (that is, information from human sources). As Yisrael Hasson, former deputy head of the ISA, stated "You can stand on the rock that the wanted person is hiding under and if you don't have an agent to tell you he's there, you'd never know." In other words, a human source is needed to confirm a target's whereabouts and to confirm his identification.[243] The people of the Palestinian Preventive Security Force assess that the ISA almost always used more than one informant to carry out a target killing operation, when the informants are unfamiliar with each other. One of the Palestinian security forces' commanders in the Gaza Strip stated that "when we arrested Akram Al-Zatme, who helped Israel target Salah Shkhade [head of the Hamas terror wing in Gaza], he said that every time he was asked to follow Shkhade, his operator in the ISA knew to tell him where the Hamas operative was and where he was headed. Al-Zatme understood that the ISA had another informant reporting to them." In another case, he adds "for the targeting of Jamal Abd Al-Razek, the commander of the Al-Aqsa Martyrs Brigades in Raffah, killed on November 2000, his nephew, Majdi Mikawi, was recruited. The nephew was arrested by the ISA, jailed, and then released. His mission was very simple: to report his uncle's movements. No one told him that Abd Al-Razek was going to be targeted. He called his operator from the ISA and reported that his uncle was leaving Raffah to pray in a Han Yunis mosque. Meanwhile, the other informant, Haider Ranem, waiting between Raffah and Han Yunis, reported that the Fatah operative was passing by the settlement of Morag. A truck blocked the way of Abd Al-Razek's car and the soldiers that came out of it shot the wanted man and another passenger to death from a close range."[244]

Targeted killing provides many advantages. First, as Moshe Yaalon, former IDF Chief of Staff, notes, it disrupts the military operatives daily routine: "He tries to avoid being seen in public, doesn't use means of communication since he understands we listen in, so he has is forced to rely on messengers and face-to-face meetings, he's persecuted."[245] In other words, targeted killing forces the terrorist operative to be preoccupied with self-survival instead of planning attacks. Furthermore, the pool of central operatives in any terrorist network is limited. Their systematic elimination reduces their ranks. It is true that new operatives quickly take their places, but their quality and level of professionalism decreases, and it takes them time to gain the experience and knowledge.[246] To all this we should add the lowering of the terrorists' morale, the accumulative affect on their operational abilities, and, of course, the higher morale on the attacker's side.

On the other hand, targeted killing has many disadvantages that include, among others: the creation of new "martyrs," evoking feelings of revenge in the Palestinian street, the controversial or frail legal grounds they rely on, and the risk to innocent civilians' lives.[247] Above all these, the targeted killing has one central weakness: the inability to produces additional information from the operation. One of the most important principles in terror prevention is "the translation of information into more information." The arrest of an operative and his interrogation lead to more arrests, which in turn lead to more exposures and arrests. Targeted killing prevents a specific attack, but it can also cut the chain of prevention by leaving intelligence void. Therefore, says Dichter, former head of the ISA, "targeted killing is a preventative measure for the poor."[248] Accordingly, any time that it is possible, the IDF prefers to carry out arrests and not targeted killing.[249]

Arrests, Raids, and Special Units. The arrest of suspected terrorists is the beginning of every thwarting activity and also its core. "The chain of prevention," as the former head of the IDF's Operations Branch and deputy chief of staff today defines it, is composed of "the arrest of terrorists, their interrogation, using the findings to arrest the rest of the members of the plot, trial and incarceration."[251] Therefore, in the chain of military action in the territories, the arrest is the key link, the heart of the matter. Almost every interrogation of an ISA detainee will supply intelligence that would result in the capture of terrorist squads and preventing attacks. The ISA's data are enlightening: Approximately 5000 Palestinians were arrested in the West Bank in 2006—that is, an average of 13 arrested every night. Among the arrested that year, the ISA counted 279 potential suicide bombers (that is, wanted men or women on their way to commit a suicide attack or that had agreed to commit one), compared to 154 the year before. There is a direct correlation between the arrests activity and the decrease in the number of Israeli casualties. In 2006, 30 Israelis were killed by Palestinian terrorism, the lowest rate since the Second Intifada commenced (55 deaths in 2005, 117 in 2004). The number of suicide bombings also decreased: 5 in 2006, as opposed to 7 and 15 in the previous two years.[252]

In addition to the production of intelligence and arrest of more network members, the arrests produce more results. A systematic arrest activity over a long period of time may neutralize a terror network for a long duration, even if all its members haven't been arrested. Those who remain at large usually suffer a blow to their morale, a lack of leadership, and lower organizational abilities. As Israel discovered, even operations that seem to be very specific and tactical such as arrests are strategically important, since due to legal or operational constraints, not all terrorists can be arrested. In Israel itself, for example, people cannot be detained for more than 48 hours without a judge's approval. There are also two operational considerations in carrying out arrests in a focused way. First, when a suspect is arrested, his partners in the terrorist network usually change their patterns of behavior or go underground, fearing that they'll be next. That is why Israel's security forces usually limit the number of arrests and wait until they have a clear intelligence picture that can allow them to either arrest or eliminate the entire terrorist cell. This tactic is not risk-free, since the operatives can carry out further attacks in the meanwhile. Second, Israel's priority in planning arrests put a special emphasis on the arrest of the activity's organizers, recruiters, and explosives makers. However, another group that has to be taken into consideration when planning the priorities is the network's facilitators. Obviously the security services are most eager to arrest the leaders of the terror organizations. They are right to see the leaders, bomb makers, and executers as targets that head the priorities lists.[253]

At the same time, one cannot ignore the facilitating factors, the transporters, and logistics operatives that are a key element for the survival of a terror organization in general or a specific terrorist plot. For example, in the Israeli case and in general, there is usually a difficulty in arresting the senior members of a terror network. However, the suicide bomber's transporters are usually more care-free in their behavior, since they do not belong to the "hard core" of the terrorists and don't view themselves as targets. Naturally then, the transporters are more vulnerable. An arrest of a suicide bomber's transporter, especially if he has been working with a certain terrorist group for a while, not only delays the execution of attacks, but forces the entire network to change its behavior. This, in turn, leads at times to an increase in its intelligence signature and the creation of new operational and intelligence opportunities. Another group of facilitators that are of interest are the courier. A terrorist cell whose members cannot communicate safely is not effective. For this reason, the arrest of the courier is especially important. Couriers usually know terrorists from several terrorist cells. Furthermore, they usually possess documents or money, which makes it easier to convict them. Accordingly, based on this weakness, it is usually easy to recruit courier and turn them into valuable intelligence sources. Furthermore, despite the problematic nature of courier use, Israel's electronic intelligence capabilities have forced the Palestinian terrorist groups to rely more and more on couriers over the years. Accordingly, the number of intelligence opportunities created for the Israeli security services grew.[254]

However, the Israeli experience suggests that a systematic arrest of operatives requires an operational free hand in the field. This doesn't mean that there's a need to remain in permanent bases in the centers of population, but the forces do need to have an operational ability to reach any place at any time.[255] This can be achieved with the use of two kinds of forces: the first, the regular or conventional arm; and the other, by operating special units. The Israeli security system makes a wide use of elite units, and their correct deployment is responsible for a large part of the Israeli success in fighting suicide terrorism.

Deploying Special Units Against Suicide Bombers. In Israel, there are a number of special units operating, of which some are involved in fighting suicide terrorism. The army operates Sayeret Matkal—a special unit very much like the American Delta Force and the British SAS. The unit carries out intelligence and commando missions and answers directly to the army's General Staff. The unit specializes in most anti-terrorism operations, including hostage situations. Two additional units called "Duvdevan" and the Border Patrol's Yamas units specialize in operations in the territories and use different undercover methods including disguising themselves as Arabs to arrest terrorists. The Israeli police operates its own counterterrorism unit, called the Yamam, which was founded in 1974 and specializes, among other things, in hostage rescue operations, although it is deeply involved with the arrest of terrorists and fighting suicide bombers.[256] This unit is one of the backbones of the Israeli war on terrorism. The fact that this is a police unit allows it to operate flexibly both in the territories and in Israel. Accordingly, in the intense years of suicide terrorism, the Yamam was involved in over 1000 operations in which it arrested or eliminated dozens of suicide bombers and senior terrorists.[257]

That being said, operating the special units in the territories during the Confrontation was not a natural process. The IDF, as Hagai Peleg, the Yamam's commander between the years 1999–2002, notes, operated in the first few months slowly and cautiously. "The army prefers using masses. It had little patience for special operations."[258] Beyond this, the military had to overcome several operational inhibitions and psychological barriers. The operation of a special unit outside of the state's borders is a complex operation that requires a set of authorizations and complex preparations. It seems that in the early months of the Confrontation, entering the Palestinian Authority territory, not to mention a special operation in the heart of a Palestinian city, was perceived in the IDF as similar to a force operating abroad, with all of the implications and operational complexities that go along with it. Accordingly, the IDF preferred not to authorize invasive activity in the Palestinian Authority territory. And when a special operation was authorized, it was preceded by many preparations. For example, the first arrest operation that the undercover Duvdevan unit carried out in Palestinian-controlled territory was planned to be a kidnapping from a vehicle. The target was a Palestinian police officer who was involved with terrorism. The operational rehearsal that preceded the operation included the participation of the Command commander and division commander themselves as the abductees to make sure that the operation was possible. Gradually the army understood that operating a special unit in the territories is less complicated than its operation in an Arab country. The concepts of time and space were completely different, as is the potential resistance of the enemy, in case the soldiers were discovered. Furthermore, as the units themselves acquired new skills such as the ability to operate in a crowded urban environment and as the number of successes grew, more and more operations were authorized.[259] As Ilan Paz, former commander of the Ramallah brigade, notes: "Slowly, the army learned that these operations were of less risk than originally perceived and lowered the authorizing level to the Command, the division, and the field. Even the units themselves learned to operate in the field both undercover and in a more aggressive manner."[260]

Apart from operating the special units in the Palestinian territories, these units are operated against suicide bombers in two other geographical circles: The first one is "on route to the target," meaning in the range between the Palestinian -controlled territories and the target in Israel. In this circle the unit is sent out after an intelligence indication has been received that the terrorist is on his way. In this situation a tactic called "Dagger" code was mostly used, which includes blocking roads in order to create traffic jams. These delay the suicide bomber and enable the intelligence services to produce more information and to direct the special unit to the transporter's car. For example, at one time the Yamam caught a terrorist sitting in traffic in the entrance to Petah Tikva (a large city east of Tel Aviv), 40 minutes after the initial warning.[261] A second geographical circle that the special units are used in is at the target itself, meaning in a chase after a suicide bomber, explosive belt, or safe house in Israeli cities. "This situation", says Zohar Dvir, Yamam commander during the peak years of the Intifada "isn't good, but these cases are rare."[262]

An arrest or elimination of a terrorist, who operates in an urban environment, especially if he is a suicide bomber, requires the use of different operative measures than those used against a regular military target. Among others, there needs to be independent intelligence collection ability, an ability to get ready and set out in a short period of time, a covert approach to the target, an ability to operate in small forces within a hostile population, and a use of a variety of combat techniques (sniper shooting, assault, undercover arrest, kidnapping, etc.). The Israeli experience in dealing with suicide bombers shows that in each of these areas the special units have notable advantages. As far as intelligence collection in concerned, a special unit has an ability to use a large number of special technological measures—to deploy covert sensors in a hostile area and to maintain them and eventually dismantle them, in a way that won't expose the intelligence collection operation. Furthermore, a special unit can usually collect intelligence while in close proximity of the target, by using different tracking techniques in the heart of hostile territory.

Another advantage of using a special unit against suicide bombers is its ability to reach the target covertly. As Lior Lotan, a former commander in an elite unit, points out, this point is especially critical, when dealing with a suicide bomber or his dispatchers "the main challenge ... is how to reach the target and less about what to do when contact is made with him, meaning that the movement to the target and the ability to concentrate an effective force near the terror target are more complicated than fighting the object himself."[263] Finally, when contact is made with the terrorist or his facilitators, the special unit has three other major advantages: first, an ability to distinguish the terrorist from his civilian surroundings; second, the ability to carry out a chase—that is, to pursue the object until he is arrested/eliminated or, alternatively, to produce additional information from the object and immediately set out on a follow-up operation.[264] A third advantage is the special unit's ability to do combat in small groups. During combat in a crowded urban environment the organic unit usually splits up to a number of smaller units. If the unit members aren't accustomed to this situation or don't plan ahead for it, many operational problems occur, such as loss of contact, lack of means, lack of information, a blow to the morale, and so on.[265]

Passive Measures

Checkpoints and a Security Fence. Alongside offensive actions, Israel has taken a number of more passive-defensive measures to deal with terrorism in general and suicide bombers specifically. These steps are of three types: the first, setting checkpoints and roadblocks; the second, building a security fence and the third; police and civilian security in the city centers—near potential targets for attack. In general, it can be said that based on the Israeli experience these steps can achieve significant goals as far as the fight of suicide terrorism is concerned: The first of these is disrupting the suicide bomber's way to his target. As Berko, who interviewed failed suicide bombers in the Israeli jail, notes, the potential suicide bombers describe a "sense of high," a spiritual uplifting, once they made the decision to die in a suicide attack, and a state of "robotic" behavior when they are on their way. However, if something disturbs the human weapon on its way to its target, there is a chance that he reconsiders or changes his mind.[266] That is the reason why checkpoints are set in the territories and the security in the Jewish centers of population was increased during the Intifada. Among others, a special security unit was founded to protect public transportation, safety measures and the security guards were increased in schools, and private security guards were placed in the entrances to shopping centers and most businesses and restaurants.[267]

A second goal that physical obstacles achieve is *piling logistical difficulties on the terror network*—the checkpoints; and especially the security fence's objective is to force the terror networks to find more roundabout complicated ways to infiltrate suicide bombers into Israel. This leads to two main outcomes: The first outcome is increasing the potential of error of the plotters. The physical barriers and the difficulty in passing around them compel the terror network to find complicated logistical solutions to get the bomber or explosive belt into Israel. Mostly, these solutions include expanding the circle of conspirators; and as former head of the ISA, Avi Dichter, notes, "the more links a chain has, the higher the chances are of prevention."[268] A second outcome of the logistical difficulty is the lengthening of the preparation time for an attack. This allows intelligence production and increases the chances of prevention.[269] Similarly, once the suicide bomber is on his way, the checkpoints and road blocks lengthen the amount of time he's on the road and allow for the production of more accurate information regarding his position. This information results many times in the terrorist's arrest in the car transporting him to the attack.[270]

Alongside this, the checkpoints and security fence achieve another important goal, which is *channeling the suicide bomber toward an electronic detection device.* As Avi Dichter, former head of the ISA, notes, fighting suicide bombers is complicated because you can't detect the means (usually hidden) and intent. Identification of the means—explosive belt or charge—is only possible using technological devices. This requires the channeling of the suicide bomber to a necessary passage- a checkpoint or road block where this kind of device can be used.[271] This is exactly what the security fence built along the border between the territories and Israel does.

Alongside the thwarting efforts, the passive defensive measures also *deter the suicide bomber.* As former head of the Yamam, Zohar Dvir, says, checkpoints and fences create a "psychological barrier" for the terrorist.[272] This barrier could cause the suicide bomber to make navigational errors as well as to reduce his determination to carry it out, thinking that he wouldn't be able to reach the target anyway. Dutter and Seliktar refer to this as "deterrence using denial", and they claim that apart from physical destruction, the denial of access to potential targets could be the "most effective way" to deter a terrorist.[273] Dichter refers to this as "fighting opportunities"[274]—meaning that to reduce the amount of opportunities, the bomber has to carry out his death wish. Beyond the effect on the bomber himself, interrogations of Palestinian detainees show over and over again that the checkpoints affect the decision-making processes within the terror networks. Terrorists incarcerated in Israel have reported that they've canceled terrorist attacks when they noticed what they perceived as increased security activity at the checkpoints, which in reality was mostly routine activity.[275]

Last but not least, the Israeli experience shows that physical barriers help *strengthen the standing ability* of the society under attack. Often, in the fight against terrorism, the dilemma of where the resources should be invested is raised—in offense or defense. If in defense, then in covert or visible defense. Cost–benefit considerations require first and foremost an investment in offensive steps aimed at dealing with the terrorist in his living environment. However, one of the most important elements of successfully handling terrorism is the public's resilience. One factor that contributes to the strengthening of the standing ability is the sense of personal security—feeling that the government hadn't abandoned the citizen. To create a sense of public safety, one must also invest in visible defense measures—checkpoints, a security fence, and security guards.[276]

Movement limitations can result in heavy prices paid by the other side. Checkpoints and fences often create inconveniences and even humiliation. These, in turn, evoke feelings of hatred and may create the next generation of terrorists and suicide bombers. For example, Muhammad Daghlas, who sent the suicide bomber to the attack in the Sbarro restaurant in Jerusalem, in which 15 Israelis were killed, claims that he decided to join Hamas to get back at Israelis, after soldiers stopped him at a checkpoint and kept him handcuffed for six hours, without him doing anything.[277] This is why, as the former IDF Chief of Staff claims, these measures have to be used "in a manner that is proportional to the threat."[278] Furthermore, the Israeli experience shows that checkpoints and passages in a security fence have a negative effect on the forces manning them and on their conduct toward the population. This factor, in the opinion of Ilan Paz, former commander of the Ramallah brigade, outweighs the advantages of these measures.[279] From a military point of view, checkpoints and roadblocks become immobile targets for terrorist organizations.

Either way, Israeli statistics show that at every place where a physical barrier was set—especially a security fence—a sharp drop in the amount of attacks and Israeli victims killed by suicide bombers was noted. The security fence around the Gaza Strip was completed even before the Confrontation with the Palestinians started. As a result, terror networks that operate in Gaza had so far managed to carry out only two suicide attacks in Israel. These attacks happened because of errors in the security checks in the border crossings.[280] In one case, suicide bombers entered Israel after hiding in double-layered shipping container. In the other case, it was two Muslim terrorists from England who made use of their foreign documentation to get by the border crossing and carry out an attack in a popular pub in Tel Aviv on March 14, 2004.

In contrast to the Gaza Strip, the terror organizations in the West Bank carried out hundreds of suicide bombings in Jewish centers of population until a fence was built in that area. After a security fence was built around these areas as well, the amount of suicide bombers launched from the West Bank's cities significantly decreased. For example, the terror networks operating in Samaria sent out 73 suicide bombers from the beginning of the Confrontation until 2003 (the time the fence was built in that area). In the years 2003–2005 these networks managed to send out only 11 suicide bombers. The number of Israeli casualties in these attacks decreased accordingly from 293 killed before the fence to 54 killed after it. The terrorists that had managed to enter Israel had gotten in either through the crossings or through the areas where the fence's construction was not yet complete.[281] That is why, as Dutter and Seliktar note, "despite the lack of will to admit to this based on international or moral law, walls can work."[282]

The security fence played an instrumental role in mitigating suicide terrorism in Israel. Nevertheless, there is still a dispute among top security officials whether the investment in the fence could have been more wisely spent in other venues. For example, the former head of Israel's Security Agency (ISA), Avi Dicter, views the security fence as the most important element in mitigating suicide terrorism, whereas the former IDF Chief of Staff strongly believes that intelligence is the most important factor by far.[283]

Security at City Centers. Once the suicide bomber is able to get past the checkpoints and security fence, he needs to overcome the activities of the police

Figure 2.5. Israel's security barrier.

forces and civilian security elements, though these may only be a complementary measure. As previously mentioned, according to the Israeli perception, most of the fight against terrorism is carried out in more distant circles (the suicide bomber's home environment and the geographic area between that environment and the city centers). This implies that in theory the police's role regarding the fight against suicide terrorism is secondary. However, in practice, as the number of attacks on Israeli cities increased, the police invested growing amounts of resources on this as well. As part of this, the police focused on two main areas: (a) prevention and (b) the response to an event. The prevention efforts included reinforcement (by the police) of the units that take part in the first response in the field: the special patrol units, the bomb squad, the forensic specialists, and so forth. Additionally, the checkpoints in the entrances to the cities and the border-line between Israel and the territories were reinforced. Another area that significantly changed was the relationship between security services—the IDF and the ISA. In the past, ego conflicts between the organizations prevented true cooperation. Once a number of different agencies had to deal with a growing threat of suicide bombers coming from the territories to Israel, cooperation became a must. The high frequency of these incidents did not allow the heads of the services to get involved in every event, so the people in field got authorization to work directly in real time with their colleagues in other organizations. The lifting of organizational barriers had a huge impact on the efficiency of the thwarting and preventive efforts.[284]

Another aspect of the prevention efforts was the increase in cooperation with civilian elements. The police operated based on the belief that an intelligent use of the civilian element could significantly increase the prevention ability. This led to the inclusion of civilian security officers working in various institutions in the prevention activity and response to events. Part of this was the creation of combined security areas that include both policemen and civilian security guards (referred to as "security territories"), mainly in crowded locations.[285]

Alongside the prevention activity, the police focused on the response to the attacks. Supposedly, after the suicide bomber blows himself up, handling the event seems unimportant. This is a misconception. A quick response after an attack can have critical implications both on saving human lives and on the chances of catching the plotters. Accordingly, the Israeli police put together organized working procedures for dealing with attacks. These determine that after an attack has been carried out the local police commander is the only one that has jurisdiction over the crime scene and all the other forces involved answer to him: medical teams, rescue teams, forensics, the ISA, and so on. Alongside the activity at the scene of the crime itself, which is aimed at saving human lives and restoring routine as fast as possible, the police also activate further response circles. This includes the deployment of roadblocks in an attempt to catch the suicide bombers accomplices, and a status report is passed on to the ISA and IDF. In a number of cases the suicide bombers' transporters were caught in roadblocks set dozens of miles away from the scene of the attack. For example, the transporters of a suicide attacker that targeted a popular restaurant called

"Sea Food Market" in Tel Aviv, on March 5, 2002, were caught in a police checkpoint near Jerusalem.[286] While they were slowing down before the checkpoint, an officer noticed that an object was thrown out of the car. A check showed that the transporters threw out the suicide bomber's identification documents. In another case, following roadblocks that were deployed after an attack on a Tel Aviv dance club in June 2001, the transporter decided to stay nearby and not to return to the territories. Shortly afterwards he was arrested based on intelligence information.[287]

Alongside the police, civilian security bodies also filled an important role, especially in the prevention of suicide attacks. The increase in attacks at shopping malls in the center of Israeli cities resulted in the employment of an unprecedented number of civilian security guards by different institutions (local municipalities, malls) and also small businesses like restaurants and pubs. For example, in 2003 the Jerusalem municipality hired 13 private security firms to protect schools and hired an additional security firm to protect other institutions. The police became a professional supervising body for these elements.[288] The efficiency of civilian security guards in institutions and businesses' entrances was doubted many times. However, the Israeli experience shows that often when suicide bombers saw a security guard at the entrance, they preferred to blow themselves up at the entrance or at a different location where security wasn't present.

On other occasions, civilian security guards managed to detect the explosive belt and physically stop the attack. This is how a serious suicide attack was prevented in a coffee shop near the American embassy in Tel Aviv on October 11, 2002. The terrorist, a male in his twenties, reached the promenade area of the city to carry out a suicide attack. He was carrying on him a charge that weighed approximately 10 kilograms that contained homemade explosives, shrapnel, and screw-nuts to increase the damage. The terrorist approached the entrance of "Café Tayelet," adjacent to the American embassy building, and meant to go in. The security guard, Michael Surkisov, stopped him at the café's entrance for a routine search. At this point, Surkisov says "the metal detector I was using on him beeped. Suddenly I saw a wire hanging out of his pocket and that's where the terrorist's hand was as well. I also immediately saw something sticking out from his shirt, around his back." After noticing the wires, he grabbed the terrorist's hands and twisted them. The terrorist managed to free himself from the grip and began running toward the promenade. A car driven by a young woman blocked the terrorist's way and he slammed into her front windshield with force. Surkisov continues: "I yelled to the [American—A.K.] embassy's security guards 'terrorist, terrorist' as I was chasing him." The terrorist recovered from the car hit and continued to run for a few more meters. Surkisov, together with the Israeli security guards of the American embassy, managed to catch him. In their effort to stop the terrorist, the embassy's security guards shot several warning shots into the air. An eyewitness said that a group of border patrol officers standing nearby joined the chase after they realized it was not a brawl, as they presumed originally. The security guards grabbed the terrorist's hands and feet so he couldn't detonate the charge he was carrying. Surkisov said that on the terrorist's right arm he noticed the detonating button. Meanwhile, the embassy's guards called the police. The first officer on the scene said that they sat on the terrorist's hands until the arrival, minutes later, of a member of the police bomb squad, who neutralized the explosive belt the terrorist was wearing.[289] Other security guards paid for the discovery of suicide bombers with their lives, since the latter prefer in many cases to blow themselves immediately after their exposure. This was the case of a security guard in the Kfar Saba central bus station who identified a suicide bomber in April 2003.[290] It also happened when another security guard identified a suicide bomber at the entrance to a Jerusalem coffee shop in September 2003.[291] In another case, two security guards were severely injured at the central bus station in Be'er Sheva after the suicide bomber they exposed blew himself up.[292]

Alongside the police and civilian security companies, the Israeli public is also a factor in preventing suicide attacks. The average Israeli, if such a generalization can be used, is very alert to suspicious packages, individuals, and actions that could pose a threat to public safety. Furthermore, Israelis are not hesitant to report their suspicions to the police. As a result, ordinary citizens play an active role in the prevention of terrorist attack and in spotting suspects. There is a high degree of public awareness, not necessarily because of an organized educational activity that the government carries out, but because of the reality of living in Israel and the large number of terrorist attacks that had happened in it in the last few decades. This is added to the fact that a large portion of the citizens have some kind of security background (in Israel there is a mandatory military service for men ages 18–21, women 18–20) or active participation in police work as volunteers.

The public awareness, says Uri Bar Lev, former commander of the Southern District of the police and the founding Commander of special undercover units in the IDF and Israeli Police Force, is more important than the work of private security guards: "a person walking around in the summer, wearing a coat with an electric wire hanging out of it, would immediately be reported to the police by civilians."[293] This is exactly what happened when a terrorist in a coat walked toward a shopping mall in the city of Netanya holding a large bag across his shoulder. The suicide bomber, who dyed his hair blond, drew the attention of several civilians. The latter reported him to the police and also drew the attention of the security guards at the entrance to the mall. Unfortunately, the bomber managed to blow himself up in the entrance and killed five people; but had the suicide bomber been able to enter the mall, dozens of additional casualties would surely have been suffered.[294] In another case, civilians reported to the police about a suspicious man walking with a bag in Afula. A detective who was on his way home took the call and managed to stop the suicide bomber.[295] In other cases the civilians themselves managed to stop a suicide bomber from reaching his destination. For example, the alertness of civilians in Netanya forced a suicide bomber for the Popular Front to explode at the entrance to a marketplace and not in its center.[296] In another case a city bus driver prevented a suspicious-looking suicide bomber from getting on the bus in the Jordan Valley[297]; and in a similar case, a bus driver stopped a suicide bomber from getting on the bus at a major intersection in Jerusalem.[298]

Public Resilience

In the war on terrorism, the home front has become the battle front. Israel's public resilience level was put in question both by its enemies and by its own leaders on many occasions. In the second Intifada, Israel's public resilience was a source for admiration. As noted, for terrorism to prosper, terrorists require a comforting environment and public setting; but on the other hand, for counterterrorism to be effective and for a country to be able to regroup and confront, it needs what is commonly referred to as "public resilience," which has become a key component in coping with and confronting terrorism.

As previously noted, physical and overt security elements enhance the public's feeling of security and serve as an important component of resilience. Another aspect of public resilience is education and public debate that leads to an understanding of the threats at hand. The Israeli public has unfortunately learned to cope with terror, by decades of actual experience. But such hindsight experience is not a must. The American public can prepare for the worst in advance. Targeted educational programs, public campaigns, and debate are just a few of the means that need to be applied in preparing the public and enhancing its resilience.

Deterring Suicide Bombers

Deterring a suicide bomber is a complex issue. The core question is, How can you deter a person who wishes to die? The answer given by the Israeli security system has three parts. The first is decreasing the suicide bomber's chances of success with a variety of physical barriers. A second element in deterring a suicide bomber was carrying out actions against his family and relatives. In the summer of 2002, at the height of the Confrontation with the Palestinians, the Israeli government authorized a line of responsive measure against suicide attacks. Some of these steps were meant to deter the suicide bomber by destroying his family's house and confiscating his personal possessions.[299] In the next two years, several hundred homes had been destroyed, but the opinions in the Israeli security system as to this measure's effectiveness differed. Both in the senior command and in the field ranks, some doubt the ability of these kinds of steps to deter the bomber. Danny Yatom, the former head of the "Mossad," for example, firmly opposes such measures.[300] On the other hand, there are people who enthusiastically support these kinds of punitive actions and support their claim with the fact that since suicide bombers' houses started to be destroyed, approximately 20 potential suicide bombers were turned in by their own families [*The Seventh War*, p. 163.]. This complex picture is revealed also when interviewing potential suicide bombers in the Israeli prison. For example, M., a Palestinian youngster from one of the West Bank villages, claimed that when asked if he would still go to carry out a suicide attack if he knew his family's home would be demolished, he answered: "I would carry out the operation as usual, I made the decision, and my family—God will help them." On the other hand, another prisoner, N., answered that he wouldn't have gone through with it because "I know the situation my family is in...."[301]

To this complex picture we should add the effect of actions considered by many to be collective punishment measures, which may have deepened the hatred and created feelings of revenge.[302] Eventually the IDF appointed a special committee to examine the issue and, after reaching ambivalent conclusions, decided to stop house demolitions because it was ineffective in deterring future attackers.[303] However, even after these steps were stopped, different factors in the security system and out of it feel that suicide bombers can be deterred with these measures if only they would have been implemented on a wider scale. For example, Yitzchak Eitan, commander of the Central Command, believes that deterrence cannot be created by "random acts of scaring families." According to him, destroying a house could perhaps be effective as a punishment, but in order to create deterrence it has to be a far-reaching act, such as "destroying 100 terrorists' houses in one night."[304] Boaz Ganor, head of the International Institute for Counter-Terrorism (ICT), also claims that these steps did not create deterrence "because we never reached the deterring dosage."[305]

The issue of dosage and the Israeli system's inability or lack of will to carry out collective punishing steps is also noticeable regarding the issue of the deportation of family members. During the Confrontation the security system decided to deport suicide bombers' families from Judea and Samaria to the Gaza Strip. The first case in which the IDF attempted to deport family members created a public outburst. In the center of it was the Ajuri family, because one of its sons sent out a number of suicide bombers to attacks in Israel. The brother and sister of the dispatcher were active partners in one of the attacks. The sister sewed the explosive belt for the bomber, and his brother helped the terrorist cell hide the explosives. After a long legal debate the Israeli judiciary system allowed the temporary deportation of the family as a deterring measure, but determined that this step could only be used against a family member who assisted with the planning and execution of attacks, and only when there's a fear that he would continue to do so in the future. The bottom line was that the court didn't allow this step to be used against family members who were not involved in attacks.[306] In light of the existing case law, deterrence cannot be achieved by deportation.

As opposed to trying to deter the suicide bomber himself, it seems that a deterring effect can be created against other parts of the chain of attack.[307] This assumption is based on the considerations of cost–benefit made by the organizations' leaderships. As the personal distress of the heads of the terrorist organizations worsens, there is an increased chance that they would change their attack policy. For example, says Cooperwasser, former head of the IDF's intelligence research department, when the Hamas fired rockets at Ashkelon (one of the major cities in the south), Israel killed one of the movement's senior members, Ismail Abu Shanab, in response. As a result, the Hamas refrained from shooting at the city for a long duration. In other words, even if the organization deploys suicide bombers, it also must maintain its own survival.[308]

RELEVANT LESSONS FOR AMERICAN LAW ENFORCEMENT

As many people know, suicide bombings are a complex phenomenon that stems from different hardships and motives: religious, political, social, cultural, and even financial. Therefore countries who seek to stop or mitigate the use of suicide terrorism should deal with the different layers that the problem is made of. The Israeli method of handling the suicide bombers is ostensibly successful: Within a relatively short time span the Israeli security system managed to bring the suicide bombings in Israeli cities to almost a complete halt. Therefore it might come across that Israel has managed to develop an efficient inclusive strategy for dealing with the suicide bombings phenomenon. Unfortunately, that isn't the case. The Palestinians still harbor a deep hatred toward Israelis; the religious extremism still ensues in wide circles of the Palestinian society and is raising the next generation of suicide bombers; the values of democracy and Western world view are still struggling to find their place among the Palestinians; and the living conditions in the territories are still harsh and they increase the despair and hopelessness among the Palestinian people. All these issues have not been solved by the Israelis or the Palestinians themselves. So it seems that the lessons Israel has to offer the Western world and the United States in particular, in regard to a fundamental-strategy of confronting suicide terrorism, is limited.

And yet, one cannot ignore the Israeli success in decreasing the number of suicide attacks, the impressive prevention percentages, and the efficiency in which all these were achieved. There is much to learned from this experience. Israel has been suffering from terrorism in its various forms for more than 60 years—since its very first day—and from suicide terrorism for a decade and a half. The Western world had been forced to face this kind of terrorism only recently—after the 9/11 attacks in 2001 and the bombings in Europe later on. Furthermore, while in the Western world only a handful of suicide bombings took place, in Israel there were hundreds of attacks or attack attempts in a relatively short period of time. When examining the "results of the experiment" it seems clear that, as previously noted, Israel has yet to put together a comprehensive solution for the core problems from which the suicide bombings stem, and perhaps it doesn't have the tools to change key elements such as Palestinian media education, and public opinion. However, the Israeli security system has managed to develop and implement a set of components and mechanism that have significantly mitigated the terror organizations' abilities and the opportunities the suicide bomber has when he wishes to blow himself up in centers of Jewish population. In that regard, Israel confronted suicide bombings as a fighting method—a tactic with special characteristics as opposed to a widespread phenomenon. This approach against suicide bombings allowed Israel to confront them with an operative-tactical level regardless of wider solutions that politicians, religious leaders, educators, and others may find. Accordingly, the lessons that could be learned from the Israeli experience are also mainly operative and tactical in nature. These lessons are drawn not only from the Israeli success, but just as much from the failures and mistakes made by Israel along the way.

Key Lessons Learned—Operative Level

Nobody Else Will Do the Dirty Work for You. For years, Israel relied on the Palestinian Authority to uproot and confront Palastinian terror infrastructures.[309] This misconception cost hundreds of Israelis their lives. Time after time the Israeli experience has shown that vital interests in a nation's national security cannot be entrusted to a third party. When Israel gave up this notion and began confronting the suicide bombers phenomenon by itself, it managed to mitigate both the number of attacks and the number of fatalities from this brand of terrorism. This cardinal rule is also relevant on a local level for law enforcement activity in a Western country: A religious or political leader of a local community, as charismatic and firm as he may be, cannot physically stop a suicide bomber or prevent the terror network from carrying out its plans, no matter how much you might pressure him. Third parties can facilitate in intelligence and means, but they cannot replace the efforts of the threatened party or serve as the backbone of confronting the suicide bombers phenomenon.

Release Psychological Barriers. The members of the Israeli security forces took many months to release themselves from different psychological barriers that hindered their ability to confront suicide bombings effectively. These barriers were associated with the political environment in which they operated and which initially defined the Palestinian Authority as a peace partner, and also with an operational environment—transferring time and space patterns from the activity in Lebanon to the activity in the territories, a projection of the characteristics of one scene of activity on the characteristics of a different scene. Another barrier comes from an occasional overestimation of the opponent's abilities or the difficulty of operating in an urban environment. For example, the Israeli army personnel avoided entering the refugee camps at different stages because of concern of heavy losses due to the terrorist organizations' abilities as well as the complexity of the environment. It is up to the security expert or law enforcement officer to try to identify these barriers, which might compromise the thwarting operation or intelligence operation that he is involved with. Is it based on a misinterpretation of the dimension of operational time, the environment he was accustomed to operating in, or various political thought patterns that might stop him from thinking of more original and different courses of action than he has previously taken.

Carrying Out Continuous and Systematic Activity Against the Hard Core of the Terror Networks. It seems that the first and most significant difficulty in dealing with suicide bombings is the feeling that there is no end to those who wish to kill themselves. Therefore any attempt to fight the phenomenon is like trying "to empty the sea with a teaspoon", or to try and empty "a bottomless barrel." And true enough, after seven years of violent Confrontation with the Palestinians, it seemed that there is still no shortage of volunteers for suicide attacks. However, the number of terrorist operatives who can assist them in carrying out their death wish is limited. Terror organizations, as threatening as they may be, will always be composed of a relatively small core of skilled operatives. These are the terror networks' leaders, those with a high organizational ability, those with the explosives expertise, ideologists, facilitators, and suicide bomber transporters. All these can be confronted. Systematic arrests and targeted killings aimed at this nucleus of operatives significantly hurts the terror networks' capabilities. True, many times the operatives arrested or killed can be replaced. However, the planning and execution of attacks requires skill, experience, and knowledge that are acquired over many months and sometimes years. Therefore the substitutes and their substitutes are bound to be less and less professional, and as a result also more likely to commit operational and intelligence errors. Furthermore, a large portion of them would be preoccupied in spending their time on their own survival rather than on plotting and carrying out attacks. It's obvious that when there is information concerning a certain person's intent to carry out a suicide bombing, he should be arrested; however, in principle, most of the thwarting activity should be carried out against the hard core of the operatives and, most importantly, in a continuous and systematic manner. Among the members of the hard core, those who are centers of knowledge should be handled in a higher priority: the head of the network, the explosives expert, or other operatives with specialist knowledge should be arrested or "disappear" from the scene first. Such counterterrorism cannot afford voids. If security efforts lax, the voids will be filled by terrorism plotting, reinforcement, training, and eventually attacks.

Operate Against the Terrorists in Their Own Home. There are various steps that can be taken against the suicide bombers and the terrorist networks who send them. However, as Israel realized, the most effective activity is the one carried out against the terrorist where they live, or "in their beds," as former IDF chief of Staff, Moshe Yaalon, puts it. In this kind of activity the initiative is with the operating force and thus involves the element of surprise. Furthermore, activity in the suicide bomber/terrorist's home environment has a psychological effect that influences his ability to function efficiently. The need to change places of hiding, to narrow down the circle of facilitators, and to be careful every time he goes out to the street makes life hard on the operative and forces him to divert personal and operational resources from plotting attacks to survival. This strategy of narrowing down the circle and closing on suicide bombers and their facilitators can also apply to their countries of origin overseas. This activity in the suicide bombers' home environment is much more focused and requires smaller resources than defending potential targets for suicide attacks.

Intelligence, Intelligence, and More Intelligence. Intelligence is the core element of confronting suicide terrorism. The more specific the intelligence is, the more successful the thwarting efforts would be. In this regard it is hard to overestimate the importance of intelligence. Successful intelligence activity against suicide bombers can usually be achieved by a combination of factors:

Cooperation Between Intelligence Agencies. This rule is obvious and always true, but its actual implementation is always harder. The Israeli experience shows that close cooperation between the various intelligence bodies is primarily created because of a sense of urgency and imminent threat. Subsequently, a series of practical steps follow:

Setting Clear Areas of Responsibility. Who is responsible for what and when?

Breaking Down Interorganizational Barriers at the Working Levels. That is, creating a situation in which a military intelligence officer, for instance, can communicate directly and in real time with an ISA intelligence officer operating in the field, and vice versa, even if they both belong to different intelligence agencies.

Flow of Information and Flexibility in Operation. Intelligence in general, and when confronting suicide bombers specifically, must first flow horizontally—that is, to the intelligence and operational factors who need it immediately. Only later is the information passed upwards or vertically. For example, an intelligence coordinator who receives information from a human source (humint) about immediate plans for an attack will immediately pass it on to the commander of the operational unit who should carry out the thwarting activity. The system on its part needs to be flexible enough to tolerate an initial activation of a unit without authorization by the highest levels.

Emphasizing Human Intelligence. This kind of intelligence is the most important when dealing with terrorist networks that are compartmentalized and are, at least on a fundamental level, aware of their opponents' technological abilities. The Israel experience shows that motives for cooperation with the enemy can be diverse and can include not only money, but also other benefits: decreasing of a jail sentence, financing medical treatments, receiving various licenses, and so forth. In this context, a positive incentive is stronger than a negative one—that is, one that is based on pressure and threats. In order to create intelligence opportunities for recruiting human sources, opportunities should be created for close contact between the target population and the intelligence officer who is recruiting sources: checking stations such as border crossings, improvised checkpoints, and patrols in the target area or any other place where you can interact with the type of people that you want to recruit a human source from. Who should be recruited? Sometimes intelligence officers have a tendency to try and focus on recruiting sources from within the terrorist network itself. These sources are the most difficult to recruit. However relevant information can be obtained from people in more distant circles—former supporters, family members, distant friends, or the terrorists neighbors. These would not always supply intelligence regarding the plotting of attacks but could provide information that could be translated into successful thwarting activity (a routine presence in a certain location, type of car in the operative's use, and so on.). In this regard, human intelligence is not complementary to technological intelligence as some think, but is instead a central and critical source of information.[310]

Immediate Connection Between the Intelligence and an Operational Unit. Intelligence produced for thwarting a suicide attack must be directed for operational activity—that is, to answer when and where a suicide bomber or his dispatcher would be tomorrow or in an hour, not where they were two days or an hour ago. When this kind of intelligence is produced, it has to be passed on a direct channel to a firing or operating factor. To do this, the factors involved in the process need first of all to be familiar with each other, at least on a theoretical and organizational level. Secondly, there has to be direct channels of communication; and, as importantly, they have to be able, technologically, to pass information and communicate with each other. This of course requires the development of an appropriate C4I system.

Creation of Ad Hoc Intelligence-Operational Work Teams. Because of the complexity of fighting suicide bombers and their dispatchers, who operate in a crowded urban environment and in short time frames, Israel realized that for effective thwarting activity against them, ad hoc combined working teams must be created. For example, during a targeted killing by an elite unit in Palestinian controlled territory, the members of the unit, the military commander of the area with his operational teams, members of the air force, intelligence officers, and members of the ISA would all need to cooperate. They would all focus their efforts on the same goal and would have direct channels of communication between each other.

Knowing How the Local Terrorist Networks Operate. Suicide bombings are usually based on the simmilar operational components. These in turn are composed of two parallel channels: The first is the bomber's channel, which recruits him/her, trains him/her, and hooks him/her up with the explosive means. The second is the operational-logistical channel, which includes the preparation or acquisition of the explosive belt/charge, locating a transporter and a target for the attack. The two channels merge shortly before the attack when the suicide bomber is hooked up with the explosive belt and sets out on his way. This basic formula is relevant for most suicide attacks. When there is a need to thwart several potential suicide attacks simultaneously, this knowledge enables the security forces to prioritize there efforts. It is likely that different nuances could come up in the process of planning and executing an attack, due to different environmental or human conditions. The law enforcer and intelligence officer must know them and ask themselves questions accordingly: Can explosives or certain chemicals for their preparation be obtained in his area of activity? Do the terrorist cells in his area have the knowledge to manufacture explosives themselves? Do they have religious training or a charismatic leader who can recruit and train a suicide bomber, or do they need to take in a suicide bomber from another area or from abroad? And so on.

A Quick Operation of Forces as a Basis for Success. Alongside the intelligence, law enforcement forces or the military, need to develop a quick operating ability. In suicide terrorism the time frames that set apart a successful thwarted attack and failure can be very brief and span between minutes to a successful few hours. Therefore, every system needs to have units that can operate in very short standing operating procedures. For example, if information is received about a suicide bomber entering a car at this very minute, a force should be able to operate immediately. This has been accomplished in Israel in several ways. First, in Israel there are several special units operating who have standby teams for a quick response. Secondly, the first response units and intelligence elements have open information channels and a mutual acquaintance, mainly regarding abilities and intelligence and operational limitations. Additionally, a large part of the quick response units have diverse abilities. They can operate in an urban setting and under civilian cover as well as fighting in the open field in a conventional military manner.[311]

Taking Steps That Would Increase the Likelihood of a Mistake by the Suicide Bomber or His Dispatchers. As part of the war on suicide terrorism, certain Israeli steps were taken to increase the likelihood that the suicide bombers or their dispatchers would make mistakes or alternatively would be forced to prolong the amount of time in which they are on their way to their target. Roadblocks and fences are an effective measure in this regard. Fences between countries are a matter handled on a national level. However, at the local level, checkpoints are an effective measure, especially if they are surprise checkpoints. These should be placed on potential exit routes from the terror network's area of activity or the entrance routes to areas with a large number of potential targets. All this is done as a part of the law enforcement forces operating routine as well as in response to intelligence warnings. There are two kinds of checkpoints in this context: The first kind is halting checkpoints—that is, checkpoints that are intended to create heavy traffic. These are placed when specific information is received about a suicide bomber who has been dispatched for attack. These checkpoints' purpose is to "buy time" for the intelligence to produce more relevant information. The second kind of checkpoints is "breathing" checkpoints. These would allow cars to come through, though in a slower pace, to allow law enforcement officers to locate suspicious passengers. Other steps in this regard are numerous patrols in varying frequencies in the terror networks' areas of activity, arrest and interrogation of the network members' families, widespread arrests for information, and more.[312]

A Fence is Part of Holistic Operational Concept. A security fence built to stop the suicide bomber from reaching his target cannot stand on its own and needs to be part of a whole defense system. This system should include: technological means, an technological research support that can produce information from these means and connect it with information from other sources (human and technological), and defense forces who can operate based on that intelligence or based on a warning that received from the different means of detection is received.

Confronting the Source that Molds the Next Generation of Suicide Bombers—The Leaders of the Terror Organizations and the Civilian Infrastructure. In addition to destroying the various operational terrorist frameworks, it is necessary to address the political as well as religious leaders and organizations which preach and spread the ideas that make acts of suicide bombing a ligitimate course of action.

Specific Lessons—Operational-Tactical Level

The Israeli method of confronting suicide terrorism emphasizes thwarting attacks, which mostly take place in the suicide bombers living environment before he sets out for the attack. A large portion of this activity is carried out be Special Forces. At the same time, quick response forces are operated after the bomber is dispatched. The Israeli experience in this regard provides several lessons:[313]

Abilities that the Special Unit Should Have. In order to deal efficiently with suicide bombers, a special unit or any other counterterrorism force must develop a wide range of abilities, first of which is the ability to move undercover in a hostile country or urban environment. The unit members need to be able to stay for long periods of time both in a rural environment and in a urban environment—in a building, with no outside signature. Additionally, the unit members need to know how to communicate in a real time situation with external intelligence bodies that are outside the unit and to develop an integrative working and processing ability with these systems: liaison officers, knowledge of the unit's abilities, limitations and collection capabilities, a common language. For example, a commander lying in an ambush with his men inside a house needs to know if he has to stay in his location for 6 hours or 36 hours until he gets the intelligence required for the operation. Of course, in order to allow effective combat in a crowded urban environment, it is important that information flows properly among all unit sections, among the combat forces in the field and between the combat level and its commanders. Alongside these, the unit members need to be skilled in a variety of combat techniques: open field, urban environment, sniper shooting, or close-range elimination.[314]

There Is a Big Advantage to a Special Unit's Stay in the Field Before the Operation for Independent Intelligence Collection By Its Members. There are two ways of operating a special unit against suicide bombers or their dispatchers. The first is the immediate activation of the unit after receiving indication that the bomber is on his way or about to do so. In this case, of course, there is no time for preparations or intelligence collection and the unit gets organized while heading for the target. The second option is as part of an attempt to arrest a suicide bomber or his sender as they are preparing for the attack. In these kinds of operations, there's usually time for intelligence collection. In this case it is advantageous to bring the special unit to the field as much time as possible before the operation and to let its members participate in the collection of visual intelligence on the target. In these cases the unit members could usually translate the collected information into better operational opportunities than if they would have received the information after it was processed, shortly before the operation. For example, information that a terrorist living with his family in a fifth-floor apartment in multi-story building. This information is focused but not focused enough for a special operation. Members of a special unit would first analyze the ability to infiltrate the city, reach the house, and reach the floor, assess what force they are facing and the means at their disposal, including firearms and booby traps; and analyze the ability to concentrate the appropriate size of force there and maybe then would set an operational plan.

Special Units and Other Units that Respond to Intelligence Warnings Must Have Operational Flexibility. In dealing with suicide bombers, time is of the essence. A primary response force needs to be able to set out quickly without a standing operating procedure. This means that all of the operations that could save on preparation time when setting out should be carried out in advance: putting on-call teams in different geographical regions, keeping outfitted vehicles, setting collection routes for unit members who are staying at their home, and so forth.

Putting Egos Aside. This principle is always wise to implement, but is particularly true when operating a special unit alongside other forces in short time frames. This is achieved, among other ways, by setting clear areas of responsibility and jurisdiction. In a special operation, each force must have a field in which it leads, for example, the army can be responsible for the territory (rescue, medical evacuation, plan authorization); the special unit can be in charge of the inner circle (the area close to the target); and the ISA can be in charge of passing specific intelligence on line.

Cooperation on Lower Levels. One of the main causes of success in thwarting suicide attacks is the cooperation between the security bodies at the field levels. In the operating-planning close circle, this principle also means lifting barriers of compartmentalization. While confronting suicide terrorism, the working ranks understood that it was important to expose the main points of the relevant information to all the operations participants, of course under the limitation of exposing sources. For example, it doesn't really matter who the source is or what his name is, but it is important for the operation that the ISA can place a human source in the relevant area. As time went by, as a result of the intense activity a common language developed and an almost blind understanding between the different forces flourished.

Debriefing Between Units. The ability to learn from the experience of others is an important element in horizontal improvement and in increasing thwarting capabilities. For this purpose every few months a convention of elite units must be held in which lessons from the everyday activities should be discussed. This allows both learning and personal acquaintance for solving common operational problems or for future operations.

During the Confrontation with the Palestinians, as the number of suicide bombings in Israeli cities grew, so did the police's involvement with preventive activity and especially with the initial response to the attack. The Israeli experience in this field can produce number of important lessons:

Identification of a Suicide Bomber. The Israeli experience illustrates that there isn't one profile for a suicide bomber. However, whether it's a teen, a youngster, or a woman, there are several suspicion-arousing signs. These mainly have to do with the suspect's behavior, his external appearance, apprehension, and the objects that he's holding or carrying. These characteristics are of course not sufficient by themselves to positively identify a suicide bomber. Furthermore, at different suicide attacks the bombers and their dispatchers made great efforts to blend the bomber into the target environment as much as possible and to make him appear like an innocent passerby. That being said, these signs could give the law enforcer or civilian a lead on the person they are facing. Many times, despite the camouflage efforts, the excitement before the act, and anxiety that some bombers feel can allow a trained security guard or even an alert civilian to identify suicide bomber on time. These signs include, first and foremost, different indications of abnormal behavior such as hesitation, apathy, uncharacteristic walk (slow and sluggish, zigzag, observation, and advance—in an attempt to avoid security guards). Other signs can indicate mental stress such as a stare, dilated pupils, nervous or hesitant mumble or speech, increased sweating, blush etc. One or more of these signs may combine with an unusual physical appearance such as dressing inappropriately for the season (mainly long-sleeved or heavy garments in warm weather), sleeves that cover the hands, and bumps under the clothing. Other physical signs could be a person who obviously cut his hair or shaved his beard recently. Alongside the physical signs, attention must be paid to the possessions the suspect is carrying, especially items that don't match the person's general appearance such as: a woman with a big backpack, a suspect carrying a large suitcase around the street, and so on. Another set of signs have to do with the item the suspect is carrying: unusual protrusions from the object, abnormal modifications made on the bag, or electrical components protruding from it toward the suspect's hands or pockets. Some of these signs may seem outdated due to the sophistication and lessons learned by terrorism organizations themselves, as was illustrated in the February 2008 suicide attacks in a Dimona commercial center, when one suicide attacker was composed enough to drink a cup of coffee at a nearby café prior to exploding himself in the midst of a civilian crowd, and a second attacker waited until the rescue teams arrived on the scene before he attempted to pull the charge he was wearing. Nevertheless, the above-noted suspicious signs should not be overlooked and law enforcement and the public as a whole should be aware and alert to them.

Dealing with a Suicide Bomber, Once Identified. Once a suicide bomber has been identified by a police officer or a civilian, the reaction time the security person has is very short. Therefore his initial reactions could make the difference between failure that would cost human lives and success. Past experience has shown that suicide bombers preferred to blow themselves up immediately after they were exposed. That means that handling the suspect must be done as discreetly as possible. First of all, the law enforcement officer must meet the person from whom the information about the suspect originated, and he must be able to identify the suspect himself. From this point on, the confrontation with the suspect should be carried out from a distance using a loudspeaker and a weapon, because the immediate goal is not to bring to the suicide bombers arrest, but to minimize the damage in lives and property. Accordingly, the law enforcer should attempt to distance the crowds and isolate the area as much as possible, all this, as much as possible, with the assistance of other civilians. In the next phase the policemen should order the civilians, using a loudspeaker, to lie down and for the suspect to raise his hands. If the latter stays standing and does things with his hands such as inserting them into a bag or pockets, he should be shot to neutralize him, out of the assumption that he is trying to detonate the charge. If the suspect did not blow up, he should be ordered to lie down and the bomb squad should be called in to examine him.[315] It should be mentioned that in some of the cases where suicide bombers were identified in Israel, policemen and alert civilians managed to neutralize him or at least caused him to explode early. In some cases these people paid the price for trying to save the lives of others with their own lives.

Stopping a Vehicle Suspected of Transporting the Suicide Bomber. Engaging a suicide bomber in a vehicle is usually a result of either specific intelligence information or general information which leads to the blocking of roads and detection of the terrorist in the traffic jam. In both cases the main principle in dealing with a suicide bomber in a vehicle and in general is to control the situation: to either stop or redirect the traffic, isolate the car, work against the suspicious car from a distance (at least 30 meters), use directions from afar to make sure the car is turned off, and remove the passengers one at a time and make them lie down on the ground. Subsequently, each person would be dealt with individually. After he is instructed to do so, each person must stand up, walk away from his possessions, and lift his shirt. At either this point or when the passengers get out of the car, most suicide bombers would prefer to blow themselves up. If no explosion has taken place so far, it is probably because of a technical malfunction of the explosive belt. However, one must take into account that the terrorist is waiting for the law enforcement officers to come closer before he detonates himself. That is why once he is identified, he must lie on the ground with his arms and legs spread out. He should be dealt with only by the bomb squad and only after his/her hands and legs are held by policemen. If the terrorist resists or tries to detonate the charge during the bomb squad's work, he must of course be neutralized. For example, in March 2003 a suicide bomber was stopped on his way to a suicide attack in Jerusalem. The terrorist was stopped by border patrolmen following general information warning of an attack, after they noticed a suspicious person. The terrorist was laid down on his stomach and was handcuffed. Interrogation of the transporter revealed that the detonation switch was near the stomach. The member of the bomb squad who arrived at the scene exposed the explosive belt and tried to disarm the detonation mechanism. The detonation switch was not accessible. The terrorist tried to interfere with his work and there was a concern that the terrorist's movements would create friction between the detonation switch and the ground and would detonate the charge. The only option left for the force was to shoot the terrorist in the head. A shot aimed anywhere else might have detonated the charge and killed the security force members around him, and the suicide bomber was indeed shot.[316]

The Response after a Suicide Attack. After a suicide bomber explodes at his target, Israeli law enforcement forces operate under a number of premises. The first is that a second suicide bomber or charge might explode, aimed at hurting the rescue forces and crowd of curious people. A second premise is that in order to handle the event efficiently, the command and control must be concentrated in the hands of one element—in the Israeli case, the police commander. A third premise is that the thwarting efforts don't end after the bomber has exploded, but continue in order to catch the transporters and supporters. That is why forces are operated not only at the scene of the attack but also in more distant geographical circles, sometimes dozens of kilometers from the center of the blast. Accordingly, a variety of many factors are involved with handling an event—some focus on rescue, some on isolating the scene and examining it, and others on chasing the transporters. All of these work, sometimes simultaneously:

Containment. Isolation of the scene of the attack and denying the possibility of carrying another attack at the scene.

Rescue. Casualty treatment. Alongside the police forces efforts to isolate the scene, actions to save the lives of casualties would be carried out. These actions would only be carried out by medical teams. This will include setting an isolated sector within the scene of the event for treating the injured, and a point would be set at which ambulances will concentrate and leave from. The local police commander will set the routes through which the ambulances would go to the hospitals and will make sure that these will remain clear for the medical forces passage, using police cars and barricades.

Thwarting Activity. The arrest of the transporters, supporters, or more bombers who fled the crime scene.[317]

Command and Control of an Event. The responsibility and command of an incident after the terrorist has carried out his attack belong to the police. It determines the course of events in the three circles of action: the scene of the attack, the close circle, and the far circle. For instance, when should the rescue forces be allowed to enter the scene, and by which routes would the injured be evacuated? To control all these, a tactical headquarters should be set up near the scene of the attack at which the representatives of all of the elements involved in the scene sit at. One part of the headquarters would run the event, the evacuation, and rescue activities. Another part will deal with blocking the close and far circles. Usually, the initial police force who started handling the event will continue handling it as long as its control is the most effective. This is based on the assumption that this force is the most knowledgeable regarding the current status, the forces operating at the scene, and the actions taken so far. Transferring command to a higher rank would take place only when this action does not hinder the crisis management and the continuity of the rescue efforts. Alongside these actions, the tactical headquarters contacts relevant civilian elements—primarily the media, hospitals, and the local municipality—and gives them the details of the event to help create a quick clear status report. The incident is declared over, based on the decision of the police commander. This is after the casualties are evacuated, the body parts collected, and it has been confirmed that there is no risk to public safety from buildings or other objects hit by the blast. The guiding principle in Israel in this case is to restore life at the damaged area to normal as fast as possible. In comparison, sites of attack in London and Spain took a long duration to return to daily routines.

SUMMARY AND CONCLUSIONS

From its very inception in 1948, Israel has been forced to fight terrorism. From the 1990s, suicide bombings have stood out as the most lethal and dominant type of attacks targeting Israeli civilians. Accordingly, the Israeli national security apparatus has had to come up with countermeasures and solutions for this type of threat, long before the rest of the world, particularly the West, realized that it must also confront this kind of attacks. The intensiveness of terror attacks in Israel and the long period of time it had to contend with suicide bombings allows those ample lessons to be drawn from the Israeli experience. These lessons can be pertinent to, among others, national security, strategy and policy-making, economics and social issues, and, of course, those entities actively and directly involved in counterterrorism and combat operations.

This chapter focused on the operational level, with the belief that this realm is of most relevance to law enforcement and homeland security in the United States and the West. Thus, despite the differences between Israel and the United States, they are both liberal democracies with very similar value systems and worldviews.

Beyond the numerous lessons mentioned in this chapter, we can perhaps point out two central lessons that go beyond the operational realm and which may be further corroborated by the Israeli experience.

The first pertains to national security. It seems that when examining the Israeli case, the most important insight for the United States to have from Israel's experience is not to lease its security to others. This is not meant say that when fighting terrorism the United States, or any nation for that matter, should not establish alliances and cooperate with other countries, but that this should be done with the fundamental, sober recognition that when push comes to shove there will be no one else to do its "dirty" work for it.

The second lesson directly relates to the principal fight against terrorism. Many experts rightly argue that in order to effectively deal with the phenomenon of terrorism, both the capabilities and motivations to commit such acts are to be addressed. This may hold true in theory, certainly when we examine cases where groups or nations eventually abandoned terrorism as a practice. However, when the security of a nation's population is hanging in the balance, there is often no time to wait for the completion of slow-paced historical or social process, and immediate action targeting the capabilities of the adversary is required instead. These must thus be neutralized *before* the threat is realized. Both deterrence and protective measures, however effective, will always remain limited. The Israeli experience clearly shows that terrorism—particularly suicide bombings—can be confronted and even defeated by focusing primarily on prevention, preemption, and mitigating the operational capabilities of the terrorists.

Finally, it is important to keep in mind that all people act in accordance with certain patterns of thinking that are molded by their respective life experiences and backgrounds. Therefore, according to the "pattern" of the 9/11 attacks, suicide-bombing attacks may be perceived (especially in the United States and the West) as grandiose acts that require complex planning and intelligence gathering over a period of years and are executed in a precise and meticulous manner. Nevertheless, the Israeli experience proves that suicide bombings can be relatively simple and quick to plan as well as execute. This is particularly true when such attacks originate in a community that is in daily interaction with the supposed "enemy"—for example, Muslim communities in cities across the United States. This close contact increases both the capability to perpetrate an attack and its effectiveness. A series of "small" suicide bombings in various places in the Unites States over a sustained period of time can have the same psychological effect as a single large-scale attack. Hence, as Israel's case shows, it is crucial to remember that beyond the 9/11 model there is a wide range of suicide attacks models for which there needs to be both a mental and operational readiness.

In sum, we must bear in mind that using terrorism against innocent people did not begin in Israel, nor did it end with the relative success Israel had in dealing with the suicide bombings that were launched against its civilians. As a guiding principle, terrorism is still a preferred course of action for many groups and countries around the world, and as such it transforms and adapts according to time and circumstances. In that regard, a measure of diffidence is called for, and to recognize that Israel's success in confronting suicide terrorism may indeed prove to be temporary and its lessons are mostly relevant and perhaps limited to the present mode of attacks, namely suicide bombings.[318]

ENDNOTES

1. Avi Shleim, *The Iron Wall: Israel and the Arab World* (Hebrew), Tel Aviv: *Yedioth Ahronot Books*, 2000, pp. 25–26.
2. Harel Amos, Ben Aloff, and Amira Hess, 24 Killed in Passover Terror Attacks, (Hebrew), *Haaretz*, March 29, 2002.
3. Oren, Amir, The Purpose of the Operation: to demonstrate the Cost of Terror, (Hebrew), *Haaretz*, March 31, 2002.
4. IDF's summary of 2004 data booklet www.1.idf.il/SIP_STORAGE/DOVER/files/4/37604.pdf
5. Interview with Yitzchak Eitan, Genral (retired), Commander of the Central Command 2000–2002, August 20, 2007.
6. Interview with Ilan Paz, Brigadier General (retired), brigade commander in Ramallah during the Intifada, who was also head of the civil administration in Judea and Samaria, August 7, 2007.
7. Harel Amos and Isacharoff Avi, *The Seventh War*, (Hebrew), Tel Aviv: Yedioth Ahronot Books, 2004, p. 219.
8. *The Seventh War*, pp. 137–138.
9. *The Seventh War*, p. 200.
10. Gal Luft, The Palestinian H-Bomb, *Foreign Affairs* **81**(4), 2, 2002, p. 1.
11. For the purpose of this chapter the term "Confrontation" relates to the armed conflict between Israel and Palestinians that erupted in 2000, which is also commonly referred to as the "Second Intifada".
12. Moshe Yaalon, "Lessons from the Palestinian War against Israel," The Washington Institute for Near East Policy, Policy Focus No. 64, January 2007, p. 2; Interview with Yitzchak Eitan, Commander of the Central Command 2000–2002, August 20, 2007.
13. Yaalon, "Lessons," pp. 2–3.
14. Amos Harel, "In Retrospect I Should Have Put a Reporter with Each Battalion," an interview with Dan Harel head of the IDF's operational section (Hebrew), *Haaretz,* April 24, 2003.
15. Amos Harel, Security Sources: There Isn't Any Military Solution to Terror, *Haaretz,* December 19, 2001; The Minister of Internal Security, Avi Dichter, who was head of the ISA at the time, also notes the feeling of helplessness that ensued within various elements of the security system regarding the wave of suicide attacks. Interview with Avi Dicter, head of the ISA 2000–2005, September 26, 2007.

16. ISA website, summary of the number of suicide attacks 2000–2007: http://www.ISA.gov.il/
17. When dealing with the Palestinian fundamentalist terrorist organizations, Hamas and Islamic Jihad, this solution should result in the destruction of the state of Israel and the establishment of an Islamic Caliphate in the entire area spanning between the Jordan river and the Mediterranean.
18. Yitzhak Ben Israel, Coping with Suicide Terrorism—The Israeli Case, in Golan Hagai and Shay Shaul, eds., *Ticking Bomb—Confronting Suicide Attacks* (Hebrew), Tel Aviv: Maa'rachot, 2006, p. 23.
19. Yoram Schweitzer, Palestinian Istishadia: A Developing Instrument, *Studies in Conflict and Terrorism*, 30(Aug 2007), p. 668.
20. Anne Speckhard, Understanding Suicide Terrorism: Countering Human Bombs and Their Senders, in Jason S. Purcell and Joshua D. eds., *Topics in Terrorism: Toward a Transatlantic Consensus on the Nature of the Threat*, Vol. 1, *Weintraub Atlantic Council Publication* 2005, pp. 1–2.
21. Steven Emerson, *American Jihad,* New York: *The Free Press*, 2002. See also Yaacov Perry [former head of ISA], *First to Strike*, Tel Aviv: Keset, 1999, pp. 171–173.
22. Dan Eggen, FBI Warns of Suicide Bombs, *Washington Post*, May 21, 2002.
23. Based on an interviews with former Minister of Justice, Dan Meridor and the minister in charge of the intelligence community 2001–2003, September 24, 2007; and Dany Yatom, Head of the "Mossad" 1996–1998, May 21, 2008.
24. Interview with Avi Dichter, Head of the ISA 2000–2005 and Internal Security Minister since 2006, September 26, 2007.
25. Interview with former Minister of Justice Dan Meridor, who also served as the Minister in charge of the intelligence community 2001–2003, September 24, 2007.
26. Immanuel Sivan, Collision Inside the Islam, in Meridor Dan and Pass Haim eds., *21th Battle—Democracies Fights Terror* (Hebrew), *the Israeli Institute for Democracy*, Jerusalem, 2006, p. 51; Dan Eggen, Terror Threat Grows Quietly, *Washington Post*, August 18, 2007.
27. Interview with former Minister of Justice Dan Meridor, who also served as the Minister in charge of the intelligence community 2001–2003, September 24, 2007.
28. Giora Eiland, A Strategice Perception for Fighting Terror, in Meridor Dan and Pass Haim eds., *21th Battle—Democracies Fights Terror* (Hebrew), Jerusalem: *The Israeli Institute for Democracy*, 2006, pp. 306–314.
29. Ganor Boaz, *The Counter Terrorism Puzzle—A Guide for Decision Makers* (Hebrew), Mifalot: *Herzelyia Inter-Disciplinary Center*, 2003, p. 139.
30. Arie Dayyn, If We Can Use Torture Why Waste Time on Other Methods? (Hebrew), *Haaretz*, January 18, 2000.
31. Ronen Bergman, Interview with Shabtai Ziv, ISA Judicial Counsler Adviser (Hebrew), *Haaretz*, April 14, 2000.
32. See, for example, Amir Oren, Low-Tech against High-Tech (Hebrew), *Haaretz*, February 15, 2001.
33. The Public Committee against Torture in Israel vs. The Government of Israel (HCJ 5100/94).
34. *The Seventh War*, p. 162; Ganor, *Counter-Terrorism Puzzle*, pp. 151–152.

35. Ze'ev Segal, "Large Room for IDF Maneuver" (Hebrew), *Haaretz*, February 15, 2001; Yuval Yoaz, The Last Verdict of Barak: IDF Can Carry Out Target Killings (Hebrew), *Haaretz*, December 15, 2006.
36. Ze'ev Schif, The Proportionality of Israel's Confrontation (Hebrew), *Haaretz*, October 8, 2004.
37. Menahem Finkelshtein, Democracy Fight Terror: Analyzes of the Practical Dilemmas, in Dan Meridor and Haim Pass, eds., *21th Battle—Democracies Fights Terror* (Hebrew), Jerusalem: *the Israeli Institute for Democracy*, 2006, p. 183.
38. Interview with Moshe Yaalon, *IDF Chief* of Staff 2002–2005, October 9, 2007.
39. Ganor Boaz, Israel, Hamas and the Fatah, in Robert J. Art and Louise Richardson, eds, *Democracy and Counterterrorism: Lessons from the Past*, Washington, DC: *United States Institute for Peace*, 2007, p. 297. It should also be noted that by safeguarding liberties, Israel frequently paid a substantial price in terms of security. For further detail, refer to "'Security and Liberty' Striking the Right Balance," http://www.ict.org.il/apage/5450.php
40. Interview with Aharon Zeevi-Farkash, head of Aman 2002–2006, August 19, 2007.
41. *The Seventh War*, pp. 139–140; Chris Quillen, Mass Casualty Bombing Chronology, *Studies in Conflict and Terrorism*, **25**(5), (September 2002), pp. 295–296.
42. The February/March 1996 attacks was at the time an unprecedented peak that many believe affected the national elections held in May 1996.
43. Schweitzer, Istishhadia, pp. 672–673.
44. *The Seventh War*, p.140.
45. Schweitzer, *Istishhadia*, p. 674.
46. Luft, pp. 2–3.
47. Based on IDF data—2004 summary booklet, www.1.idf.il/SIP_STORAGE/DOVER/files/4/37604.pdf; Also see Amos Harel, and Ben Aloff, Less Israelis Killed in Terror Attacks in 2004 (Hebrew), *Haaretz,* December 30, 2004.
48. *The Seventh War*, p. 133.
49. *The Seventh War*, p. 177.
50. *The Seventh War*, pp. 181–184.
51. *The Seventh War*, p. 186; Clive Jones, p. 281.
52. *The Seventh War*, p. 215.
53. *The Seventh War*, pp. 218–219.
54. Based on ISA data 2000–2005 summary booklet, http://www.shabak.gov.il/
55. Efrat Wies, Official Report: 60% Decrease in Casualties, Increase in Kassam Rockets (Hebrew), *Ynet*, January 2, 2006, http://www.ynet.co.il/articles/0,7340,L-3193436,00.html
56. Interview with Avi Dichter, head of the ISA 2000–2005, former Internal Security minister 2006–2009, September 26, 2007.
57. Quoted in *The Seventh War*, p. 151.
58. Amira Hess, Floating to Heaven (Hebrew), interviews by a security prisoner named Walid Daka with Palestinian youngsters caught on route to carry out suicide attacks, *Haaretz*, April 4, 2003.
59. See, for example, Anat Berko and Edna Erez, "Ordinary People" and "Death Work": Palestinian Suicide Bombers as Victimizers and Victims, *Violence and Victims* **20**(6),

2005. A. M. Oliver and P. F. Steinberg, The Road to Martyrs' Square: A journey into the World of the Suicide Bomber. New York: Oxford University Press, 2005.

60. Dan Eggen, Terror Threat Grows Quietly, *Washington Post*, August 16, 2007.
61. Interview with Ilan Paz, Brigadier General (retired), a brigade commander in Ramallah during the Intifada, who also served as the head of the civil administration in Judea and Samaria, August 7, 2007.
62. Quoted in *The Seventh War*, p.143.
63. Jonathan B. Tucker, "Strategies for Countering Terrorism: Lessons from the Israeli Experience." *Journal of Homeland Security*, March 2003. http://www.homelandsecurity.org/newjournal/articles/tucker-israel.htm#endref1
64. Data regarding suicide bombers and the attacks they carried out are based on the suicide bombings summary by the Intelligence and Terrorism Information Center: Suicide bombing terrorism during the current Israeli–Palestinian confrontation (September 2000–December 2005), January 1, 2006, http://www.terrorism-info.org.il/malam_multimedia/English/eng_n/pdf/suicide_terrorism_ae.pdf
65. *The Seventh War*, pp. 143–144.
66. The Intelligence and Terrorism Information Center: Suicide bombing terrorism during the current Israeli–Palestinian confrontation, September 2000–December 2005, pp. 24–26, January 1, 2006. http://www.terrorism-info.org.il/malam_multimedia/English/eng_n/pdf/suicide_terrorism_ae.pdf
67. A. Moghadam, (2003). The characteristics of suicide terrorist: An empirical analysis of Palestinian terrorism in Israel. Retrieved from http://nssc.haifa.ac.il/Terror/articles/profile.html
68. Anat Berko, *On the Route to Heaven—The World of Suicide Bombers and Their Dispatchers* (Hebrew), Tel Aviv: *Yedioth Ahronot Books*, 2004, p. 26.
69. For a summary of the research literatures regarding the motives of suicide bombers see Shmuel Eben and Shaul Kimchi, *Who Are the Palestinian Suicide Bombers?* (Hebrew), Memo No. 73, Tel Aviv: *Jaffe Center for Strategic Surveys*, 2004, pp. 19–29.
70. The analysis of the psychological motives of suicide bombers can be found at Speckhard, pp. 5–7. For other studies that analyze the issue from this point of view, also see E. Sarraj, Suicide Bombers: Dignity, Despair, and the Need for Hope, *Journal of Palestine Studies*, Vol. **31**(4), 71–76, 2002. J. Post and L. M. Denny, The terrorists in their own words: Interviews with 35 incarcerated Middle Eastern terrorists. *Terrorism and Political Violence*. 15, p 171–184, 2003.
71. For example, see the study by L. L. Schwartz, The Cult Phenomenon: A Turn of the Century Update, *American Journal of Family Therapy* **29**(1), p. 13–22, 2001.
72. Schiff, Zeev, Targeted Killing: From Ticking Bomb to Ticking Infrastructure—An interview with the retiring deputy of the ISA (Hebrew), *Haaretz*, September 10, 2003.
73. Berko, p. 27.
74. Interview with Lt. Col. D. H. Commander of IDF Crisis Management and Hostage Negotiation Unit, June 18, 2008.
75. Assaf Moghadam, Palestinian Suicide Terrorism in the Second Intifada: Motivations and Organizational Aspects, *Studies in Conflict & Terrorism*, **26**(2), 72, 2003.

76. Interview with Aharon Zeevi-Farkash, Head of Aman 2002–2006, August 19, 2007.
77. Quoted in *The Seventh War*, p. 142.
78. Amira Hess, Floating to Heaven (Hebrew), interviews by a security prisoner named Walid Daka with Palestinian youngsters caught on route to carry out suicide attacks, *Haaretz*, April 4, 2003.
79. Ibid.
80. Uri Blau, If I Were Palestinian—An interview with Nisim Levi (One of the ISA's Highest Ranking Officials) (Hebrew), Haaretz, December 22, 2006.
81. There is relatively a lot of literature on the Hamas. On its early years of development until the late 1990s, see: Shaul Misha and Avraham Sela, *The Palestinian Hamas: Vision, Violence, and Coexistence*, New York: *Columbia University Press*, 2006. For more contemporary literature, see: Matthew Levitt, *Hamas Politics, Charity and Terrorism in the Service of Jihad: Politics, Charity, and Terrorism in the Service of Jihad*, New Haven, CT: *Yale University Press*, 2007. Azzam Tamimi, *Hamas—A History from Within*. Northamptonm, MA: *Olive Branch Press*, 2007. Also see an online contemporary review in: http://www.terrorisminfo.org.il/malam_multimedia/English/eng_n/pdf/hamas_e0206.pdf
82. Izz Al-Din Al-Kassam was a fundamentalist religious preacher who operated in Israel during the British mandate and was killed in a battle with the British in 1935 after trying to start a violent revolt against them.
83. Moghadam, p. 79.
84. For an analysis of the activities of one of Hamas' foundations (The Bethlehem orphans foundation) and its link with terrorist activities see: http://www.terrorism-info.org.il/malam_multimedia//ENGLISH/MARKETING%20TERRORISM/PDF/JAN22_05.PDF
85. Yossi Melman, There Is a Partner for Negotiation, an interview with Shalom Sharabi, one of the ISA's high ranking officials and head of the non-Arab department in the ISA, (Hebrew), *Haaretz*, July 14, 2006.
86. For a summarized review on Hamas' financing methods see http://www.justice.gov.il/NR/rdonlyres/E4160F6C-8F86-483E-9CC0-C9C0144DDE86/0/430.doc. For a review on the movement's current civilian infrastructure and its financing see the Intelligence and Terrorism Information Center's review at http://www.terrorism-info.org.il/malam_multimedia/English/eng_n/pdf/hamas_0706e.pdf
87. For example, see the story of Hamas' Nablus terrorist network, *The Seventh War*, pp. 205–206; For a review on Hamas' Tulkarm network and its attacks see שוויצר, מעריב, April 6, 2007.
88. Ma'ariv Schweitzer, April 6, 2007. Schweitzer, *Istishhadia*, p. 681.
89. Moghadam, p. 79.
90. See the Intelligence and Terrorism Information Center's review summarizing suicide bombings in the first 5 years of the confrontation at http://www.terrorism-info.org.il/malam_multimedia/English/eng_n/pdf/suicide_terrorism_ae.pdf
91. It should be noted that although the Fatah refrained from carrying out suicide attacks until 2001 and that other forms of terrorist activities were not carried out in an official manner, there were acts of lethal violence carried out by Fatah members against Israel—for example, the shootings on IDF soldiers in September 1996 that resulted in 16 deaths.

92. *The Seventh War*, pp. 216–217; Moghadam, pp. 82–83.
93. *The Seventh War*, p. 215.
94. See, for example, the case of Abd Al-Karim Awiss, who placed an explosives charge at a major junction in northern Israel, *The Seventh War*, pp. 132–133; Luft, pp. 2–3.
95. Yossi Kuperwasser, The Causes for Terror and Its Risks: The Intelligence Assesment, in Dan Meridor and Haim Pass, eds., *21th Battle—Democracies Fights Terror* (Hebrew), Jerusalem: *the Israeli Institute for Democracy*, 2006, p. 225.
96. Quoted in *The Seventh War*, p. 214.
97. Schweitzer, *Istishhadia*, pp. 680–681.
98. Interview with Benjamin Netanyahu on April 23, 2006 and interview with Moshe Yaalon on October 9, 2007.
99. Meir Hatina, *Palestinian Radicalism: The Islamic Jihad Movement.* Tel Aviv: *Tel Aviv University*, 1994, pp. 17–18.
100. See the Intelligence and Terrorism Information Center's review summarizing suicide bombing statistics in the first 5 years of the confrontation at http://www.terrorism-info.org.il/malam_multimedia/English/eng_n/pdf/suicide_terrorism_ae.pdf
101. Amos Harel, Israel Is Worried: Hamas and Iran Are Developing Strategic Relations, (Hebrew), *Haaretz*, December 17, 2006.
102. Interview with Giora Eiland, Head of the National Security Council 2003–2006, August 8, 2007. Amos Harel, Security Sources: There Isn't Any Military Solution to Terror, *Haaretz*, December 19, 2001.
103. *The Seventh War*, p. 156.
104. See the Intelligence and Terrorism Information Center's review summarizing suicide bombing statistics in the first 5 years of the confrontation at http://www.terrorism-info.org.il/malam_multimedia/English/eng_n/pdf/suicide_terrorism_ae.pdf
105. A data sheet published by the Prime Minister's Office summarizing the conflict in 2004: http://www.pmo.gov.il/
106. Amos Harel, Say Hooray! or Get Inside the Police Car (Hebrew), *Haaretz*, March 23, 2007.
107. *The Seventh War,* pp. 156–157.
108. Interview with Yossi Kuperwasser, Brigadier General (retired), Head of the Military Intelligence's (Aman's) research department 2001–2006, July 29, 2007.
109. Schweitzer, *Istishhadia*, p. 675.
110. See, for example, a poll by the Palestinian Center for Policy and Survey Research (PSR), headed by Khalil Shikaki, at http://www.pcpsr.org/survey/polls/2004/p13a.html#fouryrs
111. Yaalon, "*Lessons*," pp. 7–8.
112. *The Seventh War*, pp. 217–218.
113. See, for example, learning material and accessories captured in Bethlehem which carry radical Islamic messages in favor of Jihad and pictures of martyrs. See the Intelligence and Terrorism Information Center's review at http://www.terrorism-info.org.il/malam_multimedia/English/Hate-Authority/PDF/june_03.pdf. Regarding the indoctrination of Islamic fundamental values in kindergartens in the territories see the Intelligence and Terrorism Information Center's review at http://www.terrorism-info.org.il/malam_multimedia/English/eng_n/pdf/kindergarten_

gaza060607.pdf, Inculcating Kindergarten Children with Radical Islamic Ideology and the Culture of Anti-Israel Terrorism.

114. Berko, pp. 28–29; *The Seventh War*, p. 42; Yossi Kuperwasser, The Causes for Terror and Its Risks: The Intelligence Assesment, in Dan Meridor and Haim Pass, eds., *21th Battle—Democracies Fights Terror* (Hebrew), *the Israeli Institute for Democracy*, Jerusalem, 2006, p. 229.
115. Moghadam, p. 72.
116. Berko, p. 28.
117. Yossi Kuperwasser, quoted from Protocol of Discussion that was held at the Israeli Institute for Democracy on October 31, 2004, in Dan Meridor and Haim Pass, eds., *21th Battle—Democracies Fights Terror* (Hebrew), *the Israeli Institute for Democracy*, Jerusalem, 2006, p. 251.
118. Amos Harel, In Retrospect I Should Have Put a Reporter with Each Battalion, an interview with Dan Harel, Head of the IDF's operational section (Hebrew), *Haaretz,* April 24, 2003. On Saddam Hussein's support of suicide bombers, see the Intelligence and Terrorism Information Center's review: How Saddam Hussein's Régime Transferred Funds from Iraq to the West Bank and Gaza Strip to Encourage Palestinian Terrorism, http://www.terrorism-info.org.il/malam_multimedia/English/eng_n/pdf/saddam8_05.pdf
119. Berko, p. 43.
120. Interview with Moshe Yaalon, IDF Chief of Staff 2002–2005, October 9, 2007.
121. 11.8.2001 http://news.walla.co.il/?w=//100540. Amos Harel, Ben Aloff, and Baroch Kara, 14 Died and 140 Wounded in a Terror Attack at the Center of Jerusalem (Hebrew), *Haaretz*, August 10, 2001.
122. *The Seventh War*, pp. 135–136.
123. *The Seventh War*, pp. 136–137.
124. Ma'ariv Schweitzer, April 6, 2007; *The Seventh War,* p. 147.
125. Berko, p. 23.
126. *The Seventh War*, pp. 148–150.
127. Quoted at Schweitzer, *Istishhadia*, p. 685.
128. *The Seventh War*, p. 150.
129. Schweitzer, *Istishhadia*, p. 683.
130. *The Seventh War*, p. 152.
131. Interview with Avi Dichter, Head of the ISA 2000–2005, and Internal Security Minister since 2006, September 26, 2007. Moghadam, p. 84.
132. Schweitzer, *Istishhadia*, p. 684.
133. Interviews by a Security Prisoner Named Walid Daka with Palestinian Youngsters Caught on Route to Carry Out Suicide Attacks, *Haaretz*, April 4, 2003.
134. Interviews by a Security Prisoner Named Walid Daka with Palestinian Youngsters Caught on Route to Carry Out Suicide Attacks (Hebrew), *Haaretz*, April 4, 2003.
135. The attack in the outdoor mall in Jerusalem in December 2001, in which 10 Israeli boys were killed; the attack at the "Moment" café in Jerusalem in March 2002, in which 11 Israelis were killed, the attack at the "Sheffield Club" in Rishon LeZion, in which 15 Israelis were killed; the attack at the Hebrew University in Jerusalem in July 2002, in which 9 Israelis were killed; the attack on a bus in Allenby Street in Tel

Aviv in September 2002, in which 6 Israelis were killed. See: Amos Harel, and Arnon Regular, The Terrorist Who Prepared the Charges for the Suicide Terror Attacks at Sbaro and Moment Coffee House was Arrested (Hebrew), *Haaretz*, March 9, 2003; Amos Harel, Hamas Operative Confess and Convicted in Committing Terror Attacks (Hebrew), *Haaretz*, June 2, 2003.

136. *The Seventh War*, p. 83.
137. Schweitzer, *Ma'ariv,* April 6, 2007.
138. Amos Harel, Experts for Munitions Charges which Were Sent to the Territories by the Hamas Headquarter in Jordan Were Arrested (Hebrew), *Haaretz*, September 13, 2001. Amos Harel, The ISA: The Hamas Trains Experts for munitions charges at Camps in Sudan (Hebrew), *Haaretz*, September 19, 2002.
139. Amos Harel, Experts for munitions charges that were trained in Iran Infiltrate to Gaza Strip (Hebrew), *Haaretz*, January 9, 2006; *The Seventh War,* p. 155.
140. Interview with Giora Eiland, head of the National Security Council 2003–2006, August 8, 2007. *The Seventh War*, pp. 131, 151.
141. *The Seventh War*, p. 128.
142. Schweitzer, *Ma'ariv*, April 6, 2007.
143. Summary by the Intelligence and Terrorism Information Center: Suicide bombing terrorism during the current Israeli–Palestinian confrontation (September 2000–December 2005), January 1, 2006, p. 58.
144. Hagai Hitron, Exaggerated Enthusiasm from the Democracy (Hebrew), *Haaretz*, February 5, 2007.
145. Quoted in: *The Seventh War*, p. 151.
146. Summary by the Intelligence and Terrorism Information Center: Suicide bombing terrorism during the current Israeli–Palestinian confrontation (September 2000–December 2005), January 01, 2006, pp. 64–65.
147. Schweitzer, *Ma'ariv*, April 6, 2007.
148. Quoted in Schweitzer, *Istishhadia*, p. 682.
149. Review by the Intelligence and Terrorism Information Center: Suicide bombing terrorism during the current Israeli–Palestinian confrontation (September 2000–December 2005), January 01, 2006, http://www.terrorism-info.org.il/malam_multimedia/English/eng_n/pdf/suicide_terrorism_ae.pdf; The ISA's website, the summery of suicide attacks 2000–2007: http://www.shabak.gov.il/
150. Based on Israel Security Agency Report 2007, http://www.shabak.gov.il/
151. Interview with Yossi Kuperwasser, Brigadier General (retired), Former Head of Military Intelligence's (Aman's) research department 2001–2006, July 29, 2007.
152. Interview with Aharon Zeevi-Farkash, head of Aman 2002–2006, August 19, 2007.
153. Giora Eiland, Quoted from Protocol of Discussion which was Held in the Israeli Institute for Democracy on 31st October 2004, in Dan Meridor and Haim Pass, eds., *21th Battle—Democracies Fights Terror* (Hebrew), *the Israeli Institute for Democracy*, Jerusalem, 2006, pp. 329–330.
154. A large part of this section in based on interviews held with Moshe Yaalon, IDF Chief of Staff 2002–2005, October 9, 2007; Avi Dichter, Head of the ISA 2000–2005 and Internal Security Minister since 2006, September 26, 2007; Giora Eiland, Head of the National Security Council 2003–2006, August 8, 2007; Yitzchak Eitan,

Commander of the IDF's Central Command 2000–2002, August 20, 2007, Uri Bar Lev, Israeli Police Commander of the Southern District, and Former Commander elite counter-terrorism units, August 23, 2007; Aharon Zeevi-Farkash, Head of Aman 2002–2006, August 19, 2007; Ilan Paz, Brigadier General (retired), a brigade commander in Ramallah during the Intifada and also Head of the civil administration in Judea and Samaria, August 7, 2007; Lior Lotan, Head of the IDF general staff's negotiations team, November 2, 2007; Zohar Dvir, Yamam commander 2002–2005, November 4, 2007; Danny Yatom, Head of the "Mossad" 1996–1998, May 21, 2008; Dan Meridor, former Minister of Justice and the minister in charge of the intelligence community 2001–2003, September 24, 2007; Hagai Peleg, Commander of Special Israeli Police Unit—Yamam, 1999–2001, August 2, 2007.

155. Interview with Moshe Yaalon, IDF Chief of Staff 2002–2005, October 9, 2007.
156. Interview with Zohar Dvir, Yamam commander 2002–2005, November 4, 2007.
157. Interview with Zohar Dvir, Yamam commander 2002–2005, November 4, 2007.
158. http://www.ynet.co.il/articles/0,7340,L-1962833,00.html
159. Amos Harel and Arnon Regular, In the Last Few Day There Was an Increase of Attempts to Carry Out Terror Attacks (Hebrew), *Haaretz*, February 7, 2003; *The Seventh War*, pp. 147–148.
160. Boaz Ganor, Israel, Hamas and the Fatah, in Robert J. Art and Louise Richardson eds. *Democracy and Counterterrorism: Lessons from the Past. United States Institute for Peace*, 2007, pp. 281–283.
161. Interview with Uri Bar Lev, commander of the Southern District of the Israeli police, founder and former commander of two elite counter-terrorism units, August 23, 2007.
162. Interview with Lior Lotan, head of the IDF general staff's negotiations team, November 2, 2007.
163. The following analysis is based on an interview with Moshe Yaalon, IDF Chief of Staff 2002–2005, October 9, 2007; an interview with Yossi Kuperwasser, Brigadier General (retired), Head of the Military Intelligence's (Aman's) research department 2001–2006, July 29, 2007; Giora Eiland, A Strategic Perception for Fighting Terror, in Dan Meridor and Haim Pass, eds., *21th Battle—Democracies Fights Terror* (Hebrew), Jerusalem: *the Israeli Institute for Democracy*, 2006, pp. 305–317; Tucker, "Lessons"; Yaalon, "Lessons," p. 7.
164. Interview with Zohar Dvir, Yamam commander 2002–2005, November 4, 2007.
165. Clive Jones, pp. 275–277; Tucker, Lessons, Yaacov Perry, pp. 23–24.
166. Yossi Melman, The ISA: More Technology, Less Human Agents (Hebrew), *Haaretz*, September 2, 2001; 23–22, 1999; Yaacov Perry, pp. 22–23.
167. The data regarding the terror attacks thwarting percentage was drawn from Ben Israel, "Coping with Suicide Terrorism," p. 18.
168. Interview with Ilan Paz, Brigadier General (retired), a brigade commander in Ramallah during the Intifada and also head of the civil administration in Judea and Samaria, August 7, 2007.
169. Uri Blau, If I Were Palestinian—An interview with Nisim Levi (one of the ISA's high-ranked officials) (Hebrew), *Haaretz*, December 22, 2006.
170. Amos Harel, For Serious People Only!! (Hebrew), *Haaretz*, May 19, 2006.
171. Yaakov Perry, First to Strike, p. 22.

172. Interview with Ilan Paz, Brigadier General (retired), a brigade commander in Ramallah during the Intifada and was also head of the civil administration in Judea and Samaria, August 7, 2007.
173. Uri Blau, If I Were Palestinian—An interview with Nisim Levi (one of the ISA's high-ranked officials) (Hebrew), *Haaretz*, December 22, 2006; Amos Harel, For Serious People Only!! (Hebrew), *Haaretz*, May 19, 2006; Yossi Melman, The ISA: More Technology, Less Human Agents (Hebrew), *Haaretz*, September 2, 2001; Yossi Melman, There is a Partner for Negotiation, an interview with Shalom Sharabi, one of the ISA's high-ranked officials and head of the non-Arab department in the ISA (Hebrew), *Haaretz*, July 14, 2006.
174. *The Seventh War*, p. 90. Interview with Yossi Kuperwasser, Brigadier General (retired), Head of the Military Intelligence's (Aman's) research department 2001–2006, July 29, 2007.
175. Yossi Melman, The ISA: More Technology, Less Human Agents (Hebrew), *Haaretz*, September 2, 2001.
176. *The Seventh War*, p. 161. Clive Jones, p. 277.
177. Oren, Amir, To Germany and Back (Hebrew), *Haaretz*, January 30, 2004.
178. Interview with Moshe Yaalon, IDF Chief of Staff 2002–2005, October 9, 2007.
179. Interview with Aharon Zeevi-Farkash, Head of Aman 2002–2006, August 19, 2007.
180. Yaalon, "Lessons," p. 14.
181. Yaalon, "Lessons," p. 14. Interview with Moshe Yaalon, IDF Chief of Staff 2002–2005, October 9, 2007. Interview with Giora Eiland, August 8, 2007.
182. Galit Yeminy, The ISA Is Hunting for computing programmer—An Interview with the Head of the Information Technology Section at the ISA (Hebrew), *Haaretz*, September 1, 2005.
183. Interview with Aharon Zeevi-Farkash, Head of Aman 2002–2006, August 19, 2007.
184. Interview with Moshe Yaalon, IDF Chief of Staff 2002–2005, October 9, 2007.
185. Interview with Ilan Paz August 7, 2007.
186. Interview with Zohar Dvir, Yamam commander 2002–2005, November 4, 2007.
187. Yossi Melman, The ISA: More Technology, Less Human Sources (Hebrew), *Haaretz*, September 2, 2001.
188. Interview with Yossi Kuperwasser, Brigadier General (retired), Head of the Military Intelligence's (Aman's) research department 2001–2006, July 29, 2007.
189. *The Seventh War*, pp. 160, 325. Clive Jones, p. 276. Oren, Amir, The Green is More Red from the Blue (Hebrew), *Haaretz*, November 22, 2002.
190. Yossi Melman, The ISA: More Technology, Less Human Agents (Hebrew), *Haaretz*, September 2, 2001.
191. *The Seventh War*, p. 177.
192. *The Seventh War*, p. 185.
193. Amir Oren, The Green is More Red from the Blue (Hebrew), *Haaretz*, November 22, 2002.
194. Interview with Yossi Kuperwasser, Brigadier General (retired), Head of the Military Intelligence's (Aman's) research department 2001–2006, July 29, 2007.

195. Interview with Moshe Yaalon, IDF Chief of Staff 2002–2005, October 9, 2007.
196. Interview with Ilan Paz, Brigadier General (retired), a brigade commander in Ramallah during the Intifada, and also head of the civil administration in Judea and Samaria, August 7, 2007.
197. *The Seventh War*, p. 128.
198. Interview with Ilan Paz, Brigadier General (retired), a brigade commander in Ramallah during the Intifada and also head of the civil administration in Judea and Samaria, August 7, 2007.
199. Yaalon, "Lessons," p. 8.
200. Interview with Yossi Kuperwasser, Brigadier General (retired), Head of the Military Intelligence's (Aman's) research department 2001–2006, July 29, 2007.
201. Amos Harel, If We Pull Back, the Situation Will Return to Its Former Status (Hebrew), *Haaretz*, October 26, 2001.
202. Amos Harel, Security Sources: There Isn't Any Military Solution to Terror, *Haaretz*, December 19, 2001.
203. *The Seventh War*, p. 112.
204. Interview with Yossi Kuperwasser, Brigadier General (retired), Head of the Military Intelligence's (Aman's) research department 2001–2006, July 29, 2007.
205. Interview with Ilan Paz, Brigadier General (retired), a brigade commander in Ramallah during the Intifada and also head of the civil administration in Judea and Samaria, August 7, 2007.
206. *The Seventh War*, pp. 223–224.
207. Interview with Hagai Peleg, Commander of Special Israeli Police Unit—Yamam, August 2, 2007, 1999–2001; Amos Harel, The Former Yamman's Commander: There Is No Significance to Territories "A" Anymore (Hebrew), *Haaretz*, February 1, 2002.
208. *The Seventh War*, pp. 223–227.
209. *The Seventh War*, p. 235.
210. *The Seventh War*, pp. 234–235.
211. *The Seventh War*, p. 238; Yaalon, "Lessons," p. 8.
212. Amos Harel, Retrospectively I Should Have Put a Reporter with Each Battalion, an interview with Dan Harel, Head of the IDF's operational section (Hebrew), *Haaretz*, April 24, 2003. The rest of the data are drawn from: Zeev Schiff, Successful Thwarting vs. High Motivation (Hebrew), *Haaretz*, May 2, 2003 and from Shay Nitzan, Fighting Terror: Legal Aspects, in Dan Meridor and Haim Pass, eds., *21th Battle—Democracies Fights Terror* (Hebrew), Jerusalem: *the Israeli Institute for Democracy*, 2006, p. 138.
213. Interview with Yitzchak Eitan, Commander of the Central Command 2000–2002, August 20, 2007; Amos Harel, Say Hooray! or Get Inside the Police Car (Hebrew), *Haaretz*, March 23, 2007; IDF data booklet of 2004 at www.1.idf.il/SIP_STORAGE/DOVER/files/4/37604.pdf
214. Interview with Moshe Yaalon, IDF Chief of Staff 2002–2005, October 9, 2007.
215. Yossi Melman, There is a Partner for Negotiation, an interview with Shalom Sharabi, one of the ISA's high-ranked officials and head of the non-Arab department in the ISA (Hebrew), *Haaretz*, July 14, 2006.

216. Apart from the name of the Head of the ISA (Avi Dichter) and the event itself—the targeted killing of Hussein Abayat—all other details in the scenario are notional. The disruption of the operational process and chain of events is based on a number of sources including: Amir Oren, Low-Tec against High-Tec (Hebrew), *Haaretz*, February 15, 2001; Amos Harel, 15 Surveillance Means Fails in the Most Crowded Areas in the World (Hebrew), *Haaretz*, June 22, 2006; Alex Fishman, The Code Name: Bell (Hebrew), *Yedioth Ahronot*, May 25, 2007; *The Seventh War*, pp. 98, 202–204.
217. Zeev Schiff, The First Targeted Killing (Hebrew), *Haaretz*, June 5, 2006.
218. Yossi Melman, In the Past, Targeted Killing Was the Last Method. Today It Is Carried Out Across-the-board (Hebrew), *Haaretz*, March 24, 2004.
219. *The Seventh War*, p. 194.
220. Alex Fishman, The Code Name: Bell (Hebrew), *Yedioth Ahronot,* May 25, 2007.
221. *The Seventh War*, p. 115, footnote 39.
222. *The Seventh War*, pp. 194–195.
223. Clive Jones, p. 279.
224. Interview with Giora Eiland, Head of the National Security Council 2003–2006, August 23, 2007.
225. Interview with Giora Eiland, Head of the National Security Council 2003–2006, August 23, 2007. Interview with Moshe Yaalon, IDF Chief of Staff 2002–2005, October 9, 2007. Tucker, "Lessons." Uval Yoaz, The Last Verdict of Barak: IDF Can Carry Out Target Killings (Hebrew), *Haaretz*, December 15, 2006; Ze'ev Segal, A Large Room for IDF Maneuver (Hebrew), *Haaretz*, February 15, 2001.
226. Interview with Moshe Yaalon, IDF Chief of Staff 2002–2005, October 9, 2007.
227. Interview with Moshe Yaalon, IDF Chief of Staff 2002–2005, October 9, 2007.
228. Zeev Schiff, Targeted Killing: From Ticking Bomb to Ticking Infrastructure—An interview with the retiring deputy of the ISA (Hebrew), *Haaretz*, September 10, 2003.
229. *The Seventh War*, pp. 189–191. Yossi Melman, In the Past, Targeted Killing Was the Last Method. Today It Is Carried Out Across-the-Board (Hebrew), *Haaretz*, March 24, 2004.
230. Interview with Giora Eiland, Head of the National Security Council 2003–2006, August 23, 2007.
231. Interview with Moshe Yaalon, IDF Chief of Staff 2002–2005, October 9, 2007.
232. Interview with Danny Yatom, Head of the "Mossad" 1996–1998, May 21, 2008.
233. *The Seventh War*, p. 210.
234. Interview with Moshe Yaalon, IDF Chief of Staff 2002–2005, October 9, 2007.
235. Interview with Yitzchak Eitan, Commander of the Central Command 2000–2002, August 20, 2007. Interview with Avi Dichter, Head of the ISA 2000–2005 and Internal Security Minister since 2006, September 26, 2007; *The Seventh War*, p. 188; Zeev Schiff, Targeted Killing: From Ticking Bomb to Ticking Infrastructure—An interview with the retiring deputy of the ISA (Hebrew), *Haaretz,* September 10, 2003. Clive Jones, p. 281.
236. Amos Harel, 15 Surveillance Means Fails in the Most Crowded Areas in the World (Hebrew), *Haaretz*, June 22, 2006. Alex Fishman, The Code Name: Bell (Hebrew), *Yedioth Ahronot*, May 25, 2007.

237. Interview with Giora Eiland, Head of the National Security Council 2003–2006, August 23, 2007; Amos Harel, 15 Surveillance Means Fail in the Most Crowded Areas in the World (Hebrew), *Haaretz*, June 22, 2006. *The Seventh War*, pp. 119–200.
238. Interview with Giora Eiland, Head of the National Security Council 2003–2006, August 23, 2007.
239. Yossi Melman, The ISA: More Technology, Less Human Agents (Hebrew), *Haaretz*, September 2, 2001.
240. *The Seventh War*, p. 205.
241. Dicter, *Israel's Lessons*, p. 9.
242. Interview with Giora Eiland, Head of the National Security Council 2003–2006, August 23, 2007.
243. Amos Harel, 15 Surveillance Means Fail in the Most Crowded Areas in the World (Hebrew), *Haaretz*, June 22, 2006.
244. *The Seventh War*, pp. 203–204.
245. Interview with Moshe Yaalon, IDF Chief of Staff 2002–2005, October 9, 2007.
246. Interview with Yitzchak Eitan, Commander of the Central Command 2000–2002, August 20, 2007.
247. Daniel Byman. Do Targeted Killings Work? *Foreign Affairs* **85**(2), 95–112, 2006.
248. Interview with Avi Dichter, Head of the ISA 2000–2005, and Internal Security Minister since 2006, September 26, 2007.
249. Interview with Giora Eiland, Head of the National Security Council 2003–2006, August 23, 2007. Alex Fishman, The Code Name: Bell (Hebrew), *Yedioth Ahronot*, May 25, 2007.
250. Interview with Avi Dichter, Head of the ISA 2000–2005, and Internal Security Minister since 2006, September 26, 2007.
251. Amos Harel, In Retrospect I Should Have Put a Reporter with Each Battalion an interview with Dan Harel, Head of the IDF's operational section (Hebrew), *Haaretz*, April 24, 2003.
252. Amos Harel, Say Hooray! or Get Inside the Police Car (Hebrew), *Haaretz*, March 23, 2007.
253. Dicter, *Israel's Lessons*, pp. 5–8.
254. Dicter, *Israel's Lessons*, pp. 5–8.
255. Interview with Yossi Kuperwasser, Brigadier General (retired), Head of the Military Intelligence's (Aman's) research department 2001–2006, July 29, 2007.
256. Lessons Tucker and Amir Oren, To Germany and Back (Hebrew), *Haaretz*, January 30, 2004.
257. Interview with Zohar Dvir, Yamam commander 2002–2005, November 4, 2007.
258. Interview with Hagai Peleg, Former Commander of Special Israeli Police Unit—Yamam 1999–2001, August 2, 2007; Amos Harel, The Former Yamman's Commander: There is No Significance to "A" Territories Anymore (Hebrew), *Haaretz*, February 1, 2002.
259. Interview with Ilan Paz, Brigadier General (retired), a brigade commander in Ramallah during the Intifada and also Head of the civil administration in Judea and Samaria, August 7, 2007. *The Seventh War*, p. 121.

260. Interview with Ilan Paz, Brigadier General (retired), a brigade commander in Ramallah during the Intifada and also Head of the civil administration in Judea and Samaria, August 7, 2007.
261. Interview with Hagai Peleg, Former Commander of Special Israeli Police Unit—Yamam 1999–2001, August 2, 2007.
262. Interview with Zohar Dvir, Yamam commander 2002–2005, November 4, 2007.
263. Interview with Lior Lotan, Head of the IDF general staff's negotiations team, November 2, 2007.
264. Interview with Lior Lotan, Head of the IDF general staff's negotiations team, November 2, 2007.
265. Yaalon, "Lessons," p. 14. Interview with Lior Lotan, head of the IDF general staff's negotiations team, November 2, 2007.
266. Berko, p. 27.
267. Boaz Ganor, Israel, Hamas and the Fatah, in Robert J. Art and Louise Richardson eds. Democracy and Counterterrorism: Lessons from the Past, Washington, DC: *United States Institute for Peace*, 2007, p. 283.
268. Interview with Avi Dichter, Head of the ISA 2000–2005, and Internal Security Minister since 2006, September 26, 2007.
269. Interview with Aharon Zeevi-Farkash, Head of Aman 2002–2006, August 19, 2007.
270. Interview with Zohar Dvir, Yamam commander 2002–2005, November 4, 2007. Ben Israel, Coping with Suicide Terrorism, pp. 36–37.
271. Interview with Avi Dichter, Head of the ISA 2000–2005, and Internal Security Minister since 2006, September 26, 2007.
272. Interview with Zohar Dvir, Yamam commander 2002–2005, November 4, 2007.
273. Dutter and Seliktar, p. 438.
274. Interview with Avi Dichter, Head of the ISA 2000–2005, and Internal Security Minister since 2006, September 26, 2007.
275. Zeev Schiff, Successful Thwarting vs. High Motivation (Hebrew), *Haaretz*, May 2, 2003.
276. Interview with Lior Lotan, Former Head of the IDF Crisis Management and Negotiation Unit, 2.11.2007; Boaz Ganor, Israel, Hamas and the Fatah, in Robert J. Art and Louise Richardson, eds. Democracy and Counterterrorism: Lessons from the Past, Washington, DC: *United States Institute for Peace*, 2007, pp. 281–283.
277. *The Seventh War*, p. 137.
278. Interview with Moshe Yaalon, IDF Chief of Staff 2002–2005, October 9, 2007.
279. Interview with Ilan Paz, Brigadier General (retired), a brigade commander in Ramallah during the Intifada and also head of the civil administration in Judea and Samaria, August 7, 2007.
280. Interview with Avi Dichter, Head of the ISA 2000–2005, and Internal Security Minister since 2006, September 26, 2007.
281. Efrat Waiss, Official Report: 60% Decrease in Casualties, Increase in Kassam Rockets (Hebrew), Ynet, January 2, 2006, http://www.ynet.co.il/articles/0,7340,L-3193436,00.html
282. Dutter and Seliktar, p. 438.

283. Avi Dicter, Head of the ISA 2000–2005, September 26, 2007. Interview with Moshe Yaalon, IDF Chief of Staff 2002–2005, October 9, 2007.
284. Pinchas Yehezkeli and Yisha'ayahu Horowitz, The Struggle of the Israeli Police Against the Terror, 2000–2004, in Golan Hagai and Shay Shaul eds. *Ticking Bomb—Confronting Suicide Attacks* (Hebrew), Tel Aviv: *Maa'rachot*, 2006, pp. 140–141.
285. Yehezkeli and Horowitz, p. 144.
286. Interview with Uri Bar Lev, Commander of the Southern District of the Israeli police, August 23, 2007.
287. Interview with Uri Bar Lev, Commander of the Southern District of the Israeli police, August 23, 2007.
288. Yehezkeli and Horowitz, p. 144.
289. Roni Zinger, Suicide Bomber that Carried a Charge of 10 Kilogram Was Stopped at the Entrance of a Coffee House (Hebrew), *Haaretz*, November 13, 2002.
290. Intelligence and Terrorism Information Center: Suicide bombing terrorism during the current Israeli–Palestinian confrontation (September 2000–December 2005), January 1, 2006, p. 106.
291. Jonathan Liss, The Security Man Dreamt to Open His Own Coffee House (Hebrew), *Haaretz*, January 9, 2006.
292. Intelligence and Terrorism Information Center: Suicide bombing terrorism during the current Israeli–Palestinian confrontation (September 2000–December 2005), January 1, 2006, p. 44.
293. Interview with Uri Bar Lev, Commander of the Southern District of the Israeli police, August 23, 2007.
294. Intelligence and Terrorism Information Center: Suicide bombing terrorism during the current Israeli–Palestinian confrontation (September 2000–December 2005), January 1, 2006, pp. 36–37.
295. Interview with Uri Bar Lev, commander of the Southern District of the Israeli police, August 23, 2007.
296. Intelligence and Terrorism Information Center: Suicide bombing terrorism during the current Israeli–Palestinian confrontation (September 2000–December 2005), January 1, 2006, p. 163.
297. Intelligence and Terrorism Information Center: Suicide bombing terrorism during the current Israeli–Palestinian confrontation (September 2000–December 2005), January 1, 2006, p. 208.
298. Intelligence and Terrorism Information Center: Suicide bombing terrorism during the current Israeli–Palestinian confrontation (September 2000–December 2005), January 1, 2006, p. 195.
299. Zeev Schiff, Israel Responses to the Terror Attacks Are Clumsy and Slow (Hebrew), *Haaretz*, August 5, 2002.
300. Interview with Danny Yatom, Head of the Mossad, 1996–1998, May 21, 2008.
301. Amira Hess, Floating to Heaven (Hebrew), interviews by a security prisoner named Walid Daka with Palestinian youngsters caught on route to carry out suicide attacks, *Haaretz*, April 4, 2003.
302. Interview with Ilan Paz, Brigadier General (retired), a brigade commander in Ramallah during the Intifada and also head of the civil administration in Judea and Samaria, August 7, 2007.

303. Yuval Yoaz, We Shouldn't Be Ashamed: This Is a Targeted Killing and It's Legal—An Interview with the Head of the International Law Department at The IDF (Hebrew), *Haaretz*, March 8, 2005.
304. Interview with Yitzchak Eitan, Commander of the Central Command 2000–2002, August 20, 2007.
305. Interview with Boaz Ganor, head of ICT, July 18, 2007.
306. Shay Nitzan, Fighting Terror: Legal Aspects, in Dan Meridor and Haim Pass, eds., *21th Battle—Democracies Fights Terror* (Hebrew), Jerusalem: *the Israeli Institute for Democracy*, 2006, pp. 124–126.
307. Interview with Ganor Boaz, Head of ICT, July 18, 2007.
308. Yossi Kuperwasser, quoted from Protocol of Discussion which was Held in the Israeli Institute for Democracy on 31st October 2004, in Dan Meridor and Haim Pass, eds., *21th Battle—Democracies Fights Terror* (Hebrew), *the Israeli Institute for Democracy*, Jerusalem, 2006, pp. 250–251.
309. This conceptual downfall was based on the ill-fated Oslo Accords, in which Israel recognized the PLO, headed by Yassar Arafat, as the official representatives of the Palestinian People and the PLO was bound to refrain from terror and confront Palestinian Terror. The late Israeli hero and Prime Minister Yitzak Rabin was of the view that the PLO could better fight Palestinian terror than Israel could because it was not hindered by the legal and moral constraints—or in his words: "They will fight terror without Bagatz [Israel's High Court of Justice] or Betzelem [a prominent human rights organization in Israel]."
310. Amos Harel, 15 Surveillance Means Fails in the Most Crowded Areas in the World (Hebrew), *Haaretz*, June 22, 2006.
311. Interview with Lior Lotan, Former Commander of IDF Crisis Management Unit, November 2, 07.
312. Interview with Lior Lotan, Former Commander of the IDF general staff's negotiations team, November 2, 2007.
313. This section is based on the authors' insight and analysis which serves as recommendations in confronting suicide terrorism. These recommendations and/or procedures are not necessarily used by any specific security apparatus.
314. Interview with Zohar Dvir, Former Commander of YAMAM, November 4, 2007. Interview with Lior Lotan, Former Commander of IDF Crisis Management Unit November 2, 2007. Interview with Hagai Peleg, Former Commander of YAMAM 1999–2001, August 2, 2007.
315. Based on workshop findings held at ICT.
316. Yehezkeli and Horowitz, pp. 146–149; Interview with Zohar Dvir, November 4, 2007.
317. Based on interview with Uri Bar-Lev on August 23, 2007 and Zohar Dvir on November 4, 2007.
318. The people interviewed for this chapter are is follows: Giora Eiland, General (retired) Head of the National Security Council 2003–2006, August 23, 2007, August 8, 2007; Uri Bar Lev, Israeli Police Commander of the Southern District, and Former Commander of elite counterterrorism units, August 23, 2007; Avi Dichter, Head of the ISA 2000–2005, and Internal Security Minister since 2006, September 26, 2007; Zohar Dvir, Yamam commander 2002–2005, November 4, 2007; Yitzchak Eitan, Genral (retired), Commander of the Central Command 2000–2002, August 20, 2007; Dr. Boaz

Ganor, Head of the International Counter-terrorism Center (ICT), July 18, 2007; H.D. Lieutenant Colonel, Head of IDF Negotiation Unit June 18, 2008; Yossi Kuperwasser, Brigadier General (retired), head of the Military Intelligence's (Aman's) research department 2001–2006, July 29, 2007; Lior Lotan, Col. (retired) Former Head of the IDF general staff's negotiations team, November 2, 2007; Dan Meridor, former Minister of Justice, and the Minister in charge of the intelligence community 2001–2003, September 24, 2007; Benjamin Netanyahu, Prime Minister of Israel 1996–1999 April 23, 2006; Ilan Paz, Brigadier General (retired), brigade commander in Ramallah during the Intifada and headed the civil administration in Judea and Samaria, 2001–2005, August 7, 2007; Hagai Peleg, Commander of Special Israeli Police Unit—Yamam, 1999–2001, August 2, 07; Moshe Yaalon, IDF Chief of Staff 2002–2005, October 9, 2007; Danny Yatom, Head of the Mossad 1996–1998, May 21, 2008; Aharon Zeevi-Farkash, Head of the Military Intelligence (Aman) 2002–2006, August 19, 2007.

3

THE EAGLE AND THE SNAKE: AMERICA'S EXPERIENCE WITH SUICIDE BOMBINGS

Yaron Schwartz

We love death. The U.S. loves life.
That is the difference between us.
—*Osama bin Laden*, November 2001

THE ORIGINS OF SUICIDE ATTACKS AGAINST THE UNITED STATES

In the minds of many in the United States and around the world, the maelstrom of large-scale suicide bombings generated by al-Qaeda in the mid-1990s and that culminated on the morning of September 11, 2001 in the most devastating terrorist attack in history, mark America's first experience with the scourge of suicide terrorism. Indeed, under the tutelage of Osama bin Laden and his chief ideologue, Ayman al-Zawahiri, al-Qaeda and its affiliated groups, cells, and networks in the Global Jihad movement have managed to paint erstwhile local political or religious conflicts as having to do with or being part of a larger global, ideological/religious war between an infidel West and a righteous Islam.

This nihilistic, fundamentalist ideology of the radical Salafi/Takfiri school of Sunni Islam explicitly pits Islam against the West or any other religion/ideology/

Suicide Terror: Understanding and Confronting the Threat. Edited by Falk and Morgenstern

worldview that is antithetical to Salafi notions of what Islam should be and what is required of its "true" believers. In that respect, Salafi Jihadis of al-Qaeda's ilk see those Muslims who do not follow this particular interpretation of Islam and its subsequent tenets as *apostates*—a status that is worse than "infidel" because it deems these Muslims as traitors to Islam, as opposed to "infidels," who are simply people of another or of no faith—and have not eschewed killing and maiming thousands of innocent Muslims at the altar of jihad. Much like Nazism and fascism, this extremist Islamic ideology is strictly dogmatic and cannot realistically be reconciled or placated while at the same time allowing the West to maintain its core values and practices—with those who carry the banner of the Takfiri ideology proclaiming as much time and again. And whereas the driving force in Nazism was racial supremacy and intolerance, *Islamofascism* is driven by a religious supremacy that cannot tolerate or give legitimacy to any other faith or secular forms of government.[1]

An ardent follower and proponent of this pernicious ideology, Osama bin Laden placed suicide bombing attacks as a pivotal tactic in his grand strategy and as al-Qaeda's mainstay operational tactic. For al-Qaeda and its like-minded disciples and sympathizers, this tactic not only had the necessary lethality for the desired effect of a large number of casualties, but has served as a weapon of defiance and a symbolic tool to show the supremacy and purity of Muslims over the decadence and weakness of their rivals. Following 9/11, where for the first time an unprecedented number of suicide terrorists were used in four simultaneous suicide missions using hijacked airliners as guided missiles, al-Qaeda and its affiliates helped spread suicide terrorism not only in the Middle East and other Muslim countries, but elsewhere as well. By July 2007 these terrorist groups executed more than 100 suicide attacks around the world, not including those carried out in Iraq since 2003 by Sunni insurgent/terrorist groups, primarily al-Qaeda in Iraq (AQI).[2] Apart from 9/11 (and actually leading up to it), al-Qaeda under Osama bin-Laden conducted deadly attacks on American targets that claimed hundreds of lives, including dozens of Americans, and wounded thousands: the August 1998 simultaneous bombings of the U.S. embassies in Kenya and Tanzania and the October 2000 attack on the USS *Cole* in Aden, Yemen.

Yet long before extremist Sunni groups like al-Qaeda adopted, perfected, and waived the banner of this deadly *modus operandi*, the tactic of suicide bombings was actually introduced by the radical Iranian-Shiite school of Islam, which first brought the idea to the test in the Iran–Iraq War. There, thousands upon thousands of radical Islamists, many of whom were teenage boys, volunteered as martyrs in the infamous human-wave attacks orchestrated by Iran's Islamic Revolutionary Guards Corps (IRGC), or *Pásdárán*. These columns of "martyrs" marched into enemy fire as cannon fodder to both draw out Iraqi forces and overwhelm them psychologically, as well as clear minefields and fortifications in their path, paving the way for the regular Iranian Army formations that followed.[3]

As the concept of using volunteers who are knowingly willing to die a "holy death" in order to carry out an attack further developed, it was initially exported to Lebanon where Hezbollah, trained by the IRGC, began to employ this tactic in suicide car and truck bombings. It was Lebanon 25 years ago that introduced America to the modern tactic of suicide bombings when on April 18, 1983 a suicide bomber detonated a truck laden with explosives just outside the U.S. Embassy in Beirut, killing 63 people, including 17 Americans (this was one of the very first truck bombs in the world). Six months later, on October 23, a truck bomb rammed into the barracks of the U.S. Marines in Beirut, killing 241 Marines and wounding over a hundred—the deadliest single attack against Americans prior to September 11, 2001 (another truck bomb simultaneously struck the French paratroopers' barracks, killing 58 French peacekeepers). With such devastating attacks—first against the Israeli Defense Forces headquarters in Tyre in late 1982 and again, in a far more lethal attack that killed 60, in November 1983 (just 10 days after the attacks on the U.S. Marines and French military barracks)—and dozens of suicide bombings throughout the 1980s, Lebanon became the locus for the advent of suicide bombings as a modern tactic, later adopted in other parts of the world.[4] Both attacks were the brainchild of Imad Mughniyah and his Iranian sponsors[5] and were carried out by the "Lebanese Islamic Jihad" (a front name for the budding Hezbollah), trained and supplied by the special al-Quds Force of the Islamic Revolutionary Guards Corps (IRGC-QF), whose intelligence agents had set up shop in Lebanon's Bekaa Valley in the early 1980s.[6]

Iran's presence in southern Lebanon and support of the large Shiite population there was partly an effort to empower the embittered and disenfranchised Lebanese Shi'a community that from the mid-1970s to the early 1980s was occupied, bullied and repressed by the forces of the PLO, which had relocated to Southern Lebanon after its expulsion from Jordan in the aftermath of its failed coup attempt of September 1970 (a.k.a. "Black September"). In addition to the plight of Lebanese Shiites and their aspiration to reclaim political influence in Lebanon and fight the Palestinian occupation of the Shiite south (which under PLO control was known as Fatah-land), the establishment of Hezbollah, originally known as "The Organization for the Oppressed of the Earth," gained further impetus by Israel's invasion of Lebanon in June 1982, the consequent expulsion of the PLO from Lebanon, and the political alliance between Israel and Lebanon's Maronite Christians. But most important, Hezbollah's creation was also an integral part of Iran's greater policy of exporting its brand of religious revolution to other parts of the world.[7]

Shortly after coming to power, Khomeini's regime began supporting radical Shiite elements in Iraq, Lebanon, Saudi Arabia and other Persian Gulf states. Foreign Shiite terrorists trained in Iran in the use of firearms, explosives and executing suicide bombings.[8] Throughout the early 1980s, the Iranian regime supported a string of terror attacks, failed coups and attempts of destabilization in countries like Qatar, Bahrain, Saudi Arabia, Lebanon and Iraq.[9] As the United States pulled its troops out of Lebanon four months after the Beirut attacks and as other nations in the peacekeeping force followed suit, the barracks bombings became a source of prestige for the burgeoning Hezbollah and an inspiration for other terror groups to emulate it. But by far the main significance of those attacks lay in their real and perceived *lethality and effectiveness: Tactically* (dozens, sometimes hundreds of casualties and the lack of effective countermeasures); *psychologically* (instilling fear in the enemy and at the same time empowering the perpetrators' constituency); and for many in the Arab and Muslim world who drew a direct line between the attacks and the American withdrawal thereafter, *politically* as well.

Finally, on December 12, 1983 the U.S. embassy in Kuwait was bombed as part of a series of attacks whose targets also included the French embassy, Kuwait's international airport, its main oil refinery, and a residential complex for employees of American defense contractor Raytheon. The terrorists were members of al-Dawa—"The Call"—an Iranian-backed Iraqi Shiite group and one of the main Shiite outfits that operated against Saddam Hussein's regime during the Iran–Iraq War. Despite Iranian denials, much like the bombings of U.S. targets in Lebanon, the hallmarks of Iran's terrorist network and its methods were clearly present even though no definite evidence of a "smoking gun" had been found.[10]

As Hezbollah grew in strength and influence, it conducted more terrorist acts against the United States, primarily the taking (and sometimes killing) of American hostages, both civilian and military, in Lebanon from the mid-1980s and into the early 1990s, but also such terrorist acts as the hijacking of airliners, most notably TWA flight 847. Hezbollah continued to hone its operational capabilities and was eager to assist other radical Shiite groups, often with full logistical and financial support from Iran, in targeting Americans using, among other tactics, suicide bombings. In 1996, Hezbollah al-Hijaz (a.k.a. Saudi Hezbollah)—a Shiite terrorist outfit based in Saudi Arabia that opposes the Saudi royal family's control of the country and particularly over the Muslim holy cities of Mecca and Medina—attacked the U.S. military facility at the Khobar Towers complex in Dhahran, killing 19 Americans and injuring 347. Iran's *Pásdárán* helped create this group and trained many of its members in IRGC camps both in Iran and in Lebanon's Bekaa Valley.[11] In 1994, the IRGC-QF instructed the group in preparation for attacks on potential American targets in the Saudi kingdom, and in the fall of 1995 the Khobar Towers were specifically earmarked for an attack. The Iranians furnished Hezbollah al-Hijaz with additional funding and the explosives used in the attack, and together with (Lebanese) Hezbollah helped the Saudi group design the truck bomb.[12] For political reasons, the United States decided not to act in retaliation for this blatant attack against its citizens, just as it refrained from retaliating following the Beirut bombings. One might argue that this kind of inaction or lack of adequate retaliatory/proactive action (as in case of the 1998 Embassy bombings and the attack on the USS *Cole* in 2000) from the early 1980s until the aftermath of 9/11 was perceived by the mullahs in Iran, Hezbollah in Lebanon, and al-Qaeda alike as indifference/apathy/lack of resolve and perhaps even fear on the part of the United States government—an impression that only further emboldened the terrorists, particularly al-Qaeda, to pursue increasingly audacious attacks on American targets abroad and finally on the American homeland itself.

Hezbollah, which has been on the State Department's list of terrorist organizations for a almost a quarter of a century,[13] has also been playing a major role in facilitating attacks on U.S. forces in Iraq, working together with the IRGC's al-Quds Force in training (including explosives, ambushes and use of rockets and mortars) and funneling weapons to Shiite extremists there since 2006—namely Shiite "Special Groups," the U.S. military's name for splinter factions of Muqtada al-Sadr's Mahdi Army militia. The United States believes the IRGC-QF has armed and financed the Special Groups: According to the Pentagon, Hezbollah instructors have trained these militias at camps near Basra in southern Iraq from mid-2006 to as recently as April 2008, when they then crossed the border into Iran after U.S.-backed Iraqi forces launched a campaign against the various Shiite militias in the Basra area.[14] Hezbollah members were also allegedly involved in planning some of the most brazen attacks against U.S.-led forces, including the January 2007 raid on a provincial government compound in Karbala that killed five American troops (in that attack, English-speaking militants wearing U.S. military uniforms and armed with American weapons stormed the building, killing one U.S. soldier and abducting four others who were later found dead and were apparently executed).[15] The allegations made by these Pentagon sources point not only to an Iranian hand in the Iraq war, but also to Hezbollah's attempts to expand beyond Lebanon and assume a broader role in the Iranian/Shiite struggle against American influence in the Middle East.[16] That seems to fit into what Iran has *de facto* been doing since 2003: waging a proxy war against the United States so as to secure a dominant political and strategic role in Iraq and the region. Hence, Iraq seems to have replaced Lebanon as Tehran's top priority in the Middle East due to its oil reserves, holy Shiite shrines, large Shi'a population, and its significance as a buffer between Iran and Saudi Arabia—a longtime ideological rival.[17]

But despite the long history, relevance, significance and continuous threat of suicide terrorism from both the Sunni fundamentalism of al-Qaeda and the Global Jihad movement on the one hand, and Shiite/Iranian fundamentalist ideology of revolutionary Jihad on the other, the most pressing challenge facing the Unites States since 2003—and where suicide bombings have been a major factor—is the war in Iraq. Whatever future developments America is to face, sooner or later, with regard to terrorism and suicide attacks by fundamentalist Shi'a and/or Sunni Islam and their respective agents, it is Iraq and the presence of suicide bombings there that has held center stage. Accordingly, this chapter focuses primarily on the war in Iraq, the suicide-bombing phenomenon, and the wider lessons that can be drawn from that experience, especially vis-à-vis U.S. homeland security.

SUICIDE BOMBINGS IN IRAQ

More than five years after the United States invaded Iraq and toppled Saddam Hussein's regime, Iraq—a country with no history of suicide bombings prior to the American-led invasion of 2003[18]—has seen an almost unprecedented campaign of terrorism and guerilla/insurgency warfare that, as it claimed the lives of tens and perhaps hundreds of thousands of Iraqi civilians[19] and thousands of American soldiers,[20] has turned that country into the terrorism and suicide bombing capital of the world. Since late 2004, Iraq has seen the largest number of suicide bombings in the world—a dubious position previously held by Sri Lanka, where between 1987 and 2007 the Tamil Tigers (LTTE) killed thousands in dozens of attacks. Another major hotspot for terrorism and suicide bombings since the mid-1990s—now overshadowed and that pales in comparison to Iraq—is the Israel–Palestinian conflict, where more than 700 Israelis died in some 150 suicide bombings between 1994 and 2005.[21] With no history of suicide attacks prior to 2003, Iraq very quickly became a Mecca for suicide bombers. Sunni insurgents—almost invariably radical Salafi/Takfiri Islamists under the umbrella of *al-Qaeda in Iraq* (AQI)—carry out suicide bombings on a regular basis, exacting a heavy physical and psychological toll on the civilian population and the Iraqi Security Forces (ISF). In 2007 the majority of the 658 worldwide attacks took place in Iraq.[22] Consequently, suicide bombings are a challenge for U.S. troops on the ground and the United States government because of (a) the political repercussions of these attacks on the overarching effort of stabilizing Iraq both security-wise and politically and (b) the geopolitical ramifications of this enterprise. With more and more suicide bombings taking place in Afghanistan, tactical lessons from Iraq are very important.

While the bulk of American and Coalition troop casualties have been incurred by improvised explosive devices (IEDs) of various types as well as mortar attacks,[23] by mid to late 2004 suicide bombings mostly targeted Iraqi civilians (primarily Shiites) and the ISF, mostly the Iraqi National Police (INP). From city streets, markets, schools, and government offices, to sports and public events, mosques, and even funerals, almost every imaginable type of public venue and settings ("soft target") have been targeted.[24] As one observer remarked, the Shiites of Iraq—viewed by many if not most Iraqi Sunnis as the ones who benefited most from the political windfall of the American invasion while usurping the Sunnis from their rightful place in power—are "walking targets."[25]

There were undoubtedly tactical reasons as to why AQI shifted its focus from American military and Coalition targets of suicide bombings to Iraqi civilians and the ISF (e.g., better armor and protection for U.S. troops[26] and the implementation of better tactics to counter the threat). That said, the main reason for the change, as suggested herein below, was the insurgents' own analysis of the emerging political reality on the ground and the changes required on their part in order to facilitate the realization of their various strategic goals for reinstating Sunni political power and/or establishing an extremist Sunni shari'a state in Iraq, starting with the end of the occupation and the withdrawal of all foreign forces.

Despite the Surge and the dramatic improvements that have been since mid-2007, maintaining security in Iraq is still a challenge.[27] The year from March 2006 to March 2007 was the bloodiest in terms of both American troop and Iraqi civilian casualties.[28] About 45% of all violent civilian deaths since the end of the 2003 U.S.-led invasion took place during those 12 months. Fatal suicide bombings (those of both vehicle-borne (SVB-IED) and person-carried explosives) as well as non-suicide roadside car bombs (VB-IED)[29] doubled in that same period (from 712 to 1476), while deadly mortar attacks quadrupled (from 73 to 289).[30] The average daily rate of civilian deaths is another somber indicator of the tragic reality in Iraq: In the 11 months immediately after the 2003 campaign, the daily death rate stood at 20; by the fourth year of the conflict it averaged 75 per day. Four times as many people were killed in the fourth year as in the first, and more than 120,000 Iraqis have been wounded since the formal conclusion of the war on May 1, 2003.[31] Geographically, the city of Baghdad and the Baghdad province—the most populated area in Iraq—suffered about five times more deaths per capita than the rest of the country (although since the summer of 2006 there has been a sharp and continuous decline in the number of violent attacks and deaths in Baghdad).[32]

All in all, 27,000 deaths were reported in 2006, a huge increase when compared to the previous three years (14,000 killed in 2005, 10,500 in 2004, and just under 12,000 in 2003).[33] As 2007 drew to a close—and even when taking into account the improvements on the ground associated with the Surge—it turned out to be the second-worst year in casualties since 2003, second only to 2006 and still almost twice as deadly as the first year after the invasion. One measure by which 2007 quickly exceeded 2006 was the number of large-scale, predominantly vehicle-borne suicide (SVB-IED) and nonsuicide bombing attacks (VB-IED) exacting more than 50 civilians deaths each (at times killing up to 150 civilians). Between January and April 2007 there were already 13 such deadly attacks, whereas in the entire year of 2006 there were 12, and as of September 2007 there have been more than 20 such attacks, claiming more than 2000 civilian victims (the worst of these attacks took place in August 2007, when more than 500 members of the minority Kurdish Yazidi sect were killed by several truck bombs[34]). Altogether, there have been more than 50 of these attacks since 2003, with a civilian body count in excess of 4500.

Although 25 countries have deployed troops to Iraq as part of the U.S.-led coalition since 2003, the total number of non-U.S. troops today is only a fraction of the number of American troops in Iraq (between February 2004 and June 2007 non-U.S. troops constituted at most 17% and as little as 6.8% of the number of American troops; since January 2005, the numbers of non-U.S. troops has averaged about 9% that of American troops).[35] And while other countries have joined the United States in deploying troops to Iraq, with large-scale United Nations involvement and that of other international nongovernmental organizations,[36] the current ongoing campaign in Iraq—that is, fighting the insurgency in order to bring about the security and stability necessary for the ongoing effort of domestic political reconciliation, nation-building, and democracy—is, for all intents and purposes, an American military and political endeavor. As such, the United States is inextricably connected to and affected by the various deadly tactics employed by the terrorist insurgents, especially suicide bombings. What has been taking place in Iraq, in that regard, has many implications on America's military and homeland security policy planning for years, if not decades, to come. More ominously, some pundits have suggested that there are dangerous parallels between the anti-Soviet campaign of the Afghan Mujahideen during the 1980s and the insurgency in Iraq, in the sense that just as the Afghan War was the breeding ground for al-Qaeda and the global jihad movement and proved to have unforeseen strategic implications for the security of the United States and the rest of the world a decade or so later when al-Qaeda came knocking on America's door, today's Iraq—a new hub of for Islamist terrorism and a hands-on training center—may, according to some experts,[37] already be playing the role Afghanistan did a quarter of a century ago as a hatchery of future terrorists.[38] Hence, the nature of what is commonly referred to as the "insurgency" in Iraq and the tactics used by the insurgents—in this case, suicide bombings—is of strategic significance to the United States. To that end, the birth, evolution, preponderance, and complexity of the insurgency must be examined.

The Insurgency: Who, Why, When, and How

Quite possibly America's most challenging military undertaking since the Vietnam War, the harsh reality of the war in Iraq—commonly dubbed as the "insurgency"—is actually a multifaceted armed conflict with centuries-old social and religious strife that run deep into the very fabric of Iraqi society, compounded by the presence of the Global Jihadis who have made it their business to "set things right" in Iraq and going about it by targeting practically anyone and everyone they deem as legitimate targets. Not unlike the chaotic implosion in the Balkans after the demise of Tito's Yugoslavia, the innate social and religious tensions and fissures within Iraqi society that were contained under the tight lid of Saddam's repressive autocracy erupted in full force once that control was gone. This is precisely what the United States initially failed to understand and which in turn led it to pursue the wrong, ineffective, and often counterproductive policies well into 2006.

At the heels of the 2003 invasion, centuries-old hostilities came into play just as the United States was set to implement its democratic vision for Iraq and, by extension, some argue, for the region. By 2004, The world's only military superpower found itself in a serious predicament, contending with three different yet interconnected conflicts[39]: *first*, an anti-occupation guerilla warfare where most targets are American and coalition troops, as well as Iraqi security forces (this specific aspect of the war is usually referred to as the "insurgency"); *second*, a Sunni-versus-Shiite sectarian conflict (i.e., the "civil war") where most casualties and targets are unarmed civilians, and the deadliest terrorist attacks are suicide bombings (with Baghdad as the main locus for this violence); *third*, Shiite-versus-Shiite violence (taking place mostly in the predominantly Shiite south, particularly the city of Basra and its environs and in the Shiite "Sadr City" sector of Baghdad) between rival Shiite militias.

The one constant interlaced with these three facets of the war and the insurgency is the crucial role of fundamentalist Islamists or Jihadis—spearheaded by al-Qaeda in Iraq (AQI)—who had been targeting everyone involved in the cauldron of ethnic violence (including erstwhile Sunni insurgents mostly with tribal/

clan affiliations who, starting in 2006, increasingly turned against the AQI and the Jihadis and instead allied themselves with the United States). These Sunni Jihadis are the overwhelming group of insurgents and terrorists to use suicide bombings as well as the various kinds of IED's in Iraq, exacting the greatest number of U.S. casualties and Iraqi civilians. AQI's significance as a player in this war lies not only in the devastating attacks it has executed and their toll, but on a deeper level: With AQI hovering above the fray ready to spoil any hard-earned progress that "threatens" to bring any measure of stability to Iraq, successes in one area end up producing setbacks in other areas (AQI's bombing of the Askariya Mosque in Samara in February 2006 immediately after the successful conduct of the Iraqi elections is a good example.) Consequently, counterinsurgency operations plans, smart and complex as they may be, are often insufficient because there is much more to the insurgency than just chasing down and "smoking out the bad guys." It is this very understanding of the nature of the insurgency that eluded America's decision-makers at the earlier stages of the conflict that unfortunately led to costly mistakes that will require more effort, time, money, and lives to undo, hopefully allowing for much more progress down the road.

It is safe to say that the lack of a clear, single leadership to both the Sunni and Shiite insurgency made it even more difficult for the United States to bear down effectively on the insurgents and defang them, particularly before the autumn of 2007 (see the section below dealing with the Surge). The question of how the insurgency in Iraq came about and transformed over time has and will continue to be the subject of ongoing analysis and study. What is already clear, however, is that at the outset the insurgency was different than what it was by 2005 and what it has been at various points since 2003.

It goes without saying that all of the groups that make up the insurgency want to see the quick exit of American and coalition forces from Iraq, since they constitute the main obstacle standing in the way of the insurgents' strategic goals. But beyond this lowest common denominator of resisting the American-led occupation and the presence of foreign troops in Iraq—which has motivated the different militias and terrorist groups to cooperate when it suited them—the diverse elements on this playing field and their respective nationalistic, religious, tribal and other goals differ greatly. [Note: For example, many insurgent groups' reasons for fighting both Coalition troops and the ISF are of a more local-political nature, and these groups are concerned with immediate and medium-term implications (e.g., Sunni tribal militias and criminal elements) rather than having their actions derived from a long-term plan for Iraq as a country (e.g., AQI and other Islamists, as well as Shiite militias).]

The Sunnis. Concentrated primarily in the "Sunni Triangle" of central Iraq, Sunnis make up about 30% of the country's population. Over the past several hundred years, they have ruled Iraq while denying the rights of and oppressing the Shiite majority and other minority groups, such as the Kurds. This longstanding social reality preceded Saddam Hussein, but his regime consolidated the inequity and oppression even further. The Ba'ath party, and to a much greater extent Saddam Hussein, afforded Sunnis with vast economic benefits from the country's immense oil and gas reserves. The huge profits generated by these natural resources provided for funds that were put into development and investments in Iraq's Sunni hub in central Iraq, while at the same time the Shiite and Kurdish populations in the south and north were completely neglected. The economic boom in the Sunni provinces and the deliberate disregard for the rest of the country galvanized Kurds and Shiites to increasingly oppose Saddam, which in turn served as an excuse for him to further marginalize these groups and exacerbate their economic difficulties with rising unemployment and poverty.

The U.S.-led invasion and the toppling of Saddam's regime brought an end to this system of Sunni minority rule over a Shiite majority that afforded the former with political, social, and economic windfall. Naturally, the abrupt end of that auspicious situation created widespread resentment and bitterness amongst Sunnis, particularly loyalists of the former regime who benefited most from their now-defunct positions. In the post-Saddam power vacuum immediately after the 2003 war and the implementation of the de-Ba'athification policy by the Coalition Provisional Authority (CPA), the complex reality that has materialized and continuously changed since March 2003 began more as a guerilla-like warfare and terrorist campaign primarily spearheaded by Sunni "nationalists" who were anything but a homogenous group. Rather, they consisted of militias of now-disenfranchised Ba'athists and members of Saddam's "old guard"—the Republican Guard, secret police (the "*Mukhabarat*"), and the Fedayeen—who wanted to get rid of the Americans as soon as possible and by any means possible so as to restore what they regarded as Iraqi sovereignty (with the Sunnis and the Ba'ath party back at the helm), or at least maintain a lingering chaos to prevent the Shiites from taking control of the country.

As some experts point out, many more of the early insurgents were average, non-Ba'athist yet deeply proud nationalist Sunnis who served in the Iraqi army and were now disbanded and out of a job as a result of de-Ba'athification and who yearned to take up arms in resistance against a foreign power that had defeated their military, occupied their country, and then denied them of an income to boot.[40] Others still were tribal insurgent groups who vied for power in their respective locales—power that was denied to them by Saddam. Added to the mix were criminal elements that, as a result of a deliberate, large-scale release of criminals from Iraqi prisons by Saddam shortly before the collapse of his regime (many of whom serving life sentences for serious, violent crimes), were now free to engage in their heretofore-dubious activities. Many hired themselves out for attacks on American and Coalition forces, launched mostly by AQI and its allies. For example, in November 2003, the U.S. Army's 4th Division's "Taskforce Iron Horse" tellingly reported that 70–80% of the people they arrested for attacking U.S. forces were paid between $150 and $500 to do so and that many of them turned out to be criminals[41] (when compared to the $500 billion the United States has spent on the war so far, further pause must be given to the asymmetrical nature of this war and its impact on America's future campaigns).

Soon, however, radical Salafi Sunnis came to the forefront of the insurgency as the influx of foreign Jihadis coming to join local Iraqi Islamist insurgents soared. By late 2004, Iraqi and foreign Islamists had taken the reins of the insurgency, carrying out spectacular, vicious attacks—most and foremost suicide bombings—to rally sympathizers from across the Arab and Muslim world for the cause of Jihad against the "occupying infidels." Videos of attacks on American and other foreign troops posted on numerous Jihadi websites[42] proved very effective not only in galvanizing and radicalizing many Muslims harboring knee-jerk anti-American sentiment, but also in drawing hundreds if not thousands of volunteers to Iraq to take part not only in the fight against the American occupiers, but also in the fight for Sunni-Salafi religious primacy.

Yet even the Jihadis were not part of a homogeneous group. Rather, they made up several groups that did not see eye to eye on many points but rallied together only for their utilitarian purpose of fighting U.S. and Coalition forces and wreaking as much sectarian and internecine violence as possible, with the overall goal of curtailing, if not preventing, the effort to create a new, democratic and stable Iraq. And because of the initial widespread grassroots support for resistance to the occupation, the various Iraqi armed groups did not need the infrastructure and networks emblematic of classic guerilla wars and insurgencies such as Vietnam or Algeria because the society these militants came from was supportive of their efforts to begin with.

An important element glaringly overlooked by the United States was the tribal component of Saddam's power structure, which proved to have a great impact on the advent of the insurgency. In that regard, the importance of family and tribal connections in rural areas was pivotal: Where these are very prevalent, such as in the western al-Anbar province, the principle of *fiz'a*, or solidarity, gives strong motivation for joining the struggle in defense of an honored relative or important person. Some experts have also suggested that similar social and cultural motivations for participating in violence are based in *intiqam*, or revenge, whereby insurgent killings are viewed through the prism of retaliation for the transgressions, whether real or perceived, of the American occupation.[43] Iraqi insurgents were therefore not hindered by one of the main challenges of guerilla warfare: winning the "hearts and minds" of the people.[44] Furthermore, these militants received both moral and logistical support from those segments of Iraqi society—particularly the hundreds of thousands of former regime loyalists in Saddam's military and security apparatus (now unemployed overnight and armed to the teeth with weapons they took home with them) and their families, bitter and resentful—who, by-and-large, never accepted the demise of the Ba'ath regime and the placing of a U.S.-backed, Shiite-dominated government in its stead. Moreover, shortly after the fall of Saddam, tribal leaders and clerics in places like Fallujah and other spots in central Iraq as well as the al-Anbar province had independently set up a civil affairs council, which actually brought about a fair degree of stability during the period of riots and looting in major Iraqi cities on the heels of the 2003 invasion.[45] These tribal sheikhs and religious leaders had a lot of influence on stirring up popular opposition to American presence following several intense counterterrorism operations in 2004 carried out by U.S. forces in these areas.

As noted above, the Sunni insurgent groups did not constitute a monolithic network with a single leadership and thus constituted both an advantage and a disadvantage for the conduct of successful U.S.-led counterinsurgency (COIN) operations. But at the core, the majority of Sunni insurgents shared a set of basic political, social, and economic motivations. The resurgence of Shiite political power and fear of further disenfranchisement of the Sunnis at best—and, at worst, vengeance—helps explain Sunni insurgent attacks against both Iraqi security forces and civilians, as both are intended to jolt the political system and delegitimize it by demonstrating to the general Iraqi population that the Shiite-led government and its American allies cannot protect the civilian population, much less effectively curb the insurgency. Many Sunni Iraqis were motivated to join the insurgency as a result of (a) widespread embitterment over the loss of power, with its social and political perks, (b) the end of the economic boon Sunnis enjoyed under the Ba'ath party coupled with high unemployment and a steep drop in the standard of living, and (c) civil chaos and rampant crime immediately after the 2003 invasion.[46]

In addition to suicide bombings against civilians and Iraqi security forces for the sheer tactical efficacy that these attacks generate, Sunni insurgents have also attempted to create a reality whereby it is necessary for Coalition forces to take security measures so tough as to make life increasingly difficult for the average Iraqi, with the goal of making the United States and the Iraqi government bear the brunt of the people's frustration instead of the insurgents.[47] This idea has been quite successfully stymied by the realignment of Sunni tribal groups and their decision to side with the United States and against AQI and its affiliates, in what is known as "The Awakening" movement that has swept Iraq's Sunnis starting in 2006.[48] Today, while still a very destabilizing element that can by no means be written off as a threat, AQI is at its weakest in large part because of growing popular Sunni rejection of its conduct and ideology.[49]

Al-Qaeda in Iraq and the Jihadis. Perhaps the single most destabilizing element in Iraq has been the role of the Islamists and global Jihadis, with their devastating attacks—including suicide bombings—against Iraqi civilians, Iraqi security forces, and U.S. troops.

Al-Qaeda in Iraq is the largest of the Sunni insurgent groups. Ostensibly founded by Abu-Musab al-Zarqawi, its overarching goals are to overthrow the Iraqi government and establish an Islamic state in Iraq by forcing out the U.S.-led coalition. Elements of the Islamist group Ansar al-Islam and indigenous Sunni Iraqi Islamists constitute its ranks. Prior to the Sunni "Awakening" movement, AQI derived most of its domestic support from Iraqi Sunnis and focused its attacks on Shiite Iraqi civilians and the Iraqi National Police. AQI perpetrated the bombing of the Shiite Askariya shrine in Samara in February 2006 and again in June 2007 (see herein), which set off the deadliest reprisal killings between Sunnis and Shiites. In addition to the frequent attacks in Iraq, the group claimed responsibility for the 2004 bombing of three hotels in Amman, Jordan that left scores of dead and injured. The group has claimed responsibility for attacks on American troops and ISF, often taking credit for several attacks in one day. AQI employs a variety of guerilla and terrorist tactics that include RPG (rocket-propelled grenade) attacks against U.S. and ISF armored vehicles, guerilla-style attacks and ambushes, and suicide bombings, as well as the kidnapping and beheadings of foreigners. Abu Ayyub al-Masri replaced al-Zarqawi as leader of the group after the latter was killed by Coalition troops in June 2006. In January of that year the group was one of six insurgent organizations to unify under the *Mujahideen Shura Council* (see below). As of this writing, all attacks perpetrated by AQI are claimed in the name of the Council.[50]

The Mujahideen Shura Council or "Freedom Fighter Consultation Council," is an umbrella organization made up of several insurgent groups, including AQI. The Council first appeared in the spring of 2005, when it claimed the kidnapping of an Australian citizen contracting for Bechtel and demanded the departure of all foreign forces in Iraq.[51] In January 2006, several Sunni insurgent groups announced that they were joining to form the Mujahideen Shura Council. They claimed to unite so as to continue the struggle against the occupation and drive out the "invading infidels." The Jihadi groups that formed the Council included AQI, Jaish al-Taifa al-Mansoura, al-Ahwal Brigades, Islamic Jihad Brigades, al-Ghuraba Brigades, and Saraya Ansar al-Tawhid. The goal of the Council is to unify insurgent efforts in Iraq against both government and Coalition forces. Its creation may also be aimed at mending a rift between various Sunni insurgent groups but is clearly an attempt to unify disparate groups with a radical Sunni ideology striving to destroy the Iraqi government and its international support. The formation of the Council was possibly also meant to shore up support for the insurgency by distancing itself from the extremely violent tactics and targeting of innocent Iraqis by AQI. To that end, the Council allows AQI to continue its suicide attacks but without claiming responsibility. It is now the Council claiming attacks by its member factions. Thus, the goals of the Council and its members are the same: the removal of U.S. troops from Iraq and the formation of an Islamist Sunni-Salafi government in place of the current, predominantly Shiite government.[52]

Ansar al-Sunna ("Followers of the Tradition") is a group of Iraqi Jihadis vying to establish an Islamic state with Shari'a law in Iraq, akin to Afghanistan under the Taliban, by defeating the Coalition forces in the Iraq. To them, waging Jihad in Iraq is the duty of every Muslim and they regard anyone opposed to Jihad as an enemy. The group's membership is varied, including radical Kurds in the Ansar al-Islam, as well as foreign al-Qaeda members and other Sunni terrorists. Ansar al-Sunna has targeted Coalition troops, ISF and governmental institutions, and other political establishments in Iraq the group treats as puppet regimes of the United States. The group has claimed many attacks, many of which have not been substantiated. The group reportedly teamed up with the banned Arab Socialist Ba'ath Party and AQI, pledging to continue and increase attacks on U.S., Coalition, and Iraqi government forces.

The Islamic Army in Iraq conducts a brutally violent campaign against foreigners within Iraq—specifically anyone believed to be cooperating with the U.S.-led coalition. The group has been implicated in several gruesome beheadings. It aims to drive all U.S. and related Coalition forces, both military and civilian, out of Iraq. But the group does not limit its attacks to such targets: It has also murdered French journalists, Pakistani contractors, an Italian journalist, and Macedonian citizens working for an American firm. According to previous statements by its leader, there are thousands of terrorists in its ranks. Its leader also claims that the group is predominantly Iraqi, not foreign-born—a claim that also cannot be substantiated. Statements released in November 2004 declared that the Islamic Army in Iraq has collaborated with Ansar al-Sunnah and AQI.

Of all the Islamist and Jihadi groups, al-Qaeda in Iraq is the largest and most active and has been the focus of America's COIN and counterterrorism operations in Iraq. Per suicide terrorism in Iraq, AQI leads in the numbers of attacks it has launched and for which it claimed responsibility. Therefore, it is the one organization that is the most destabilizing because it targets U.S. and foreign troops, Iraqi security forces, Iraqi Shiites, and Sunni Muslims. About 90% of AQI's members are Iraqis, yet it is the foreign fighters who dominate the leadership echelons[53] and are overwhelmingly present among the suicide bombers (again, about 90%). Since AQI's members are mostly Iraqis, it has been argued that AQI is not part of the Global Jihad movement and hence the war in Iraq is not part of the global war on terror against bin Laden's al-Qaeda "central." This argument goes on to say that the United States should be fighting AQI through precise, intelligence-driven, SOF (special operations forces) attacks and precision air strikes on key leaders, and not through the deployment of large conventional forces, which only stirs resentment in Muslim countries and creates more terrorists. In reality, AQI is part of the global al-Qaeda Jihadi movement, despite great efforts on AQI's part to portray itself as an Iraqi organization.[54] Ideologically, it lies on the extreme end of Salafi-Takfiri Islam and by targeting Iraqi civilians has actually killed many more Muslims than Americans. Its preferred weapon is suicide car or truck bombing (SVB-IED) aimed at public places with large numbers of Iraqi civilians, especially Shiites—considered "apostates" because they do not follow the Sunni branch of Islam.[55] Although Zarqawi pursued attacks on Western targets (he was involved in the 2002 murder of Lawrence Foley, a USAID official, in Jordan, as well as the bombing of the United Nations office in Baghdad in August 2003), he focused his attention on targeting the Shiites of Iraq. His constant efforts—via deadly suicide bombings that killed dozens of people—to provoke a widespread Shi'a backlash against Sunnis finally succeeded with the February 2006 bombing of the Askariya Mosque in Samara. Coalition troops eventually killed Zarqawi not long after the Samara bombing, but his successors continued to attack Iraqi Shi'a, even as they began to attack Iraqi Sunnis. In this regard, AQI has been even more extreme in its ideological and religious views than al-Qaeda "central." But as Frederick Kagan has pointed out, although AQI behaves like al-Qaeda in almost every respect, it is not just a local franchise of the global al-Qaeda movement: Its leaders participate in the development of the global ideology, as Zawahiri's exchanges with Zarqawi and al-Masri demonstrate; it sends aid to the global movement and asks for and receives aid from it; most importantly, AQI has received tens of foreign fighters each month, recruited by al-Qaeda throughout the Muslim world, helped in their training and travel, and, once in Iraq, controlled by AQI.[56] Finally, AQI's non-Iraqi top leaders were part of the global al-Qaeda movement *before* coming to Iraq. Therefore, there should be no question regarding AQI as a vital and central part of al-Qaeda.

The movement against AQI began as the group tried to solidify its position in the al-Anbar province by marrying some of its senior leaders to the daughters of local tribal leaders, just as al-Qaeda has done in South Asia. When the sheikhs resisted, AQI began to attack them and their families, including the assassination of prominent sheikhs. This led to the rise of numerous clashes between AQI and tribal families, and the viciousness of AQI's retaliation and the relative weakness of the tribes as a fighting force put these locals in a predicament from which they were rescued by the determined work of Coalition and Iraqi security forces. By refusing to cede al-Anbar's capital and major population centers to the Jihadi insurgents, and at the same time reaching out to the local population increasingly disenchanted with AQI, American troops helped bring about a major change. The Surge did not create this shift but did accelerate it, because the general troop increase allowed U.S. commanders to keep their forces in al-Anbar instead of shifting them in order to protect Baghdad, as had happened in previous occasions.[56] The increased U.S. presence and the more aggressive operations of American forces and ISF made possible a faster, more widespread turn against AQI. By late spring 2007, all of the major tribes in al-Anbar had sworn to oppose AQI and began sending their sons to volunteer for service in the Iraqi army and the INP. By summer, thousands of these recruits began patrolling the streets of their own cities and towns against AQI. U.S. commanders now benefited from the population's support and were in turn inundated with intelligence about the presence and movements of AQI fighters. By August 2007, AQI was driven out of all of al-Anbar's major towns, and its attempts to regroup in more rural areas have not been successful. According to Frederick Kagan, it was the combination of local disenchantment with AQI's religious extremism, lack of cultural sensitivity by the Jihadis themselves, and effective COIN operations by Coalition forces aimed at protecting the population that turned the tide in al-Anbar, which in turn sowed the seeds of "The Awakening" movement that has swept Iraq ever since.[56]

The Shiites. Constituting about 65% of Iraq's population, the Shiites' increasing role in the violence portrays the complexity, fluidity, and permutations of the insurgency and the difficult challenge it presented. As early as 2004 and much more since 2006, Shiite militias entered the fray and have not only conducted attacks on Sunnis but also targeted U.S. forces using Iranian-supplied Explosively Formed Projectiles (EFP) such as RPG's and anti-armor rockets, as well as military grade explosives for use in various IED's—exacting a substantial toll on U.S. troops in terms of casualties[57] (remarkably, no suicide bombings to date have been carried out by any Shiite group in Iraq, and this remains an exclusively Sunni-Islamist tactic).[58]

The power vacuum that emerged after the invasion unfortunately lingered for almost two years under the weak interim government and the United States' less-than-effective strategy. This vacuum resurrected longstanding internal conflicts between various religious and political groups, and Islamist opposition factions that were kept under control during the Ba'ath era began to pursue their political preferences through public protests or through acts of violence against their domestic and foreign political opponents. During the short-lived relative calm that prevailed at the outset of the U.S. occupation in mid-2003, there was widespread rallying of Shiites who were now free to assert their position as Iraq's majority group.[59] At the same time, new rivalries developed in urban communities between clerics attached to local sheikhs who employed private militias and attained control of public utilities and social services. In addition, with the collapse of Saddam's regime the Shiite holy cities in the south saw enormous growth in the local economy spurred by commerce and construction of hotels and housing built to accommodate large numbers of pilgrims from Iran and the greater Shiite world. Despite tensions between the native Iraqi Shiites inhabitants and recent (and richer) Iranian pilgrims-turned-residents, a large portion of the local population benefited from the economic boon generated by this religious tourism and the influx of people into southern Iraq. As the influence of more moderate clerics such as Ayatollah Sistani waned in the first year of the occupation, political radicals like Muqtada al-Sadr came to the forefront.[60] At the same time, Iranian influence and activities within Iraq steadily increased, as Iranian intelligence reinstituted and created new networks of intelligence gathering and contacts with Iraqi militant Shiites that facilitated the flow of cash, supplied arms, and provided training for Shiite insurgents groups.

The Mahdi Army's prominence as an actor began in April 2004 when it fought against U.S. forces in East Baghdad and in Najaf. In addition to fighting U.S. and Coalition forces, al-Sadr's militia has been actively involved in a long-running territorial battle with AQI and other Sunni groups for control of Baghdad, especially in what has become known as "Sadr City." It was responsible for much of the sectarian cleansing in the Sunni neighborhoods during 2005–2006, which, as previously mentioned, was mainly spurred by AQI's bloody suicide bombings against Iraqi Shiites and Shiite shrines. During that same period the Mahdi Army also managed to effectively infiltrate units of the ISF, particularly the Iraqi National Police.

At its heart, this intra-Shiite violence is driven by the political/religious rivalry and the claim to leadership of Iraqi Shiites between al-Sadr and his political faction and the other Shiite parties. The contest between al-Sadr and Ayatollah Sistani, for example, played out in violent clashes between their respective militias that, since 2006, has further destabilized southern Iraq so much so that by the end of that year the U.S. military concluded that sectarian violence by Shiite militants had surpassed AQI and the Jihadi insurgency as the principal threat to Iraq's overall internal stability. Moreover, those elements within al-Sadr's militia that turned against him after he declared the August 2007 unilateral ceasefire have only increased the level of instability and chaos. That ceasefire followed an outbreak of violence in Karbala that revealed fissures within al-Sadr's own movement and splinter elements he no longer controlled. These splinter/rogue groups, dubbed "Special Groups" by the Pentagon, have attacked U.S. forces, ISF, and the Mahdi Army. In addition, they continue to carry out extra-judicial killings of Sunni civilians, as well kidnappings and other criminal activities. Supported by Iran via the IRGC-QF, they continue to be targeted by Coalition forces and the ISF, particularly since the completion of the Surge.

The influence of al-Sadr and other radical figures further grew by the Coalition Provisional Authority's (CPA) announcement in February 2004 that a new Iraqi constitution that included any principles from the Quran or that would set up an Islamic state would not be accepted.[61] Thus, by April 2004, significant elements of the Shiite community suddenly began attacking Coalition forces as Muqtada al-Sadr unleashed his militia in the holy cities of Najaf and Karbala.

The United States found itself in a position of having to deal with a second insurgency led by a faction of its potential Shiite allies. Sadr's militia, the rather poorly trained Mahdi Army (estimated at 3000 to 10,000 fighters), was largely made up of disgruntled and unemployed young Shiites who were enticed to attend Friday sermons where they began their entry into the movement. As his militia grew stronger and larger, Sadr's electrifying sermons began to condone, if not call for, violence against U.S. forces. In March 2004 the CPA decided to shut down al-Sadr's weekly paper and arrested one of his deputies. Consequently, Sadr concluded that the United States was now planning to go after him and decided to preempt it by calling out his supporters, arming them, and throwing them into battle against Coalition forces with two major campaigns in Najaf and East Baghdad.

According to Ahmed Hashim, the United States failed to understand that al-Sadr, despite his clerical credentials and status, is essentially a political, populist entity rather than a religious one.[62] His revolt has been primarily an internal struggle within the Shiite hierarchy between Iraqi Shiites such as himself and returning exiles (heavily influenced and financed by Iran) for control of the Shiite population and, by extension, Iraq. As such, al-Sadr's insurgency is socially driven, aiming to accommodate the young, disgruntled, unemployed elements in Shiite society.

When the United States launched an assault in August 2004 intended to tighten the noose around Sadrist strongholds, there was an increase in violence in the country and sharp political criticism even within the interim government and elements of the Iraqi security forces. For a brief period, al-Sadr actually started to cooperate with Sunni insurgents in staging attacks on U.S. forces, but this cooperation was not very serious in its scope and did not last more than a few months.[63] Religious and tribal leaders within the Sunni and Shiite communities endorsed attacks against each other and thus escalated the historical tensions that existed between the two. Only after deadly attacks on mosques were there periodic calls for unity and collective Iraqi opposition to the presence of foreign troops.[64] But the situation worsened once more following the January and

December 2005 parliamentary elections that established a Shiite-led government. This reversed the longstanding Sunni political supremacy and created a new source of conflict that has fostered more sectarian strife and terrorist attacks.

After experiencing heavy losses to American forces and seeing his militias' strength diminish, al-Sadr declared a ceasefire in August 2007. While that helped hold down U.S. and Iraqi casualties, it arguably did more for al-Sadr, bolstering his importance as a political player at the same time the various Iraqi factions compete for power.

Iran's Involvement. Iranian preparations for its intervention in Iraq go back to as early as 2002, and starting in 2003 it worked to create a vast network to supply Shiite insurgents with Iranian weapons. Iranian and Hezbollah agents began to recruit, train, and supply groups of Shiite militia members, including the Mahdi Army, in 2003. These groups are known as "special groups" or "secret cells." The creation of training camps in Iran in 2005 increased the number of Special Groups and their lethality. Today, there are three training camps outside Tehran where Iraqi Shiite insurgents train for four to six weeks.[65] Iranian-made EFPs, rockets and mortars flowed into Iraq via the Special Groups' supply routes and, in May 2006, the IRGC-QF[66] and Hezbollah reorganized the Special Groups in Iraq along a Hezbollah-like model. While the precise aims of the IRGC-QF are unclear, the results are not. The Quds force developed a Hezbollah-modeled secret cell network dependent on Iranian support that could operate within the umbrella of government institutions to undermine or replace the elected government of Iraq. Thus, Special Groups have actively undermined the Maliki government ever since its creation in May 2006.[67] They have targeted important government figures, Coalition forces, and the ISF. Special Groups have also kidnapped or assassinated Iraqi government officials, individuals working for the government (including the November 2006 mass kidnapping of employees from the Ministry of Education), and U.S. soldiers.

The American response has been both military and diplomatic. Coalition and Iraqi forces have taken actions to stem Iranian influence by disrupting Special Groups networks and interdicting the flow of illegal weapons; targeting Secret Cell and rogue Mahdi Army leaders; assisting the Iraqi Army in fighting rogue Shi'a militias; and engaging Iran in direct discussions about security in Iraq. Offensive operations targeting Special Groups in Iraq have increased; this has been made possible by the cooperation of the Maliki government and by surge forces. Coalition and Iraqi forces have been redeployed to disrupt Special Groups' communication and supply routes east and south of Baghdad.

The implications of Iranian involvement in Iraq are already visible. Iranian weapons and training have made Special Groups increasingly lethal. Iranian funding, training, and arming of Special Groups will continue to have destabilizing effects on the Iraqi political situation, including the ability of the central government to control the provinces and prospects for negotiation among different Shiite groups. In addition, Special Groups have been exerting pressure on Sunni populations in mixed provinces at least since early 2006. As a result, some formerly Sunni cities like Mahmudiya have become Shiite strongholds, and mixed areas in Baghdad have become more homogeneous. Consequently, it is AQI that has benefited from this development, which it helped to produce, posing as the defender of the Sunni against the Mahdi Army even as it terrorized Sunnis into supporting it. AQI's hold on Iraq's people cannot be broken without addressing the pressure of Shiite extremists like the Special Groups on these Sunni communities, as well as defending the local population against AQI attacks.

Coalition and Iraqi forces are now actively targeting Special Groups in order to mitigate the effects of Iranian funding, training, and arming of Special Groups. Despite repeated Iranian denial, there is little doubt Iran's IRGC-QF and its Lebanese proxy, Hezbollah, are directly involved in training, arming, and funding Special Groups in Iraq. The IRGC-QF and Hezbollah have sought to destabilize the government of Iraq since its creation, ferment sectarian violence, and target Coalition and Iraqi forces. As a result of increasing Iranian financial and material support, the number and quality of attacks by Special Groups have grown, particularly since 2007.

The Role of Suicide Bombings

Although Iraq has witnessed an unprecedented number of suicide bombings in only a few years, despite the vast media coverage and the conventional public impression the media perhaps had inadvertently help create, suicide bombings are not the insurgents' main weapon of choice, particularly when it comes to attacking American and coalition forces. Unmanned (mostly remote-controlled) car and truck bombs (VB-IED), as well as various types of roadside IEDs, are far more ubiquitous than suicide bombings and account for the overwhelming majority of U.S. casualties.[68]

Aside from the much greater destructive power of car/truck bombs compared to that of individuals wearing explosive belts or vests, the insurgents prefer vehicle-borne bombs because they are also easier to assemble—all that is required is the vehicle and the explosive (whether military-grade or improvised ones), which can be put together quickly in response to changing intelligence on the targets. Vehicle-borne devices can also pass through checkpoints relatively easily by hiding the explosive or bomb under stacks of cargo.

Of suicide bombings, vehicle-borne attacks are much more common than those where one or several suicide bombers wear a vest or carry a backpack of explosives on their persons. Vehicle-borne suicide bombings are also mostly used to target Iraqi civilians in populated gathering places, whereas Iraqi police/security installations and checkpoints are targeted by human-borne suicide attacks.[68] Despite the facts on the ground, the widely held impression created by the media is that suicide bombings are the preferred weapon of the insurgents. This misrepresentation is partly the result of shallow journalism: Reporters at times do not make a real effort to ascertain the finer details of an attack and end up lumping together car bomb explosions with suicide bombers who detonate or ram explosive-laden vehicles into their targets.[69]

The sheer statistics of suicide attacks in Iraq are mind-boggling: At least 407 suicide bombings that killed thousands of people, mainly Iraqi civilians, took place just between May 1, 2003 (the day President Bush carried his "Mission Accomplished" speech aboard the USS *Abraham Lincoln*) and January 30, 2006, with dozens more since.[70] The targets of these vicious acts (particularly SVB-IED, several of which take place every month) run the gamut of almost every conceivable kind of civilian venue, showing that the terrorist perpetrators do not lack imagination when it comes to killing.

With so many bombings thus callously executed—in numerous cases causing tens, sometimes hundreds, of dead and maimed innocents—there are many examples that demonstrate both the scale of the carnage and the effectiveness of the suicide attack as a tactic, not just in sheer casualties but as a *psychological* weapon of unnerving terror. The incident that took place on August 31, 2005 illustrates this latter point: That day, approximately one million Iraqi Shiite pilgrims en route to Baghdad's Kazemiya shrine stretched out along the city's streets from the mosque and onto the bridge across the Tigris River. At some point, rumor spread (deliberately perhaps) of a suicide bomber somewhere in their midst. The panicking mass of people caused a stampede that killed more than a thousand, many of them women and children.[71] As one scholar points out, this tragic incident demonstrates the extent to which the insurgents, using suicide bombings, were able to promulgate fear among Iraqi Shiites. It also showed the limited ability of the ISF to deal with the violence and the terrorists' calculated, cold, and spiteful attitudes toward the Shiites—an approach deeply rooted in the religious chasm between Shi'a and Sunni Islam and the Jihadi insurgents' interpretation of their religious duty. Another example of the chilling statistics of Iraq's unmatched standing in the sphere of suicide attacks is the fact that of the nine most lethal suicide bombings worldwide (in terms of overall death toll) between May 1, 2003 and June 30, 2006, five took place in Iraq.[72] In addition, there have been more than a dozen bombings in Iraq since the overthrow of Saddam in 2003 where more than 100 people were killed in each of the attacks.

As an epicenter of insurgent violence, the Baghdad province has accounted for almost 30% of all insurgent attacks in Iraq between February 2005 and February 2007, and more than any other place in Iraq it is in Baghdad where VB-IEDs are more commonly used by the insurgents, both Sunni and Shiite, to target American and other foreign forces.[73] Of the total number of car bombings reported by June 2007, over half were set off by remote control and less than half were suicide bombings (SVB-IED). As for suicide attacks, most have taken place in the Baghdad and Diyala provinces, as well as in the Al-Anbar province.

Interestingly, two aspects of suicide bombings in Iraq are both distinct and perplexing. First, and contrary to pedestrian views, it is *foreigners, not Iraqis, who are the perpetrators in the overwhelming majority of suicide attacks.*[74] Although the insurgency is largely made up of and led by Iraqis (either Islamists or nationalists and former Ba'athists), the overwhelming majority—at least 80% and perhaps up to 90%—of suicide bombers have come from abroad, mostly from

Saudi Arabia, Kuwait, Syria, Algiers, Morocco, and Yemen, with a few coming to Iraq from non-Muslim/Arab countries (Europe).[75] Reuven Paz of the GLORIA Center has found, for example, that of the suicide bombers who died in Iraq in the second half of 2004, about 70% were Saudi nationals.[76] Kuwaiti and Syrian nationals also account for many of the suicide bombers. Second, while suicide bombings in other Muslim countries typically target foreign soldiers and/or non-Muslims, the main targets of suicide attacks in Iraq, as mentioned above, are not U.S. troops or coalition troops but rather the ISF (primarily the police) and, most tragically, Iraqi civilians (mostly Shiites). Another noteworthy fact, as mentioned earlier, is that *Shiite insurgents have not launched any suicide bombing attacks to date*, though that may very well change.

The Pentagon asserts that the insurgents' shift from targeting U.S. and coalition forces in the immediate aftermath of the invasion on Iraq (suicide attacks from March 2003 to mid-2004 targeted only U.S./Coalition forces) was "... especially true since Coalition and Iraqi security forces developed tactics and deployed better armor and equipment to protect themselves from attacks."[77] Notwithstanding the fact that by mid to late 2004 American troops were more prepared to counter suicide attacks—particularly car and truck bombs—through various tactical measures that were taken (e.g., better armor and more secure checkpoints) and that this was an important contributing factor, a more in-depth examination would suggest that the shift to targeting Iraqi civilians and police was mainly the result of a deliberate, thought-out change in strategy on the part of the various Jihadi groups (particularly AQI, which at the time was the largest insurgent/terror group) stemming from their reading of the official policy goal of the Coalition's effort in Iraq: to create a viable government and bringing security and stability to the country.

In AQI's case, its purported founder Abu-Musab al-Zarqawi apparently saw only one strategic option in light of the prospect for an American victory (i.e., the establishment of a democratic, stable government in Iraq): Exploit Iraq's Shiite–Sunni divide by killing as many Iraqi Shiite civilians as possible. Zarqawi counted on the Shiites to respond to these terrorist attacks with retribution against the Sunnis, thus igniting a vicious cycle of "sectarian war." He believed that this religious war would rally Sunni Arabs to support AQI in general, and Iraqi Sunnis to more actively back his own organization. "This war against the Shiites," he wrote, "must start before the Americans hand over sovereignty to the Iraqis."[78] Clearly, the other insurgent groups came to see things much as Zarqawi did, making the strategic decision to undermine the American endeavor as best they could by deciding to place their crosshairs on Iraqi Shiites.

Beyond demonstrating their unrelenting tenacity to fight for as long as necessary, the Sunni insurgents' goal in attacking the Shiites, according to some scholars, was to show to the Shiite community, the Iraqi government, and the United States that Iraq's Shiites are not safe anywhere in the country and that neither ISF nor American troops could protect them—with the hope that America would forsake the Shiites as a political ally and thus facilitate the Sunnis' return to power.[79] Hence, the callous bombings and targeting of Iraqi Shiite civilians was from the outset aimed at destabilizing Iraq's already fragile and fragmented social/religious fabric. In an insidious, calculated strategy rather than as an ambitious gamble, the Jihadis—particularly Zarqawi's AQI—managed to garner spectacular success well into 2007, pitting Sunnis and Shiites against each other by setting in motion a series of deliberate bombing attacks against Shiite holy shrines and civilians, thus provoking retaliation by Iraqi Shiite militias, mainly Muqtada al-Sadr's Mahdi Army, against Iraqi Sunnis.[80] The attacks on the Askariya Golden shrine in Samara in February 2006 and June 2007 that killed some 250 civilians, along with the ensuing series of deadly reprisal killings between Sunnis and Shiites, is only one example of the this vicious cycle and the tenuous reality it has helped to sustain and exacerbate.[81] Some observers argue that the February 2006 attack was a watershed event, in the sense that prior to the attack AQI was a foreign-led group not at all supported by most Iraqis. The Askariya bombing, the argument goes, "… was followed by months of violent reprisals by Shiites against Sunnis … al-Qaeda in Iraq, virulently anti-Shiite, became a refuge for aggrieved and beleaguered Sunnis" and thereinafter AQI has been led by Iraqi Sunnis.[86]

The Suicide Bombers. The overwhelming majority of suicide bombers in Iraq have been foreigners (mostly Arab Muslims), who heeded the call for martyrdom and Jihad against America (as was in the case of Afghanistan during the Soviet occupation and the call for Jihad that rallied the likes of bin Laden), when the focus of America's military operations shifted from Afghanistan to Iraq in 2003. These foreigners are mostly members of several insurgent groups, many of whom either have changed names, were absorbed or united with other groups, or have ceased to operate altogether.[83] Of these, *al-Qaeda in Iraq* and three other insurgent/terrorist groups (*Ansar al-Sunnah, The Islamic Army in Iraq*, and *Mujahideen Shura Council*) are most noteworthy.[84] They carry the banner of religion (Salafism) and/or Iraqi nationalism and/or Sunni grievances in their fight against U.S. and coalition forces and the Iraq government. Apropos the participation of foreigners and as noted herein above, Iraqis head most of the Sunni insurgents groups whereas the suicide bombers have been primarily foreigners.

Two types of foreign Jihadis have flocked into Iraq. The first are trained and experienced soldiers and officers from Arab countries such as Jordan, who had been discharged from their unit because they harbored extremist views. The second are young novices from Arab, Muslim, and even European countries who are eager to join the "holy war" but who have no skills coming in. The latter cadre of recruits has been a concern for groups such as AQI, precisely because these recruits are fairly young and unknown. As such, they are viewed with suspicion as potential spies and are therefore isolated and prepped-up for suicide attacks as quickly as possible. This may explain why the majority of suicide bombers in Iraq have been foreigners.

An important aspect of suicide bombings in Iraq that is becoming more and more prevalent is *the role of women bombers*. To date, roughly 15% of suicide bombers worldwide have been women[85] (from 2000 to 2005, about 50 women carried out suicide attacks), most of them belonging to the Tamil LTTE or the Turkish PKK (almost two-thirds of the PKK's suicide bombers were female). In both of these terror groups, their charismatic leaders assured the female bombers that by participating in a suicide attack they would support the cause while proving that they are as brave as their male peers. Until recently, female suicide bombers were unique to the LTTE, PKK, and other nonreligious terror organizations, but this trend has changed in the past few years because some religious leaders have sanctified women's participation in such acts under their "liberal" interpretation of Islamic tradition (ironically, these same men claim "strict" readings of the Koran to justify terrorism). Thus, the Palestinian Hamas and Palestinian Islamic Jihad, as well as Chechen separatists, have been increasingly using female bombers. Most significantly, these organizations have been operating in very conservative and traditional societies where women do not enjoy equal rights with men.

Accounting for a growing portion of suicide bombings in Iraq, women bombers have carried out dozens of attacks since 2003.[86] Following the U.S.-led invasion of Iraq, Zarqawi began using Muslim women to conduct attacks against both Iraqis and members of the Coalition forces. Although the first two attacks occurred in 2003–2004, it was not until 2007 that this trend gained pace: With eight suicide bombings involving women, that year marked a new record for both number of attacks carried out by women and the death toll they incurred. It is interesting to point out that in contrast to male suicide bombers in Iraq, almost all of the women bombers have been Iraqis, aged 15 to 35. Also, as with most suicide bombings in Iraq, almost all the attacks involving women have been attributed to AQI. The growing use of women in such attacks suggests that AQI sees great utility in using female bombers.[87]

Between January and the end of June 2008, 20 such bombings were carried out, suggesting that this upward trend will continue unless U.S. military and Iraqi security officials can develop effective and creative countermeasures aimed at women recruits (e.g., new tools for potentially profiling these women) and/or delve into and try to identify and potentially stymie the root causes for the increased role of women in such acts. American and Iraqi policymakers have noted the phenomenon but have offered differing explanations as to whether it signals the weakness of militant groups or whether it is simply yet another terrorist tactical innovation. This has led to a push within the Pentagon and the U.S. intelligence community to understand the motivations and psychology of female bombers and, if possible, prevent or disrupt the recruitment of women for such attacks. In the meantime, initial steps are being taken in an effort to deal with this phenomenon: The U.S. military and the Iraqi government, recognizing this disturbing trend, recently completed specialized training for some 600 female guards ("Daughters of Iraq").[88] With a monthly salary of $300 they work in pairs, frisking female visitors for weapons and explosives at schools, hospitals, banks, and government offices.[89]

In general, the potential susceptibility of a Muslim woman to appeals made to her by terrorists is affected by her personal, familial, organizational, and societal role in her larger Muslim community. In addition, outside influences and pressures imposed on Muslim women, such as military occupation, may also affect their decision to choose violence—albeit the vast majority of women have not chosen this path, even with their basic freedoms and rights compromised. As for more specific possible reasons for the trend in Iraq, research conducted over the past 10 years on women suicide bombers shows that most of the psychological factors that have stimulated such acts in the past are present in Iraq and that certain psychological motivators are particularly relevant[90]:

1. *Revenge Driven by Loss.* A cathartic desire for revenge appears to have motivated several mothers who had lost children to sectarian violence to become suicide bombers. As in most Muslim cultures, the importance assigned to sons in Iraqi culture is a particularly strong driver and the women who become suicide bombers often have lost close male relatives—a husband, a son or a brother—who died fighting Coalition and/or Iraqi forces.
2. *Xenophobia and Nationalism.* During Saddam's regime, many women were given firearms training to protect their families against the threat of an Iranian invasion, particularly during the Iran–Iraq war. Many of these women have since assumed responsibility for protecting their families (i.e., after a husband and/or the eldest son have died or have been detained) against sectarian attacks and the perceived threat of U.S. troops. Some of these women are also devoted nationalists.[91] Also, in the early days of the conflict, many other Iraqi women expressed a primal fear of being ruled by an external force and were thus willing to condone otherwise unthinkable acts of violence.
3. *Religious "cleansing."* Like many Muslim women, Iraqi females guard their chastity and when their chastity is perceived blemished by sinful acts, women can be manipulated to perceive violence as a way to "free" or cleanse themselves of such moral offenses.[92]
4. *Exploitation.* The abhorrent suicide bombing in March 2008 by two women with Down's Syndrome involved a more simple and common abuse of vulnerable women by terrorist groups. This is not a new trend, as Iraqi hospital administrators have occasionally "exchanged" physically and mentally disabled patients for money—with these individuals sometimes ending up in the custody of insurgents. However, the insurgents' exploitation of handicapped Iraqi women reveals a level of despair not often present in other conflicts.

In sum, continued sectarian killings and the social and cultural trends in Iraq—which increasingly restrict women's opportunities outside the home—provide more incentives for women to take part in suicide attacks. The percentage of suicide bombings carried out by women will probably continue to rise as long as the war in Iraq continues, with more women available and willing to commit suicide attacks. Can an American withdrawal decrease the level of violence by women in Iraq? Some scholars argue that a withdrawal may not necessarily restore women's rights but that the United States can play a leading role in helping Iraqi women rebuild their lives by providing security, economic opportunities, educational freedom, and other social reforms.[93]

Lastly, perhaps the most disturbing development hitherto is the recruitment of children and other innocents for suicide bombings. In May 2008 two teenage bombers were responsible for two deadly attacks on the same day: In the first, a 12-year-old boy blew himself up inside a funeral tent in a town west of Baghdad, killing at least 23 people and wounding dozens more in an attack against the relatives of a U.S.-backed police chief and former insurgent who had turned against his onetime insurgent comrades. The second bombing took place south of Baghdad, where a teenage girl blew herself up and thereby killed one Iraqi soldier and wounded seven others (AQI issued a statement claiming responsibility for both attacks).[94] This, together with the frighteningly cynical use of two women with Down's Syndrome in such bombings, makes it clear that as far as the Jihadi terrorists are concerned, there are no limits or moral boundaries in their campaign of violence and bloodshed because they see their cause as divine and therefore one that trumps anything and everything else.[95]

Assessing the Surge

In the time since General David Petraeus was called to implement the "Surge," there has been a remarkable improvement in the security situation, as well as more political cooperation and dealings amongst local political leaders in the north, west, and many areas in central Iraq (including important sectors in Baghdad). This progress has in turn made it possible to jumpstart local economies for renewed growth. Everyday life in Iraq is regaining some of its erstwhile normalcy, especially when recent improvements are measured against the grimmer reality of previous months and years since the 2003 invasion. That being said, Iraq remains a mixed bag. The progress that has been made is neither sufficient nor conclusive, because other new challenges have presented themselves. Iraq's central government is still a prisoner to recalcitrant Shiite chieftains and militia leaders who are less than interested in a compromise and coexistence either with Iraq's other ethnic groups or, as it happens, even with one another. This intra-Shi'a power struggle has led to a growing instability in the Shiite-dominated south at the same time as real improvements have taken place in central and western Iraq. The violence in southern Iraq spiraled downward in the winter/spring of 2008 and still remains one of the biggest security challenges in Iraq: Warlords fight incessantly while the two main Shiite militias—the Mahdi Army and the Badr group—compete for territorial gains and influence among local power brokers (the Mahdi Army itself has seen increasing infighting and is losing much of its cohesiveness) with the central government holding very little sway, if any, on these factions. Since 40% of Iraq's population lives in the south and almost 70% of Iraq's oil production and most of its reserves are to be found there, this is indeed a serious problem that must be addressed. In addition, the showdown between al-Sadr's Mahdi Army and the Iraqi government's forces in the Basra area and in Baghdad's "Sadr City" neighborhoods in the spring of 2008 attest to fissures not only between Shiites and Sunnis but also within the Sunni (the Sunni "Awakening" movement against AQI) and Shiite communities, respectively. Of course, longstanding Sunni–Shiite hostilities are far from being resolved and are only exacerbated as AQI and other Islamic extremist and nationalist Sunni groups continue to stoke the fire of this sectarian strife by attacking mostly innocent civilians using lethal suicide bombings; "Special Groups"—extremist Shiite militias—armed, trained, and funded by Iran, are a serious cause of destabilization and lethal attacks on U.S. forces; the INP is still heavily infiltrated by Shiite extremists; and finally, Iraq's national economy has been slow in making major headways, seeing high unemployment rates that have proven quite resilient to change.

On the positive side, the raw data on U.S. casualties and the extent of the rampant violence in Iraq since the deployment of additional troops as part of the surge show a remarkable improvement and reduction in violence compared to previous years:

1. By October 2007, when the number of U.S. troops in Iraq was at its peak, the average monthly death rate for both American troops and Iraqi civilians was down by almost 60% and has continued to drop to the levels of early 2005. For example, the monthly rate of fatalities in June 2007 (when all of the additional U.S. troops had been deployed) stood at 100, whereas five months later it was below 40.[96]
2. More specifically, when compared to the situation at the end of 2006, civilian deaths in early 2008 had decreased by about 75%, down to the levels of late 2005; From January to December 2007, sectarian attacks and deaths decreased by more than 90% in the greater Baghdad area (in June 2007 there were about 1200 insurgent/terror attacks in Baghdad, whereas in June 2008 about 110 attacks took place.)[97] In addition, Coalition forces found and cleared some 6900 weapons caches in the course of 2007, which is more than twice the number (2662) for 2006.
3. As for al-Qaeda in Iraq, its capabilities have diminished with thousands of Islamist operatives captured or killed in 2007—including hundreds of key AQI cell commanders, as well as national-level leaders and operatives—albeit it remains a threat and source of destabilization.
4. ISF and Concerned Local Citizens (CLC) groups continue to grow and to develop capabilities, and they now provide more security for the Iraqi people. The ISF grew by more than 100,000 in 2007 and are more than 500,000 strong. These CLC groups, or "Sons of Iraq"—which are local security militias with tribal affiliations (paid by the United States)—continue to play a key role in the decreasing degree of violence and in improving stability across Iraq. More than 130 different CLC militias,

numbering some 100,000 active members, have volunteered to help maintain security in their respective neighborhoods since "The Awakening" began.

5. Following clashes with government forces loyal to the rival Shiite party Islamic Supreme Council of Iraq, Muqtada al-Sadr declared a ceasefire in August 2007, which has helped curtail much of the violence against Sunnis. This in turn allowed U.S. forces to be more efficient and effective by focusing on AQI and other Sunni radicals as well as breakaway Shiite elements from within the Mahdi Army that opposed the ceasefire, helping U.S. troops to curtail more potential violence.
6. The Iraqi government's decision to confront al-Sadr's forces in southern Iraq and in Baghdad in the spring of 2008 may actually prove to have positive national political effects: When seeing the Shiite-led government forces and U.S. troops going not only after AQI and other Sunni militias but also after Shiite militias, Sunnis will perhaps feel less concerned about a U.S.–Shiite conspiracy to disenfranchise them and see this as a genuine campaign to consolidate Iraq's national authority and curb all independent militias and armed forces that are not part of the ISF. At the same time, Iraqi Shiites may appreciate Sunni resistance to AQI and the Jihadis as part of the Sunni Awakening movement, along with the CLC groups' cooperation with American forces, not as a newly forged alliance between the United States and the Sunnis at the political expense of the Shiites but rather as a genuine attempt to fend off elements that are dangerous and destabilizing to the Iraqi people as a whole (this, of course, remains to be seen).
7. Notably, the general population today has greater access to such basic necessities as potable water and electricity compared even to the Saddam era.[98]
8. Finally, oil production has also increased and now stands at prewar levels of approximately 2.5 million barrels a day, bringing in greater revenues that have been set aside for infrastructure and reconstruction projects.

As noted above, one of the most important outcomes associated with the surge has been the curbing of Shiite militants, not just AQI and Sunni insurgents. Much like what Hezbollah had done in Lebanon, Muqtada al-Sadr came very close to establishing a state within a state inside Iraq by 2006. He asserted his power by violently intimidating rival clerics, agitating against the U.S. occupation, and using force to establish de facto control over Baghdad's Sadr City and large swaths of southern Iraq. In the areas under his control he set up extrajudicial Shari'a courts to administer justice against Iraqi Shiite "heretics." Many found guilty were punished by death under the Sadr theocracy. Taking the place of Iraq's army and sparsely deployed Coalition troops, the Mahdi Army also established its own security checkpoints in Baghdad and across the south. The main reason for Sadr's ability to augment his power during that period was the absence of security in Baghdad, which left the Shiite community completely vulnerable to the unrelenting deluge of terror attacks by Sunni insurgents and AQI. In hindsight, this was the apex of Sadr's influence. As his realm expanded, it also generated popular resentment. Ordinary citizens were vexed by the harsh version of Islamic law imposed by his minions, not to mention the brutality and widespread corruption of Sadr functionaries manning checkpoints and patrolling the streets. Sadr's hold on the larger Shiite community was actually quite tenuous at that point, held together almost exclusively by the overarching fear of Sunni insurgents and AQI's brutal suicide bombings. As the new U.S. counterinsurgency strategy of the Surge began to pay large dividends against AQI and Sunni insurgents[99] in 2007 and, as a result, Sunni attacks against Shiite civilians declined, so did the rationale for Sadr's authority amongst Shiites, rendering his movement increasingly vulnerable, scrutinized and divided, and subject to actions by ISF.

At the same time, it is also patently clear that much of what has been taking place in Iraq since mid-2007 is complex and often inconsistent. As noted above, casualty rates of both Iraqi civilians and U.S. troops fell sharply since the summer of 2007 (number of Iraqi civilians killed stood at around 2000 in July, falling to about 720 in November),[100] but as it drew to a close, these lower numbers reflected the monthly casualty rates of the end of 2005.[100] Hence, although lower than in most of 2007 and that of 2006 (prior to the bombings of the Shiite mosque in Samara), the number of casualties has by no means been minimized to pre-2003 levels and still constitutes a serious element of insecurity for the average Iraqi. The example of multiple-casualty attacks, both suicide bombings and IEDs, illustrates this as well: Although the monthly number of Iraqi fatalities in attacks that killed more than three people each had dropped significantly from the time the Surge began to about eight months later (from around 700 in February 2007 to about 120 in November 2007)[101], that lower number represents the monthly number of Iraqi fatalities from such attacks in early 2005.[102] Certainly, this level of attacks and casualty rate is not sustainable, nor can it be sufficient for attaining long-term security for Iraq.

Another point is the locus of the insurgency. While the increase in U.S. troops clearly improved the security situation in Baghdad and its environs and has helped restore a semblance of normality to everyday life in the capital, the violence associated with the insurgency—particularly that of the Jihadi terrorist variety—has moved elsewhere, mainly to northern Iraq and the city of Mosul, which has become AQI's last major stronghold in Iraq. And while U.S. casualties were down in the latter half of 2007, they were still higher than those of 2006. Since the surge began in February 2007, some 30,000 additional troops were deployed to Iraq, bringing the total number of American troops there to 160,000. Due to troop withdrawals that began in the Spring, that number stood at about 130,000.

There are several ways the surge has indeed worked and proved its projected efficacy to its military and political proponents. As mentioned above, the most significant is the marked improvement in the security reality in Baghdad. From the outset, the focus of the surge was put on the capital and its environs, with a belt of military roadblocks and checkpoints on the roads leading into the city, inspecting and controlling incoming traffic and the flow of militants and arms. The U.S. military also stepped up its cooperation with local residents and tribes to help American troops cleanse areas of insurgents, thus creating safe zones where trade and normal life could resume, in turn creating a new sense of restored security and safety that Iraqis have been yearning for these last few years. At the same time, the overarching goal of the surge was to bring about an environment, through enough improved security, that is conducive to political compromise amongst the various players in Iraq, particularly the Shiite-controlled government and the Sunnis, which would have reduced violence even further and potentially pave the way for real Shiite–Sunni reconciliation. In that regard, the surge has not been very successful.[102]

Another purported achievement of the surge is the anti-Islamist shift that began in Iraq's western al-Anbar, Diyala, and Nineveh provinces, where AQI and its allies encountered growing resistance from locals starting in 2006 that culminated in violent confrontations between these Jihadi insurgents and local tribal militias, now aligned with the United States. However, the real cause for this dramatic shift, which helped establish these new alliances between U.S. troops and tribal militias, was the growing grassroots distaste for the Jihadis' misconduct of the population and the ultra-radical Salafi brand of Islam they were attempting to impose and enforce. Eventually, AQI began to murder Sunnis who refused to cooperate and/or protested the group's religious ideology in violent willy-nilly, often public executions in broad daylight. This was a poor miscalculation on the part of AQI, because it drove many Iraqi Sunnis who had initially hosted the Jihadis to reject them. As with most insurgencies sustained by the support of critical segments of the population, this bode ill for AQI's operations and influence in that part of Iraq and later throughout the country. In addition, America was also able to garner local Sunni support in al-Anbar by simply paying local leaders and militias to cooperate with it against the Jihadis. The monetary windfall coupled with the prevalent popular misgivings toward AQI and the opportunity for these local Sunni militias to be armed and supplied by the U.S. military, made it quite beneficial for these groups to now side with the United States.

One must bear in mind that the dramatic improvements in the al-Anbar and other provinces have nothing to do with the ethnic/religious Sunni–Shi'a strife. As previously mentioned, that historical religious conflict transcends Iraq itself and is at the heart of the most fundamental disagreements within Islam. But even in the context of Iraq, the Sunni–Shi'a rivalry and the "civil war" component of the insurgency pertain to central and southern Iraq, not to the al-Anbar province which is almost exclusively Sunni. It is therefore a mistake to use this example as a model for the rest of Iraq and be unduly excited about the prospect of curtailing the violence in other parts of the country just as quickly and effectively. Rather, to best tackle security challenges in Iraq, military and policy planners must use a variety of strategies, each tailored to the province and population in question—the Shiite south, the Sunni-dominated central and western parts of Iraq, the mixed Baghdad province and the capital itself, and the predominantly

Kurdish north—and the nature of the violence present in each of these areas. Furthermore, this new Sunni–U.S. alliance is no guarantee for ongoing, stable, long-term cooperation. It was local Sunni interests, not some newly found affinity to American ideals and the administration's vision for Iraq, which drove the tribal leaders closer to the United States and away from the Islamists. As such, the alliance with the Sunnis is a risky endeavor—one that may truly be necessary but is precarious nonetheless. Moreover, close cooperation with the Sunnis could be construed by the Shiites as American preference of one ethnic group over another, potentially driving them further against the United States down the road—and/or seeing that the Sunnis have a firm hold over the center and western parts of the country—to decide to escalate the fight for the Shiites' own regional self-determination, much like the Kurdish autonomy in northern Iraq. Ironically, that kind of fragmentation along ethnic and religious fault lines flies in the face of what is the stated American policy of keeping post-Saddam Iraq united, sovereign, and democratic.

On the Shiite side, there has been growing violence in the south where rival Shiite criminal gangs vie for control and are intimidating the local population. With British forces handing over security control to the ISF in late December 2007, the ability to contain the violence and effectively deal with these gangs and militias will probably diminish, potentially requiring the involvement of U.S. troops later on. That would be a daunting task, particularly once the surge is over. More noteworthy is the ceasefire declared by Muqtada al-Sadr in August 2007, which held until March 2008 and clearly accounted for much of the decrease in Sunni-Shi'a violence. It is very hard to connect al-Sadr's decision to maintain that ceasefire with the surge. Instead, it was likely spurred by cold calculations of realpolitik: Having been challenged by opposition elements from within the Mahdi Army, al-Sadr needed to contend with these forces so as to consolidate his leadership amongst the Shiites before proceeding to take on the government, the Sunnis, or U.S. forces for that matter. As for the Iraqi government led by Prime Minister al-Maliki, it has proven itself quite unable and/or unwilling to take the difficult, albeit necessary, political steps, succumbing instead to factional infighting that prevents it from doing what is best for the Iraqi people. Though entirely bankrolled by the Unites States, economically the government has functioned poorly, with very little to show for except high unemployment, inflation, and scarcity of supplies like gasoline as well as lacking infrastructure services, which have become the hallmarks of the Iraqi economy. Rampant corruption continues to be a serious problem and, compounded by factional strife, often prevents various government ministries and agencies from working with each other, making many earmarked development/reconstruction projects either impossible or very slow in progress. Since the government is Shiite-led, this kind of procrastination is seen by many Sunnis as discrimination on the part of the government and does not contribute to help foster goodwill and reconciliation.

One of the major developments taking place before and during the surge has been Iran's increasing meddling and destabilizing role in Iraq. As previously mentioned, Iran and its Lebanese proxy Hezbollah have actively supported Shiite and even Sunni insurgents groups since 2003, providing arms and training so as to target U.S. forces, Coalition forces, and the ISF and foment sectarian violence. Iran has been arming Shiite militias fighting the forces of the Shiite-led government, while at the same time politically supporting that very government—all part of Iran's hedging its bets on the various political forces within Iraq's Shiite community. By 2008, roughly half of all attacks on U.S. forces were attributed to Shi'a insurgent groups armed and supported by Iran (see above). In response to mounting Iranian intervention, Coalition forces conducted an increasing number of special and conventional military operations targeting Iranian-backed secret cells. All the while, Iran continues to deny any involvement in promoting violence in Iraq.

The goal of training and preparing the Iraqi Army and the Iraqi National Police (INP) to increasingly take over the role of security has made some headway since 2006—mostly with the army but very little when it comes to the INP, which is still highly ineffective—but there remain several core issues that must be taken into account in considering American troop reductions: First, in spite of the relative progress in training the Iraqi Army, both it and other Iraqi security forces (mainly the INP) are far from being ready and capable in numbers large enough to constitute a critical mass of an effective security apparatus that can contend with the challenges of the insurgents and their terrorist tactics. Second, there are serious problems associated with many of the rank-and-file of the INP (and to a much lesser extent the Iraqi Army) when it comes to their loyalties. Many have proven loyal to Shiite and even some Sunni insurgent militias rather than to the Iraqi government. With this in mind, can local Iraqi security units work effectively if U.S. troops are no longer patrolling neighborhoods? And as the U.S. pulls out of various parts of Iraq, will Sunni-Salafi extremists make a comeback and undo all that has been achieved?

If places like the Anbar, Diyala, and Nineveh provinces are good indicators and are able to maintain the required stability in the near future, there is a relatively good chance that Iraq will be able to weather the coming reduction in American forces. But as things stand now, it is really anyone's guess how well and for how long Iraq's security forces—and, by extension, the government—can really perform once the United States actually pulls out its troops in large numbers. (Whether or not a few thousand U.S. troops are left in Iraq for the long-term—say, for further training of Iraqi forces, as a base for counterterrorist operations in the region, or perhaps to secure and ensure the flow of Iraqi oil—is irrelevant, since that size of an American force would not be strong enough to make a difference when it comes to the overall security in the country and the challenge of the various insurgent groups and militias.) It is also quite possible that once periodical troop withdrawals begin in earnest and consequently the security situation does deteriorate, those redeployments may stop as quickly as they started. We may indeed see new "surges" in the future, dictated by the security and political realities on the ground more than anything else.

With the end of the surge, we must look at the potential ramifications of a diminishing U.S. troop presence. The most crucial long-term predicament is the question of what happens once American troops begin to pull out of Iraq in great numbers (the timing of which is anyone's guess). First and foremost, such troop withdrawals—contrary to what many politicians may claim—can be complicated both operationally and logistically, particularly in such dangerous settings as Iraq. Although they could take place quite quickly, with the operational security required for a safe pullout of U.S. troops from Iraq, the task could turn out to be a hard one for the military to accomplish if asked to do so in a very short time.

A fairly less-examined aspect of the insurgency is the effect U.S. troop presence has on the insurgents with regard to the extent that the insurgents are affected by new information about the United States' sensitivity to casualties and the pressure to withdraw from Iraq. Recent analysis provides evidence that seems to imply that the insurgents are actually very much attuned to media reports and U.S. public opinion polls and react strategically: Two Harvard researchers found that in periods following intensified criticism of the Iraq War by the American public and in the U.S. media, insurgent violence in Iraq increased by 7–10%. The authors found that insurgents do not just randomly wreak havoc, but react to developments in American politics. By using data on attacks and variation in access to international news across Iraqi provinces, the researchers identified an "emboldenment" effect by comparing the rate of insurgent attacks in areas with higher and lower access to American news media after public statements critical of the war were reported. Their most important finding is that in periods after a spike in statements in the media critical of the war and American involvement in Iraq, insurgent attacks increase by 7–10%. In addition, the authors found that following new information about the United States' sensitivity to costs, insurgents shifted attacks from Iraqi civilian to U.S. military targets, resulting in more U.S. fatalities. At the same time, the perception of a drop in American resolve seems to "embolden the enemy" to unleash more destructive attacks, thereby not only increasing the cost of fighting for the U.S. military in terms of lives and resources but also tipping the scales toward a quicker withdrawal, or so the insurgents hope. These disquieting findings suggest that there is a small but measurable cost to public debate in the media in the form of higher attacks in the short-term and that Iraqi insurgent groups—even those motivated by religious or ideological goals—are strategic actors that respond rationally to the expected probability of a U.S. withdrawal. The authors conclude that a "systematic response" in the form of "emboldenment" is evident among the insurgents; and given that they appear to respond rationally to incentives set by the policies of pro-government forces to achieve their goals, a shift in counterinsurgency strategy away from a focus on search-and-destroy and toward deterrence, incapacitation, and inducements is advisable.[103] It is the author's assertion that unless the insurgency is dealt a heavy blow and is rendered almost nonexistent (at least operationally) through effective COIN operations and successful Iraqi political reconciliation or some sort of long-term coexistence and *ad hoc* cooperation, there is a chance that once American troops start pulling out in large numbers and with the wide media coverage of the withdrawal, the remaining insurgents will likely attempt to step up their attacks—including suicide bombings—so as to both inflict heavy casualties on the columns/convoys of vehicles and troops (now much less protected compared to being stationed in fortified bases)—but, much more importantly, to project an image to the world, particularly to Muslims, of being the ones who defeated America and pushed it into leaving Iraq with its proverbial tail between its legs, much like the Afghan Mujahideen did to the Soviet Union and the way al-Qaeda's founders and recruits perceived it. Such a possibility would constitute an invaluable PR stunt for the Global Jihadis and radical Islamic movements that would re-attract many new recruits and supporters in the Muslim and Arab world, but also amongst Muslims in Europe and the West, including the United States.

Once American troops pull out of Iraq completely (perhaps to Kuwait or to autonomous Kurdistan in the north), the ISF would be the ones required to maintain security and law and order. As of this writing, the chance of seeing an effective Iraqi army and INP capable of doing just that and preventing the country from spiraling down to full-blown civil war and ethnic cleansing seems distant enough to mandate prudence with regard to just how quickly the United States can actually pull large numbers of its troops out of Iraq. At this point in time, it is more likely than not that a substantial American force will be in Iraq for years to come.

The achievements of the Surge largely rest with General David Petraeus, who changed the nature of the war and U.S. counterinsurgency strategy in Iraq. The change he created was not so much military as political, particularly the process of rapprochement he began with local tribal Sunni leaders. The first phase of the U.S. counterinsurgency—from the beginning of the Iraqi insurgency in mid-2003 until the surge in early 2007—consisted of a three-way civil war, in which the United States and the ISF, the Sunni insurgents, and the Shiite militias fought each other. Petraeus reshaped the battle by observing tensions also existed within both the Iraqi Sunni and the Shiite communities.[104] His strategy was to exploit these tensions, splitting the enemy and forming new alliances with some of the insurgents. The most important thing Petraeus did was reduce the cohesion of America's enemies by recognizing they were not in fact a uniform entity and moving forward on that basis. Thus, the U.S. effort began purely on a military track and continued on a political track that Petraeus brought to bear as the surge was implemented in 2007.

Notwithstanding these achievements, there are many other burgeoning factors of uncertainty and risk in the post-Surge era and beyond. Consider the matter of Iraqi refugees, for instance: Since 2003, more than 3 million Iraqis became refugees (about 1.2 million have been displaced within the country and another roughly two million in neighboring countries, primarily Jordan and Syria).[105] As the security situation began to improve dramatically beginning in the fall of 2007, some of these refugees have been returning to Iraq. While this is encouraging, there is the question of the refugee effect in the months and years to come: Struggling for income, will they be able to find work and reclaim their lives? Will they be able to reclaim their properties? Will they seek vengeance against those who had expelled them from their homes and/or those who have taken over their homes in their absence? How would that affect the level of disenfranchisement and embitterment toward the Iraqi government that could once again provide fertile recruiting ground for terrorists and insurgent groups, only this time along socioeconomic lines as well? In addition, if many Iraqis remain refugees in their host countries, what would be the extent of this negative development? Consider the Palestinian example: Hundreds of thousands of them still live in teeming refugee camps-turned-neighborhoods/towns in several Arab countries and denied full rights or repatriation. In recent years, Jihadis (Fatah al-Islam group) linked to al-Qaeda began to thrive in the Palestinian refugee camps of Lebanon and in autumn 2007 fought a bloody three-month battle with the Lebanese Army in the Naher al-Bared refugee camp in northern Lebanon. It is not too farfetched to assume that if the status of the more than 2.5 million Iraqi refugees currently living in Syria and Jordan is not resolved—preferably by their return to Iraq—they could become a ready source of disgruntled terrorists that is "ripe for the picking" by radical Islam.

If the Jihadis and some of the Sunni sheikhs decide to realign once again and regroup, the fabric of political relations could unravel very quickly and the military situation would change dramatically. The real issue at hand for the United States is its strategy for preventing this kind of dismal scenarios of widespread instability and bloodshed—signs of which are arguably present already. Whether or not future "surges" will be necessary rests partly on the formulation of such a strategy. The United States will have to make the fundamental decision of whether it is leaving the Iraqis to their own devices by pulling out of Iraq within a predefined timeframe no matter what the reality on the ground is, or whether it is making a long-term commitment in Iraq (and by extension the region) even in times of minimal progress or worse.

But whatever shape America's strategy in Iraq takes after the surge, it is, at the end of the day, up to the Sunnis and Shiites (and to a certain extent the Kurds) to truly want to reconcile and come together in working to keep Iraq united and safe if long-term stability and real prospects for economic progress and peace are to be possible. This may indeed prove to be naïve, but both groups must come to grips with their new status in Iraq and go beyond the oft-painful burdens of the past. Until that happens, nothing the United States does can alter the fundamental ethno-societal problems of Iraq and along with them the challenge of an insurgency.

The Impact of Iraq and Its Lessons

By the spring of 2008 the number of American troops killed in Iraq reached 4000.[106] When compared to the monthly rate of casualties as well as the total number of GI's killed in Vietnam and World War II, this number is substantially low.[107] Yet, if the history of other conflicts is of any indication, it is quite striking that so many people are actually taken aback by the prospect of American military involvement in Iraq for years or perhaps even decades into the future. After all, the United States stayed in Vietnam for more than a decade (including five years *after* Nixon made the strategic decision that the United States should get out of Vietnam) and American forces are still present on the Korean peninsula almost 60 years after the Korean War (not to mention U.S. troop presence in Europe during and even after the Cold War and for several years in post-World War II Japan).

It is purportedly in the nature of great powers, particularly the United States, to take on long-term commitments with tenacity. But in contrast to such undertakings as World War II—when more than 400,000 American servicemen died and some 16 million Americans served in uniform with the entire U.S. economy and industrial complex mobilized—the Iraq War and the "global war on terror" have so far seen the involvement of a few hundred thousands troops and a fraction of the casualty rate of past wars (much of it thanks to advances in medicine) even though they have already lasted longer than World War II. Of course, any attempt to compare these two military campaigns or any other, for that matter, is, to a large degree, pointless, since the fundamental social, political, military, and technological differences between the two eras are so profound. What the American people deemed as acceptable, often necessary, sacrifice 60, 50, or even 20 years ago—the willingness to "bear any burden" as it were—seems no longer tolerable, and any current or future military undertakings will likely be pursued by the powers that be in accordance with the prevalent zeitgeist.

As previously mentioned, prior to the Surge the lack of a single leadership to the al-Qaeda-led insurgency made it even more difficult for the United States and the ISF to conduct effective COIN operations beyond the purely tactical level of trying to eliminate as many insurgents as possible. Without a clear leadership as a target, the policy of decapitating the top echelons of such groups (as has been done effectively in some cases elsewhere)[108] is highly unlikely to affect any real change in Iraq due to the ethnic, religious, and political chasms that are the main cause of instability. Experience in Iraq as well as in other conflicts shows that in many cases new leaders step in to take command of the organization and proceed to bring about a so-called "boomerang effect," whereby they ratchet up the campaign of terror and guerrilla attacks—often with the additional drive for vengeance used to execute more daring and deadly attacks—as well as serving as an instrument for garnering sympathy and recruiting new members.[109] The old adage of "drying the swamp rather than trying to kill the mosquitoes" holds very true in Iraq.

Another noteworthy corollary of the Iraq War that, as previously mentioned, could potentially become a source of future challenges/threats is that of Iraqi refugees. In order to prevent the likelihood of a protracted state of political and humanitarian limbo that can stoke another cauldron of instability in the region, the United States must work diligently with the Iraqi government, Iraq's neighbors, and other Arab nations, as well as with international relief agencies and NGOs, to make sure these refugees—particularly those who desire to do so—are able to return to Iraq as quickly as possible (preferably to their erstwhile homes) and receive the necessary economic and social assistance they need to reclaim their lives.

Perhaps the first lesson of the war in Iraq is that in a world where media coverage, public opinion and policy-making are inextricably connected, the American public's sensitivity and aversion to sustaining casualties make such a price in lives untenable for long periods of time. This is something America's political and military leaders need to keep in mind.

One basic insight of counterinsurgency warfare from Iraq is that maintaining the progress that has been made entails recognizing that *political progress depends fundamentally on security*. This lesson—which has brought about the progress made against the Sunni insurgency, AQI, and the Mahdi Army since the summer of 2007—is perhaps the main *strategic* lesson that America has learned in more than five years of fighting in Iraq.

On the *tactical/operational* level, there is more to the U.S. military's experience in Iraq than the oft-reported strain on the troops and their equipment, or the overstretching of American forces that makes it much more difficult for the Unites States to respond to other crises and military contingencies elsewhere in the world. First off, *the U.S. military has been able to learn and absorb valuable lessons vis-à-vis COIN operations and irregular warfare*: Many of the Islamic radicals groups have opted to engage in terrorist acts and irregular warfare; and, in that regard, fighting in Iraq has made U.S. troops much more prepared to fight in such wars than ever before, with conventional forces now carrying out COIN operations with such ability and success that many experts, including in the military itself, thought was either impossible or would require much more time to achieve.[110] The flip side of this improvement in capabilities is the possible (and more long-term) deterioration in America's conventional war-fighting abilities, as the emphasis on irregular warfare and counterterrorism take precedence. The 2006 Israeli war with Hezbollah can serve as an excellent example of a modern, state-of-the-art military that over a period of several years became very good, if not the best in the world, at waging irregular warfare and counterterrorism operations while at the same time allowing its conventional war-fighting abilities take the back seat and inexorably deteriorate. Although the United States and Israel exist and operate under different circumstances (for one, Israel's army is a popular military with compulsory service, whereas the U.S. has a professional military of volunteers and career officers), it is important for the U.S. military and defense establishment to ensure that improvements in the ability to fight such irregular wars such as COIN do not come at the expense of the conventional capabilities (which are still critical and may indeed prove to be even more important in armed conflicts and future international crises).[111]

Second, *the Iraq experience has greatly improved the capabilities of the military's various counterterrorism (CT) units* in intelligence gathering, developing and implementing new tactics, training, and so on. These special units are almost exclusively manned by individuals with a wealth of real world combat counterterrorism experience, something only a very small percentage of personnel in America's military CT units had before September 11, 2001.[112]

There are also operational lessons to be gained from the American experience in Iraq—including suicide bombings—as well as broader, strategic insights that are of paramount importance for understanding current and future threats and effectively planning for such contingencies, especially pertaining to U.S. homeland security:

The Iraq experience could portend future tactics used in attacks against the United States, be it a military, civilian, economic, or otherwise symbolic target, either stateside or overseas. If anyone plans to attack American targets in the future, Iraq has clearly demonstrated how to go about it, be it IEDs, mortars, RPGs, or suicide bombings. The Department of Defense, Homeland Security, as well as state and local law-enforcement agencies, need to seriously think about these threats and how to best contend with them on an ongoing basis in preparation for future scenarios and contingencies.

Local law-enforcement, Homeland Security agencies and First Responders can greatly benefit from analyzing suicide attacks in Iraq and how U.S. forces confronted that challenge. As there have been numerous suicide attacks in Iraq using primarily explosives but also firearms, these attacks ranged from very simple (gunman opening fire and blowing himself up) to highly complex attacks (such as simultaneous bombings in different sites) that involved significant preparation, surveillance, coordination, and multiple participants in different capacities. Local law enforcement can benefit from this experience by learning the lessons and instilling them independently because, on the one hand, these units are the ones handling a high volume of the most sophisticated CT challenges a domestic agency will likely face, yet there are many friction points between local police departments and access to military CT units (though combined training is done on occasion).

The military's experience in Iraq is very relevant to the threats at home. Combat operations in Iraq run the gamut of tactics, from mobile to guerrilla to terrorist warfare and ranging from incompetent to highly competent and complex attacks. It is far better to learn from this already-acquired experience than attempt to learn after domestic contact with similar complexities.[112]

Many of the traits of the terrorists in Iraq are similar to what the United States may confront at home. The U.S. military in Iraq faced terrorists ranging from local incompetents to highly experienced international terrorists trained by foreign special operations units and intelligence (e.g., Iran's IRGC-QF). It would be logical to assume that these terror organizations would send only their most competent operatives for missions inside the United States, probably assisted by cells of "domestic" Jihadis.

Terrorists in Iraq were able to adapt to many of countermeasures that were implemented and develop new methods of attack and tactics. Those terrorists that have survived possess a lot of tactical experience (particularly in bomb-making) and have become very clever (for example, they make significant efforts in surveillance and operational intelligence gathering on potential targets and have even documented U.S. capabilities in the field.[112]

Proactive pursuit of this knowledge base and skills on the part of local law-enforcement/SWAT/CT is probably the best (and often the only) way for these units to attain this useful knowledge reasonably quickly. To best simulate the lessons of Iraq (and Afghanistan) so as to mitigate the level of threats and associated risk in America's cities, high-quality training on a national level must be available to police departments, but that requires commanders of these police forces to have the initiative and actively seek out this training.[112] Local police departments wishing to access these lessons and capabilities should seek out individuals with the desired experience and attempt to recruit them, even if only as a temporary arrangement. (Even prior to 9/11 but increasingly in its wake, a growing number of police departments—like the NYPD and LAPD—have sent officers abroad to train with foreign special operations units and even established permanent liaison posts overseas; and, as a result, their own initiatives have actually created CT capabilities that, in some instances, surpass those of the federal government.) When taking into account the sheer size and level of bureaucratic red tape that exists in the federal government, as well as the traditional friction points between the military, the intelligence agencies, Homeland Security, and state and local agencies (which pose formidable hurdles for time-sensitive needs) waiting for this knowledge to eventually trickle down from the top will most probably take years or may not even happen at all. Therefore, it is would quicker and more effective for local law-enforcement agencies to train to acquire such skills and capabilities. Practically any and every lesson and skill learned in these conflicts is now available on the commercial market for a savvy commander who has the insight to build a competent program and turn to the right service providers for each element of training that is needed.[112]

There is a large disparity in quality/capability of SWAT teams in the U.S. Some units are minimally capable in intelligence and reconnaissance training, SWAT operations and practically no CT capabilities. Some police departments (like the NYPD) have highly evolved CT capability, but they are all too few. Also, very few police patrol officers—who are far more likely be on scene of a suicide bombings than SWAT and special units—have adequate suicide terrorism training and need effective, high-quality training specific to this threat as much as, if not more than, SWAT/CT units.[112]

There is a big difference between the tactical capabilities needed to fight serious crime and those needed for the realm of suicide terrorism. Very few police departments in the United States have sufficient personnel and/or screening/selection processes to man a high-quality SWAT team in addition to attempting to build a high-quality dedicated CT team in tandem. It may therefore be better to man a few national, dedicated CT teams and encourage the better, larger, and well-resourced police departments to acquire these capabilities.[112]

An old adage in the CT community points out the three things vital for effectiveness in fighting terrorism: intelligence, intelligence and intelligence. Puns aside, acquiring quality intelligence is perhaps the most important element in fighting terrorist cells and networks, foiling plots, and thwarting attacks.[113] The latter requires real-time intelligence and developing these capabilities. The federal government should thus focus its resources more on efforts to improve and expand intelligence gathering and analysis capabilities (for instance, to prevent the legal or illegal entry of suspects into the country to begin with); increase offensive operations capabilities at home and abroad; and improve cooperation between the various intelligence agencies, Homeland Security, and other relevant national security entities—something that has been called for repeatedly after 9/11 but has yet to be addressed in a truly comprehensive and effective way.

Notwithstanding the great need for the creation of capabilities to contend with terrorism and suicide attacks, adding large numbers of personnel and units willy-nilly can be counterproductive. Quality over quantity is key, because if the government bankrupts itself with multi-billion-dollar expenditures on dead-end or ineffective solutions (e.g., much of the Transportation Security Agency), it would have succumbed to terrorism without firing a single shot. Instead, funds must be made available for creating a high level of capability with existing units and personnel.[114] In addition, the federal government should advocate educational policies that redirect U.S. citizens toward self-reliance and vigilance.[115]

FUTURE THREATS AND SUICIDE TERRORISM

In the age of hi-tech weaponry, navigation, and real-time communications, the suicide bomber stands out as a simple, perhaps even crude, yet very sophisticated and effective weapon: a mobile, thinking "guided missile" that is almost impossible to neutralize and rarely misses its target. When we look back at the advent, spread, popularization, and evolution of suicide attacks over the past quarter century, it is logical to presume that until this phenomenon can be successfully dealt with and/or countered, we can expect suicide bombings to be used more frequently than not. America's military planners and political leaders must be ready for this contingency and take the necessary steps to adequately prepare the military, intelligence, and various federal, state, and local homeland security agencies for this reality. If future wars or U.S. involvement in foreign conflicts and hotspots are to have a better chance of making a difference, then this is an imperative task.

As far as implications for homeland security, suicide attacks are understandably a major source of concern. The increasingly ubiquitous nature of this phenomenon suggests that it will rear its ugly head again on American soil and existing conditions do not present a real challenge to it: With long, mostly unobstructed borders in the north and south—and the extensive criminal networks of guns, narcotics, and human trafficking operating on the U.S.–Mexican border—it is not unlikely that at some point in the not-too-distant future foreign Jihadis or other terrorists (e.g., agents of Iran/Hezbollah) will enter the United States to carry out attacks, including suicide bombings, against civilian "soft targets" as well as establish new terrorist cells and/or join cells and networks of sympathizers that may already exist inside the United States. Furthermore, it is important to keep in mind that although the template of a 9/11-type attack—one that is spectacular in scale, planning, preparation and execution—is what most Americans and many in U.S. law enforcement tend to think of in terms of suicide attacks on the homeland, America's own experience in Iraq and Afghanistan, as well as the experience of other countries with terrorism and suicide bombings (particularly where the bomber wears the explosives on his/her person), shows that cheap-and-fast "off the shelf" attacks can be brought to bear quickly and effectively executed.[116] Although they individually do not inflict the same level of carnage and damage as large-scale, elaborate attacks, when launched either simultaneously and/or continuously over an extended period of time, the resulting psychological effects of fear and confusion of such attacks can be just as detrimental on a national scale.[117]

The marriage of terrorism (including suicide attacks) and nonconventional weapons/WMD presents a singular danger.[118] Whereas conventional suicide bombings can kill and maim tens or hundreds[119] of people, CBRN (chemical, biological, radiological, and nuclear) terror attacks can potentially kill upwards of thousands, if not tens of thousands. With lacking controls and inadequate safeguards, accessibility to and trafficking of illegal nuclear materials,[120] and the knowledge to build such weapons in several countries, as well as intentional leakage and weapon technology proliferation on the part of rogue states, the chance of this know-how and material finding it way to terrorists has undoubtedly increased since the end of the Cold War. When we take this unsettling reality into account and add to it the preponderance of nihilist terrorists of al-Qaeda's ilk—willing not only to die for "the cause" but to see it as their *divine duty* to do so, even if it means killing untold numbers of innocents, including Muslims—we can conclude with a good measure of confidence that this threat constitutes the greatest danger to America (and the world) in the medium and long term and cannot and should not be ignored.

In sum, it is more likely than not that suicide terrorism will be a greater rather than lesser threat for America in the coming years, both home and abroad. Whatever the financial costs, America's leaders would do well to plan and invest the necessary resources to prepare for this harsh possibility. Furthermore, even if the United States and the rest of the world become wholeheartedly committed to fight terrorism in general and radical Islamic terrorists in particular by targeting their attack capabilities and members, a long-term "victory" can only be achieved by winning the hearts and minds of Muslims—something that, realistically speaking, can only be achieved by the Muslims themselves. Assuming this willingness within the Muslim world takes hold and brings about a real change vis-à-vis curtailing the spread of radicalism and terrorist lore, it will undoubtedly require years, if not decades, before we get there. Until then, the extremists within Islam, whether al-Qaeda and the Global Jihad movement or radical Shi'a ideology, will fight with unrelenting zeal both (a) America, its allies, and the West and (b) "moderate" and "wayward" Muslims. This will be a long and bloody campaign. The United States, along with the rest of the world, must come to terms with this reality and prepare for the hard trials and challenges ahead.

ENDNOTES

1. For more on the ideological parallels between fascism, Nazism, and fundamentalist Islamic terrorism, see Paul Berman's *Terror and Liberalism*, NewYork: w.w.Norton, 2003.

2. The phenomenon of suicide bombings in Iraq is examined separately herein below.
3. Kenneth M. Pollack, *The Persian Puzzle*, New York: Random House, 2004, pp. 190–191; see also Library of Congress Country Studies, Iran, The Iran–Iraq War, The War of Attrition: http://lcweb2.loc.gov/frd/cs/irtoc.html
4. The United States was again targeted when a car bomb exploded outside the now-relocated American embassy annex in Beirut on September 20, 1984, killing 23 people.
5. Kenneth M. Pollack, *The Persian Puzzle*, New York: Random House, 2004, p. 203.
6. Although Hizballah and Iran have denied any involvement in these attacks, two U.S. District Court judges determined in 2003 that Hizballah was indeed responsible for the Beirut attack: Judge John Bates of the U.S. District Court in Washington, D.C., awarded $123 million to 29 victims and family members of those Americans killed in the bombing of the U.S. embassy, while Judge Royce Lamberth (U.S. District Court in Washington, D.C.) determined that the bombing was carried out by Hezbollah with the approval and financing of Iranian agents.
7. Kenneth M. Pollack, *The Persian Puzzle*, New York: Random House, 2004, pp. 199–200.
8. Kenneth M. Pollack, *The Persian Puzzle*, New York: Random House, 2004, p. 198.
9. Kenneth M. Pollack, *The Persian Puzzle*, New York: Random House, 2004, pp. 198–199 and 280–281.
10. Seventeen members of al-Dawa were consequently caught by the Kuwaitis, one of them the brother of Imad Mughniyah.
11. Kenneth M. Pollack, *The Persian Puzzle*, New York: Random House, 2004, p. 282; The 9/11 Commission also found there was strong evidence for Iran's involvement in the attack; see *Report of the 9/11 Commission*, p. 60.
12. A member of Lebanese Hezbollah was later indicted for his pivotal role in designing the truck bomb used in the bombing; see Daniel Byman, *Deadly Connections*, p. 85.
13. The U.S. Department of State's list of Foreign Terrorist Organizations (FTOs) available at http://www.state.gov/
14. Indications that Hezbollah was playing a role in Iraq first surfaced in July 2007 when the U.S. military announced the arrest of Ali Musa Daqduq, a Lebanese-born Hezbollah operative who allegedly trained Shiite militias in Iraq.
15. Michael R. Gordon, Hezbollah Trains Iraqis in Iran, Officials Say, *New York Times*, May 5, 2008; Patrick Quinn, US: Hezbollah Training Iraqi Shiite Extremists in Iran, The Associated Press, May 5, 2008.
16. Michael R. Gordon, Hezbollah Trains Iraqis in Iran, Officials Say, *New York Times*, May 5, 2008; Patrick Quinn, US: Hezbollah Training Iraqi Shiite Extremists in Iran, The Associated Press, May 5, 2008. Iran and Hezbollah's have denied giving any support to Shiite extremists in Iraq.
17. For more on Iran's role in Iraq, see herein subsequent notes below.
18. The Brookings Institution's Daniel Benjamin in testimony to Congress, July 31, 2007. http://www.brookings.edu/views/testimony/benjamin/20070731.htm
19. The United Nation reported that in 2006 alone some 35,000 civilians were killed. http://www.nytimes.com/2007/01/17/world/middleeast/17iraq.ready.html
20. By the spring of 2008 the American death toll had reached 4000 troops.
21. Rohan Gunaratna, Suicide Terrorism: A Global Threat, *Jane's Security News*, October 2000; Jonathan Lyons, Suicide bombings—weapon of choice for Sri Lankan rebels,

Reuters, August 20, 2006; "Tending to Sri Lanka", The Washington Times, August 20, 2006; http://www.intelligence.org.il/eng/eng_n/pdf/suicide_terrorism_ae.pdf; see also Chapter 2 of this book.

22. Robin Wright, Since 2001, A Dramatic Increase In Suicide Bombings, *Washington Post*, April 18, 2008.
23. From May 2003 to January 2007, IEDs (mainly roadside bombs) have accounted for 36.5% of U.S. troops killed, whereas car bombs (including suicide car bombings) have accounted for only 4.2%. See http://www.brookings.org/iraqindex
24. Ibid. Most suicide attacks in 2003–2004 targeted American and coalition troops. This changed dramatically by late 2004; and since then, suicide bombings almost invariably target Iraqi security forces and Shiite civilians, though Coalition forces still face sporadic suicide attacks.
25. Jeffrey Gettleman, Shiite Pilgrims Are Walking Targets in Sectarian Conflict, *New York Times*, March 19, 2006. http://www.nytimes.com/2006/03/19/international/middleeast/19iraq.html
26. For example, the introduction of better-armored Humvees and later the sturdier MRAP (Mine-Resistant Ambush-Protected) vehicle, which gradually replaced the vulnerable conventional Humvee.
27. For more on this please refer to the section on the surge herein below.
28. The Iraq Index, http://www.brookings.org/iraqindex, May 31, 2007.
29. The U.S. military uses the acronyms SVB-IED for Suicide Vehicle-Borne Improvised Explosive Device and VB-IED for Vehicle-borne Improvised Explosive Device.
30. The Iraq Index, The Brookings Institution, http://www.brookings.org/iraqindex, May 31, 2007
31. The Iraq Index, The Brookings Institution, http://www.brookings.org/iraqindex, May 31, 2007.
32. James Glanz, Civilians Death Toll Falls in Baghdad, but Rises Across Iraq, *New York Times*, September 1, 2007, http://www.nytimes.com/2007/09/02/world/middleeast/02iraq.html
33. The Iraq Index, The Brookings Institution, http://www.brookings.org/iraqindex, May 31, 2007.
34. Toll Rises Above 500 in Iraq Bombing, *New York Times*, August 21, 2007; http://www.nytimes.com/2007/08/22/world/middleeast/22iraq.html
35. The Iraq Index, http://www.brookings.org/iraqindex, June 28, 2007.
36. On August 19, 2003, the UN headquarters in Baghdad was attacked in a suicide truck bombing. At least 20 people were killed in the worst attack on a UN facility/personnel in the organization's history: http://www.cnn.com/2003/WORLD/meast/08/19/sprj.irq.main/. Today the UN provides extensive humanitarian aid to Iraqis through UNAMI (United Nations Assistance Mission for Iraq), which includes some 18 UN and affiliated agencies. UNAMI also oversees NGO activities in Iraq; see http://www.uniraq.org/default.asp
37. See The Brookings Institution's Daniel Benjamin in testimony to Congress, July 31, 2007.
38. British authorities have uncovered a strong connection between the plotters of the bungled attempt to blow several car bombs in London and Glasgow in June/July 2007, and al-Qaeda in Iraq. The first such case of an overseas attack that heavily

involved AQI aiding a terror cell could portend a future trend: Raymond Bonner et al., British Inquiry of Failed Plots Points to Iraq's Qaeda Group, *New York Times*, December 14, 2007.

39. Defense Secretary Robert Gates and General Peter Pace, DoD News Briefing, February 2, 2007; http://www.defenselink.mil/Transcripts/Transcript.aspx?TranscriptID=3879
40. Ahmed S. Hashim, The Sunni Insurgency in Iraq, Middle East Policy brief, August 15, 2003; http://www.mei.edu/scholars/editorial/sunni-insurgency-iraq
41. Anthony H. Cordesman, The Current Military Situation in Iraq, CSIS, November 14, 2003, p. 25; http://www.csis.org/media/csis/pubs/031114current.pdf
42. Michael Moss and Souad Mekhennet, *New York Times*, October 15, 2007; http://www.nytimes.com/2007/10/15/us/15net.html
43. Nir Rosen, *In the Belly of the Green Bird: The Triumph of the Martyrs in Iraq*, New York: Free Press, 2006, pp. 84 and 140.
44. In Vietnam, for example, the Vietcong's very existence rested on their ability to maintain and consolidate local support. Conversely, the very social structures of Iraq from which the insurgents came and operated became the foundation for a long, armed resistance movement.
45. Nir Rosen, *In the Belly of the Green Bird: The Triumph of the Martyrs in Iraq*, New York: Free Press, 2006, pp. 140–141.
46. Saddam released some 200,000 criminals from Iraq's prisons just before the invasion. Many of these criminals joined the various insurgent groups augmenting violence with widespread lawlessness. The intimidation, extortion, and violence against middle-class Iraqi Sunnis and the Coalition's inability to nip it in the bud essentially drove most Sunnis further away from becoming potential U.S. allies early on (more on this is available in Ahmed Hashim's *Insurgency and Counter-Insurgency in Iraq* New York: Cornell University Press, 2006).
47. Bruce Hoffman, Plan of Attack, *Atlantic Monthly* (July/August 2004).
48. For more on this see the section in this chapter dealing with the Surge.
49. Robert Burns, Commander: Al-Qaida in Iraq Is at Its Weakest, Associated Press, May 21, 2008.
50. Multinational Force in Iraq website: http://www.mnf-iraq.com/
51. The hostage, Douglas Wood, was eventually rescued by Iraqi forces in an operation in June 2005 after being held for several weeks.
52. Multinational Force in Iraq website: http://www.mnf-iraq.com/
53. Its current leader, Abu Ayyub al-Masri, is Egyptian and his predecessor, Abu Musab al Zarqawi, was Jordanian.
54. Frederick W. Kagan, Al-Qaeda in Iraq, *The Weekly Standard* **12**(48), September 2007.
55. Ayman al-Zawahiri argued with Zarqawi on this point in a series of letters that became public. He argued that al-Zarqawi was wrong to attack Iraqi Shiites, who should instead be enticed to join the larger movement Zawahiri wanted to create. His arguments were more tactical and strategic than ideological, because he has no problem killing "unfaithful" Muslims but has been eager to focus the Global Jihad movement on what he calls the "far enemy," that is, America and the West.
56. Frederick W. Kagan, Al-Qaeda in Iraq, *The Weekly Standard* **12**(48), September 2007.

57. The Pentagon claims Iranian-supplied explosives and weapons account for about 20% of all U.S. casualties in Iraq.
58. As of this writing (Spring 2009), this fact still holds true.
59. Nir Rosen, *In the Belly of the Green Bird: The Triumph of the Martyrs in Iraq*, New York: Free Press, 2006, pp. 10–13.
60. Nir Rosen, *In the Belly of the Green Bird: The Triumph of the Martyrs in Iraq*, New York: Free Press, 2006, pp. 17–18.
61. Nir Rosen, *In the Belly of the Green Bird: The Triumph of the Martyrs in Iraq*, New York: Free Press, 2006, pp. 111 and 129–131.
62. Ahmed Hashim, testimony to the US Senate Foreign Relations Committee, April 21, 2004, http://www.senate.gov/~foreign/testimony/2004/HashimTestimony040421.pdf
63. One could argue that this short-lived cooperation and sympathy for each other's fight with the Coalition is proof that there is genuine Iraqi nationalism in both Shiite and Sunni communities.
64. Nir Rosen, *In the Belly of the Green Bird*, New York: Free Press, 2006, pp. 108, 135.
65. Kimberley Kagan, Iran's Proxy War Against the United States and Iraq, The Institute for the Study of War, http://www.understandingwar.org/
66. The current commander of the IRGC-QF, Brigadier General Qassim Sulleimani, serves on Iran's national security council and answers solely to Iran's Supreme Leader, Ayatollah Ali Khameini.
67. The Special Groups sometimes operate alongside the Mahdi Army and other militias, but many of these Special Groups have broken away from the Mahdi Army and do not adhere to Muqtada al-Sadr. According to U.S. General David Petraeus, the primary distinction between the Mahdi Army and Special Groups is that the latter "have had extra training and selection" by the IRGC-QF and Hezbollah.
68. Multinational Force in Iraq website: http://www.mnf-iraq.com/ and The Brookings Institution's Iraq Index, http://www.brookings.org/iraqindex
69. The author suggests this impression could also be the result of a deliberate attempt by the media and/or the authorities to create public outrage and aversion towards the insurgents and their tactics, so as to attempt to minimize the support amongst Sunni Iraqis for suicide attacks in particular and the insurgency in general.
70. Figures based on data from: The Iraq Index, http://www.brookings.org/iraqindex; The Terrorism Knowledge Base (http://www.tkb.org/); and Yoram Schweitzer and Sari G. Ferver, Al-Qaeda and the Internationalization of Suicide Terrorism, The Jaffe Center for Strategic Studies (November 2005).
71. Vali Nasr, *The Shia Revival*, pp. 196–197.
72. Robert Windrem, NBC News, http://www.msnbc.msn.com/id/20287932/
73. Department of Defense News Briefing, June 6, 2007; Tim Kilbride, Power, Influence Dictate Patterns of Violence in Central Iraq, American Forces Press Service, August 20, 2007.
74. The Brookings Institution's Iraq Index, http://www.brookings.edu/iraqindex
75. These were second-generation immigrants from Tunisia and Algiers, and the number of North-African insurgents is substantial: about 20% of insurgents caught in Iraq came from North-African countries.
76. See Reuven Paz, Arab Volunteers Killed in Iraq: An Analysis, The GLORIA Center, PRISM Occasional Papers (March 2005).

77. Multinational Force in Iraq website: http://www.mnf-iraq.com/
78. Ely Karmon, Al-Qaida and the War on Terror after Iraq, *MERIA Journal* (Spring 2006).
79. Vali Nasr, *The Shia Revival*, New York: Henry Holt, 2006, pp. 202–204.
80. Interestingly, anti-Shiite bombings, including suicide attacks, invariably take place during the day while Shiite reprisal killing have targeted Sunnis in their homes and neighborhoods at nighttime.
81. Dan Murphy, *The Christian Science Monitor*, June 14, 2007; "Over 100 Killed After Mosque Attack" and "Iraqi Clerics Appeal for Calm," CBS/AP news, February 24, 2006; Alissa J. Rubin, *New York Times*, June 20, 2007; Mark Santora, *New York Times*, February 13, 2007; Robert F. Worth, *New York Times*, February 25, 2007.
82. Bill Marsh, The Ever-Changing Iraq Insurgency," *New York Times*, January 7, 2007.
83. It is therefore hard to substantiate today the exact number of insurgent groups operating in Iraq, as is the ability to confirm that a certain group claiming to execute an attack was indeed the one to have perpetrated it. The identity of the suicide bombers is therefore also almost impossible to validate.
84. Multinational Force in Iraq website: http://www.mnf-iraq.com/
85. Ophir Falk and Yaron Schwartz, The Suicide Attack Phenomenon, The Institute for Counter-Terrorism (ICT) Herzliyah, January 10, 2005.
86. Iraq is not alone: Pakistan has seen a similar trend in increasing number of female suicide bombers since 2006–2007.
87. According to Iraqi officials, several of the bombings involved women wearing vests that were detonated by remote control. Since there are two ways to detonate a suicide vest—a button the bombers can push themselves and a remote control detonation—these bombings involving women suggest that the latter was used, perhaps because there was concern that the bomber would be afraid to press the trigger herself and her operators decided do it for her.
88. Alexandra, Zavis, Daughters of Iraq: Women Take on a Security Role, *Los Angeles Times*, June 4, 2008.
89. Even though the response from women has been enthusiastic, the program has faced resistance from tradition-bound community leaders who believe that fighting the insurgents is men's work. Ibid.
90. Mia Bloom, Female Suicide Bombers: A Global Trend, Daedalus, winter 2007; Farhana Ali, Muslim Female Fighters: An Emerging Trend, *Terrorism Monitor* **3**(1), November 3, 2005; Farhana Ali, Rocking the Cradle to Rocking the World: The Role of Muslim Female Fighters, *Journal of International Women's Studies* **8**(1), November 2006.
91. For example, the first suicide attack by two women in March 2003 involved bombers who asserted that it was their duty to save their country from the U.S.-led occupation.
92. This form of atonement through bloodletting sometimes echoes the exhortations of certain Iranian-trained female Shi'a scholars, particularly during religious holidays such as the day of Ashura, which features both mourning and vows of atonement. During this period, Iranian female religious leaders have sometimes directed vitriolic speeches at both U.S. military forces and Iraq's Sunni population. These enjoinders

could tip sentiment among some Shiite women—those who suffer from strong feelings of religious guilt or shame—toward suicide attacks.

93. Farhana Ali, Rising Female Bombers in Iraq: An Alarming Trend, the Counterterrorism Blog website, April 22, 2008.
94. Ernesto Londo and Uthman al-Mukhtar, Child Bomber Kills 23 in Iraq, *Washington Post*, May 15, 2008.
95. For U.S. troops this is a rather daunting state-of-affairs: Countering the threat of suicide attacks by Iraqi women who are fully aware of their actions is difficult enough for U.S. and Iraqi security forces; preventing bombings by handicapped people who do not understand how they are being exploited is even more challenging.
96. The Iraq Index website http://www.brookings.edu/iraqindex.html and the Multinational Force in Iraq website http://www.mnf-iraq.com/
97. Multinational Force in Iraq website: http://www.mnf-iraq.com/
98. The Brookings Institution's Iraq Index website.
99. Attacks dropping to 2005 levels and Iraqi deaths due to ethno-sectarian violence declining 90% from June 2007 to May 2008.
100. http://www.icasualties.org/ and http://www.brookings.edu/iraqindex.html
101. http://www.brookings.edu/iraqindex.html and http://www.mnf-iraq.com/
102. http://www.icasualties.org/ and http://www.brookings.edu/iraqindex.html
103. The researchers also point out that the implied costs of open, public debate must be weighed against the potential gains. Radha Iyengar and Jonathan Monten, Is There an "Emboldenment" Effect? Evidence from the Insurgency in Iraq, *National Bureau of Economic Research*, February 2008. http://www.nber.org/papers/w13839.pdf
104. George Friedman, Stratfor report, May 6, 2008, http://www.stratfor.com/
105. James Palmer, Displaced Iraqis Contend with a Grim Existence, *San Francisco Chronicle*, October 17, 2007; UN urges help for Iraqi refugees, BBC News, April 17, 2007; and The Iraq Index website.
106. http://www.icasulaties.org/ and http://www.mnf-iraq.com/
107. In World War II the average monthly death toll was approximately 9000, whereas in Vietnam between 1963 and 1975 the average monthly U.S. death toll was around 400.
108. Israel exercised a policy of killing the leaders of several terrorist groups, with a mixed record of success. For example, the 1995 assassination of Fathi Shekaki, the head of the PIJ, proved detrimental to the organization, which never really recovered from the blow. On the other hand, the targeted killing of Hezbollah's chief Abbas al-Musawi in 1992 proved to backfire badly for Israel. See Boaz Ganor, *The Counter-Terrorism Puzzle*, p. 130; and Anthony H. Cordesman, *Perilous Prospects*, pp. 117–119.
109. For an excellent in-depth discussion on the dilemma of military attacks on terrorist and insurgency groups, particularly the policy of targeted killings and the "boomerang effect", see Boaz Ganor, *The Counter-Terrorism Puzzle*, pp. 38–44, 121–135.
110. Interview with a retired U.S. Army Lt. Colonel who served 15 months in Iraq in intelligence operations from 2006 to 2008.

111. Anthony H. Cordesman, *Lessons of the 2006 Israel–Hezbollah War*, CSIS Significant Issues Series, CSIS: November 2007.
112. Interview with Chris Graham: A CT consultant, former USMC Captain, Anti-Terrorism Task Force commander and three-tour veteran of Iraq.
113. Such as locating and targeting bomb-making labs, weapons caches, etc., where bombing and suicide bombings attacks are launched from and targeting them would curtail the operational capabilities of the terrorist cell.
114. Interview with Chris Graham see note 112.
115. Take Israel, for example: Decades of living with terrorism have taught Israelis to be conscious and vigilant, in many cases saving lives by alerting first responders to examine suspicious packages, vehicles, people, and so on. There is no better way to defend the public than educating it to help itself.
116. Two elaborate terrorist plots thwarted in 2006/2007—one targeting JFK airport's fuel supply system and the other an attack on Fort Dix, NJ—are but two recent examples of the vulnerability and attractiveness of "soft targets" inside the United States.
117. For more on this see Chapter 2 of this book.
118. Yaron Schwartz and Ophir Falk, Chemical–Biological–Radiological–Nuclear Terrorism, The Institute for Counter-Terrorism (ICT) Herzliyah, March 15, 2003.
119. The 9/11 attacks were a rare exception, killing thousands in the eventual collapse of the World Trade Center towers.
120. See the case of A. Q. Khan, the father of Pakistan's nuclear bomb, and his network of trafficking nuclear materials and technological know-how to Iran, North Korea, and Libya, among others. See William Langewiesche, "The Point of No Return" and "The Wrath of Khan" available at http://www.theatlantic.com/

BIBLIOGRAPHY

Atkinson, Rick, "Left of Boom," *The Washington Post*, September 30 and October 1–3, 2007. http://www.washingtonpost.com/wp-dyn/content/article/2007/09/29/AR2007092900751.html

Ali, Farhana, Rising Female Bombers In Iraq: An Alarming Trend, the Counterterrorism Blog, April 22, 2008. http://counterterrorismblog.org/2008/04/rising_female_bombers_in_iraq.php

Ali, Farhana, Rocking the Cradle to Rocking The World: The Role of Muslim Female Fighters, *Journal of International Women's Studies* **8**(1), November 2006.

Ali, Farhana, Muslim Female Fighters: An Emerging Trend, *Terrorism Monitor*, **3**(21), November 3, 2005.

Allawi, Ali A., *The Occupation of Iraq: Winning the War, Losing the Peace*. New Haven, CT: Yale University Press, 2007.

Baram, Amatzia, Neo-Tribalism in Iraq: Saddam Hussein's Tribal Policies 1991–96, *International Journal of Middle East Studies*, No. **29**, 1997.

Benjamin, Daniel, The Iraq War and the New Terrorist Threat Facing the Middle East and U.S. Testimony before the House Armed Services Committee, Subcommittee on Oversight and Investigations, July 31, 2007. http://www.brookings.edu/views/testimony/benjamin/20070731.htm

Berman, Paul, *Terror and Liberalism*, New York: W.W. Norton & Company, 2004.

Bonner, Raymond, Perlez, Jane, and Schmitt, Eric, British Inquiry of Failed Plots Points to Iraq's Qaeda Group, *New York Times*, December 14, 2007.

Byman, Daniel, *Deadly Connections*, New York: Cambridge University Press, 2005.

Cordesman, Anthony H., *Lessons of the 2006 Israel-Hezbollah War*, CSIS Significant Issues Series, CSIS, November 2007.

Cordesman, Anthony H., The Military Situation in Iraq, Center for Strategic and International Studies, November 14, 2003, http://www.csis.org/media/csis/pubs/031114current.pdf

Cordesman, Anthony H., *Perilous Prospects*. New York: Perseus Publishing, 1996.

Department of Defense News Briefings: February 2 and June 6, 2007, http://www.defenselink.mil/Transcripts/Transcript.aspx?TranscriptID=3879 and http://www.defenselink.mil/transcripts/transcript.aspx?transcriptid=3982

Diaz, Tom and Newman, Barbara, *Lightning Out of Lebanon*. New York: Presidio Press / Ballantine Books, 2005.

Dobbins, James, McGinn, John G., et al., *America's Role in Nation-Building from Germany to Iraq*, The RAND Corporation, 2003, http://www.rand.org/pubs/monograph_reports/MR1753/

Falk, Ophir, and Schwartz, Yaron, *The Suicide Attack Phenomenon*, The Institute for Counter-Terrorism (ICT) Herzliyah, January 10, 2005.

Ganor, Boaz, *The Counter-Terrorism Puzzle: A Guide for Decision Makers*. Piscataway, NJ: Transaction Publishers, 2005.

Gettleman, Jeffrey, Shiite Pilgrims are Walking Targets in Sectarian Conflict, *New York Times*, March 19, 2007.

Gunaratna, Rohan, Suicide Terrorism: A Global Threat, *Jane's Security News*, October 20, 2000; http://www.janes.com/security/international_security/news/usscole/jir001020_1_n.shtml

Gordon, Michael R., Hezbollah Trains Iraqis in Iran, Officials Say, *New York Times*, May 5, 2008

http://www.geopolitique.com/terrorisme/

Hafez, Mohammed M., *Suicide Bombers in Iraq: The Strategy and Ideology of Martyrdom*. Washington DC: United States Institute of Peace Press, 2007.

Hashim, Ahmed, *Insurgency and Counter-Insurgency in Iraq*. Ithaca, NY: Cornell University Press, 2006.

Hashim, Ahmed, testimony to the U.S. Senate Foreign Relations Committee, April 21, 2004, http://www.senate.gov/~foreign/testimony/2004/HashimTestimony040421.pdf

Hashim, Ahmed, The Sunni Insurgency in Iraq, The Middle East Institute, August 15, 2003. http://www.mei.edu/scholars/editorial/sunni-insurgency-iraq

Hoffman, Bruce, Plan of Attack, *Atlantic Monthly*, July/August 2004, http://www.theatlantic.com/

The Institute for the Study of War website: http://www.understandingwar.org/

Interview (June 2008) with Chris Graham—CT consultant, former USMC Captain, Anti-Terrorism Task Force commander and three-tour veteran of Iraq.

Interview (April 2008) with a retired U.S. Army Lt. Colonel who served 15 months in Iraq (from 2006 to 2008) in intelligence operations.

Iyengar, Radha, and Monten, Jonathan, Is There an "Emboldenment" Effect? Evidence from the Insurgency in Iraq, National Bureau of Economic Research, Working Paper 13839, February 2008, http://www.nber.org/papers/w13839.pdf

Kagan, Frederick W., Al-Qaeda in Iraq: How to Understand It. How to Defeat It, *The Weekly Standard* **12**(48), September 2007.

Kagan, Kimberly, Iran's Proxy War Against the United States and Iraq, The Institute for the Study of War, http://www.understandingwar.org/

Karmon, Ely, Al-Qaida and the War on Terror after Iraq, *Middle East Review of International Affairs* (MERIA) *Journal* **10**(1), Spring 2006.

Kilbride, Tim, Power, Influence Dictate Patterns of Violence in Central Iraq, American Forces Press Service, August 20, 2007, http://www.defenselink.mil/news/newsarticle.aspx?id=47121

Langewiesche, William, The Point of No Return, *Atlantic Monthly*, January/February 2006.

Langewiesche, William, The Wrath of Khan, *Atlantic Monthly*, November 2005.

Library of Congress Country Studies, Iran, The Iran–Iraq War, The War of Attrition, http://lcweb2.loc.gov/frd/cs/irtoc.html

Londo, Ernesto, and Al-Mukhtar, Uthman, Child Bomber Kills 23 in Iraq, *Washington Post*, May 15, 2008.

Lyons, Jonathan, Suicide Bombers: Weapon of Choice for Sri Lankan Rebels, *Reuters*, August 20, 2006.

Marsh, Bill, The Ever-Changing Iraq Insurgency, *New York Times*, January 7, 2007.

Moss, Michael, and Mekhennet, Souad, *New York Times*, October 15, 2007; http://www.nytimes.com/2007/10/15/us/15net.html

Nasr, Vali, *The Shia Revival: How Conflicts within Islam Will Shape the Future*. New York: W. W. Norton & Company, 2006.

O'Hanlon, Michael E., and Campbell, Jason H., *The Iraq Index*, The Brookings Institution, http://www.brookings.edu/iraqindex, January 9, 2004, June 21, 2004, January 19, 2005, August 29, 2005, January 30, 2006, and June 29, 2006, August 31, 2006, January 29, 2007, May 31, 2007, June 28, 2007, September 27, 2007, November 29, 2007, December 21, 2007, January 31, 2008, February 28, 2008, and March 31, 2008.

O'Hanlon, Michael E., and Pollack, Kenneth M., A War We Just Might Win, *New York Times*, July 30, 2007.

Operation Iraqi Freedom—Official Website of the Multinational Force in Iraq: http://www.mnf-iraq.com/

Palmer, James, Displaced Iraqis contend with a grim existence, *San Francisco Chronicle*, October 17, 2007.

Paz, Reuven, Arab Volunteers Killed in Iraq: An Analysis, The GLORIA Center, PRISM Occasional Papers, Volume 3, No. 1, March 2005. http://www.e-prism.org/images/PRISM_no_1_vol_3_-_Arabs_killed_in_Iraq.pdf

Pollack, Kenneth M., *The Persian Puzzle: The Conflict Between Iran And America*. New York: Random House, 2004.

Quinn, Patrick, US: Hezbollah training Iraqi Shiite extremists in Iran, *The Associated Press*, May 5, 2008.

Report of the 9/11 Commission: Final Report of the National Commission on Terrorist Attacks Upon the United States, (Washington, DC: U.S. Government Printing Office, 2004; http://www.9-11commission.gov/report/index.htm

Rosen, Nir, *In the Belly of the Green Bird: The Triumph of the Martyrs in Iraq*. New York: Free Press, 2006.

Schwartz, Yaron, and Falk, Ophir, Chemical–Biological–Radiological–Nuclear Terrorism: Assessing the Threat, The Institute for Counter-Terrorism (ICT) Herzliyah, March 15, 2003.

Schweitzer, Yoram, and Ferver, Sari G., Al-Qaeda and the Internationalization of Suicide Terrorism, The Jaffe Center for Strategic Studies, Memorandum No. 78, November 2005; http://www.tau.ac.il/jcss/memoranda/memo78.pdf

Senor, Dan, and Martinez, Roman, Whatever Happened to Moqtada? *Wall Street Journal*, March 20, 2008.

The SITE Institute website: http://www.siteinstitute.org/

Suicide Bombing Terrorism During the Current Israeli–Palestinian Confrontation, Intelligence & Terrorism Information Center, January 1, 2006; http://www.intelligence.org.il/eng/eng_n/pdf/suicide_terrorism_ae.pdf

Tavernise, Sabrina, Iraqi Death Toll Exceeded 34,000 in '06, U.N. Says, *New York Times*, January 17, 2007.

Tending to Sri Lanka, *Washington Times*, August 20, 2006.

The Terrorism Knowledge Base (http://www.tkb.org/) of the Memorial Institute for the Prevention of Terrorism (http://www.mipt.org/).

The U.S. Department of State, list of Foreign Terrorist Organizations (FTOs) available at http://www.state.gov/

The United States Institute of Peace, The Iraq Study Group Report, December 2006; http://www.usip.org/isg/iraq_study_group_report/report/1206/iraq_study_group_report.pdf

The UN Assistance Mission for Iraq website: http://www.uniraq.org/default.asp

UN urges help for Iraqi refugees, BBC News, April 17, 2007; available at http://news.bbc.co.uk/2/hi/middle_east/6562601.stm

Wright, Robin, Since 2001, a Dramatic Increase in Suicide Bombings, *Washington Post*, April 18, 2008.

Zavis, Alexandra, Daughters of Iraq: Women Take on a Security Role, *Los Angeles Times*, June 4, 2008.

4

THE INTERNATIONALIZATION OF SUICIDE TERRORISM

Ophir Falk and Hadas Kroitoru

Suicide Terrorism did not make its international debut on September 11, 2001. The carnage, destruction, and images of people jumping to their death from the World Trade Center's skyscrapers were, however, unprecedented. The fact that 19 unarmed terrorists could turn four different civilian airliners into missiles and themselves into human bombs elucidated a shift in the terrorism paradigm. In terms of assessing the capabilities and intentions of potential perpetrators, there is now little room for speculation as to whether terrorists can consciously set out to kill thousands of innocent people. To a shocked and horrified public, evil seemed limitless in the wake of the attacks.

A once far-away phenomenon struck home for Americans on that now infamous date. For the British, that date was July 7, 2005, as four suicide bombers attacked London's public transportation system, the first act of suicide terrorism in a country with decades of experience combating local terrorist groups. Bali, Indonesia felt the devastating effects of large-scale suicide terrorism in October 2002 and again in October 2005 as al-Qaeda's affiliate group Jemaah Islamiyah attacked night clubs and restaurants in the tourist hub. Egypt joined the growing list of countries that have experienced suicide terrorism with suicide attacks in April 2005 and in July 2005. Kenya, Tanzania, Morocco, Saudi Arabia, Pakistan,

Suicide Terror: Understanding and Confronting the Threat. Edited by Falk and Morgenstern

Russia, Iraq, Spain, Israel, Turkey, and Tunisia have also established their place on the list, all of which have been victim to suicide attacks perpetrated by international terrorist organization al-Qaeda, its affiliate groups, or local networks.

The strategic tactic of suicide terrorism has itself spread along with the organization's Jihadist ideology, resulting in an apparent "internationalization of martyrdom." As an organization with global reach, al-Qaeda's adoption of suicide terrorism in its attacks around the world has helped spread the tactic to regions previously untouched by suicide terrorism.

Yet, while widely perpetrated and dispersed by al-Qaeda, the suicide tactic does not have its origins with the global network, nor is it limited to al-Qaeda activities alone. In fact, the suicide attack tactic has been adopted by over 30 different organizations around the world, with targets in more than 30 countries and four different continents.[1]

In the 25 years that have elapsed since one of the first acts of modern suicide terrorism—the 1983 Hezbollah attacks on the U.S. Embassy in Lebanon—suicide terrorism has spread to become an international phenomenon. What was once a local tactic, limited to a certain geographic sphere and on-ground conflicts, has spread internationally through the open borders and flow of ideas that characterize globalization.

The process of globalization entails a certain level of widened—and in many cases, unprecedented—access to communications, transportation, and information. Such a process provides advantages to society, and terrorist groups are no exception. Advanced communication technologies allow terrorist organizations to become linked, to draw inspiration from each other, and to adopt the tools and tactics of the other.[2]

As suicide terrorism has spread, groups in different geographical areas have employed various tactics to further develop the classic model of a suicide attack. Hezbollah focused on the use of explosive-laden cars and trucks, a tactic made famous by its double suicide attack of the U.S. Marine Base and French forces in Beirut in October 1983. Using explosive-laden trucks and simultaneous attacks subsequently became an al-Qaeda hallmark style. Sri Lanka's Liberation Tigers of Tamil Eelam (LTTE) redefined suicide terrorism in the maritime arena, introducing an exploding boat steered by a suicide bomber and used for naval attacks. LTTE also introduced the explosive belt, which was adopted heavily by Palestinian groups, as well as the Chechens and the PKK.[3]

The organization that initiated the tactic, Hezbollah, physically spread its use by conducting an "extraterritorial attack," leaving its base in Lebanon to launch a suicide attack in Argentina in 1992. Terrorist organizations also draw inspiration from the successful tools and tactics of their counterparts. They adopt the suicide tactic for their own purposes and justify its use through their own causes and ideologies. The spread of ideas through propaganda, the mass media, and the internet encourage such adoption and imitation. The ability to physically cross borders, receive external training, and, finally, cooperate at some level with outside terrorist groups are also factors of this growing phenomenon. The tactic

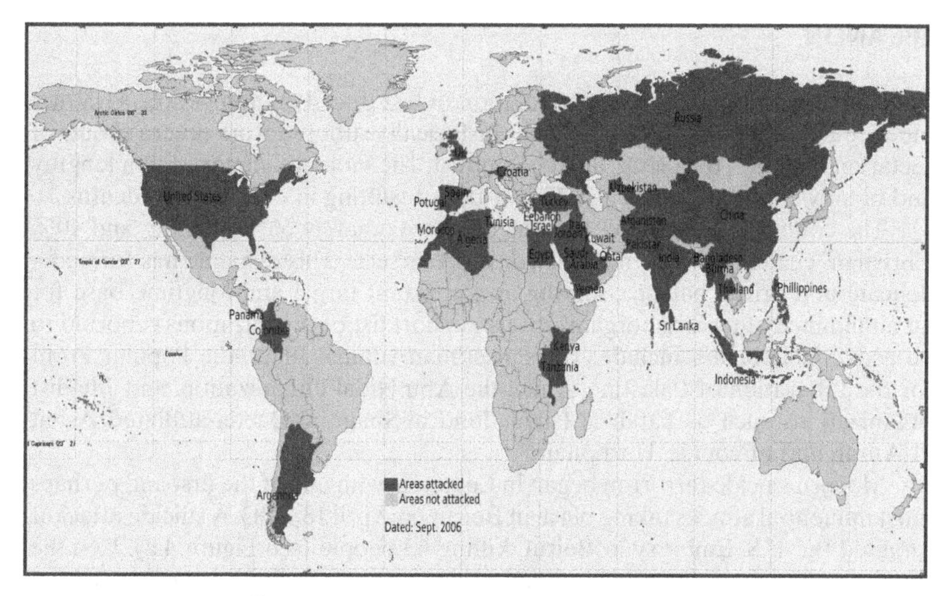

Figure 4.1. Suicide attacks around the world.

of suicide terrorism has gradually spiraled to become a force of its own, carrying with it unique meaning and advantages to the groups that employ it.

In the 1980s, suicide terrorism was witnessed in Lebanon, Kuwait, and Sri Lanka. In the 1990s, it spread to Israel, India, Panama, Algeria, Pakistan, Argentina, Croatia, Turkey, Tanzania, and Kenya.[4] As the 21st century commenced, suicide terrorism was practiced in Russia, Chechnya, Afghanistan, Iraq, and the United States (see Figure 4.1).[5] Many groups, like al-Qaeda, have extended the international reach of suicide terrorism by conducting attacks outside of their base countries or areas of conflict. These "extraterritorial attacks" include Hezbollah's attacks in Argentina, the LTTE attack in India, Egyptian militants in Croatia, and Pakistani acts in Egypt.[6]

This chapter will discuss how suicide terrorism is an international phenomenon with far-reaching implications. By exploring infamous incidents of suicide terrorism around the world—in addition to the organizations that perpetrated them and the conflicts that framed them—the chapter will present context and key lessons for law enforcement practitioners, providing a comprehensive review of the adoption and evolution of the suicide tactic. While other groups worldwide have perpetrated suicide attacks, the campaigns of the six aforementioned organizations—Hezbollah, the LTTE, Palestinian organizations, the PKK, al-Qaeda, and the Chechens—account for more than 80% of worldwide suicide attacks.[7] As such, an analysis of these most adversely affected regions and most influential organizations appears chronologically within this chapter.

LEBANON

Lebanon is a small state with a population of less than 4 million people. Granted independence in 1943, the country has historically suffered from severe spouts of sectarian violence, terrorism, and civil unrest. The violence climaxed to a lengthy and bloody civil war between 1975 and 1990, resulting in over 150,000 deaths.

The country's population consists of approximately 60% Muslim[8] and 40% Christian[9] citizens with 17 recognized religious sects. This dynamic has formed a delicate and brittle balance, serving as a constant target and longtime base for an abundance of terrorist organizations. A short list of organizations reported to be active in Lebanon include: The Palestinian Islamic Jihad, the Popular Front for the Liberation of Palestine, Amal, the Abu Nidal Organization, and Jihadist organizations such as Fatah al-Islam, Jund al-Sham, al-Qaeda-affiliated Asbat al-Ansar, and of-course, Hezbollah.[10]

Modern suicide terrorism began in Lebanon, with one of the first and perhaps most influential attacks taking place in Beirut on April 18, 1983. A suicide attacker targeted the U.S. Embassy in Beirut, killing 63 people (see Figure 4.2).[11] At the time considered a rather small organization, a group of Shiite terrorists funded by Iran named Hezbollah (The Party of God) perpetrated the attacks, consequently elevating themselves into international terrorism notoriety.[12] Another attack followed soon after, this time targeting the U.S. Marine headquarters in addition to the French Army barracks in Beirut (see Figure 4.3). As a result of the nearly simultaneous attacks, U.S., British, Italian, and French Peacekeeping forces all withdrew from Lebanon.

The success—from Hezbollah's point of view—of triggering a withdrawal of "foreigners" from Lebanon was that it sowed the seeds for other suicide attacks in the country. The small nation experienced approximately 50 suicide attacks between 1983 and 1999.[13] Shiite organizations—Hezbollah and Amal—were responsible for about half of the attacks. The other half can be attributed to five additional groups of nonreligious, nationalist-oriented ideology.[14] In fact, there were four pro-Syrian, Lebanese, and Syrian political parties engaged in suicide terrorism in the 1980s, responsible for about 25 suicide attacks in Lebanon, although figures are difficult to verify. These groups, which are not active today, include the Natzersit Socialist Party of Syria, the Syrian Nationalist Party, the Lebanese Communist Party, and the Baath Party of Lebanon.[15] It is interesting to note that aside from Hezbollah, these groups were primarily motivated by national and ideological aspirations and were not always directly linked to Islamic militancy.[16]

The suicide terrorism tactic first introduced in Lebanon would soon be adopted by militant Muslims in nearby regions, eventually spreading to almost every corner of the world.

Hezbollah

Of all the terrorist organizations based in or targeting Lebanon, Hezbollah has represented the greatest danger. Hezbollah was formed in 1982 as an umbrella

CASE STUDY

U.S. EMBASSY BOMBING IN BEIRUT

Date of Incident: April 18, 1983
Location: Beirut, Lebanon
Target: U.S. Embassy in Beirut
Type of Incident: Suicide Attack
No. of Attackers: 1

Organization: Hezbollah or a Shiite group affiliated with it

Description: On April 18, 1983, a reconnaissance vehicle was deployed to observe the physical security of the U.S. Embassy. Subsequently, the vehicle drove a few blocks and flashed its lights to a truck awaiting the signal. As the truck, laden with explosives, sped toward its destination, embassy staff—including the entire intelligence division—had no idea what was about to happen. Within minutes, 63 people were dead and hundreds more were injured as the truck rammed into and exploded at the U.S. Embassy compound.

Consequence: 63 people killed and hundreds injured

Analysis: The success of this strike encouraged enemies of the U.S. peace mission in Lebanon to carry out additional attacks, including more fatal strikes six months later that eventually led to the withdrawal of U.S. and other peacekeeping forces from Lebanon. This attack also served as an inspiration and a catalyst for terrorist organizations around the world, making Hezbollah a pioneer in the use of the suicide attack. The perpetrators of this attack managed to obtain 2,000 pounds of explosives, triggering an explosion as the truck rammed into a strategic target in the heart of Lebanon's capital. The attackers showed ingenuity in gathering information on the target and coordinating the attack between all the relevant participants. The ability to recruit an attacker willing to sacrifice his life was not a trivial task at the time and eventually led to a watershed of similar attacks by various organizations.

Figure 4.2. Case Study: U.S. Embassy bombing in Beirut.

organization for various radical Shiite groups. It was established as an organizational body for Shiite fundamentalists, led by religious clerics who saw, in the adoption of revolutionary Iranian religious doctrine, a solution to the Lebanese political malaise.[17] Hezbollah seeks to create a Muslim fundamentalist state using Iran as a model, and it strongly opposes Israel. In the latter part of 1982, Iran sent fighters from its "Iranian Revolutionary Guards" to assist in the establishment of a revolutionary Islamic movement in Lebanon with the intention of participating in the "Jihad" against Israel.[18]

Hezbollah plays a significant role in the Lebanese political scene, and it provides social services for thousands in the general population. It also operates the al-Manar satellite television channel, which has been designated by the

CASE STUDY

U.S. MARINE HEADQUARTERS AND FRENCH BARRACKS

Date of Incident: October 23, 1983
Location: Beirut, Lebanon
Targets: U.S. Marine Headquarters and French Army Barracks
Type of Incident: Two almost simultaneous suicide truck bombing attacks
No. of Attackers: 2
Organization: Hezbollah

Description: At around 6:20 A.M. on October 23, a large yellow delivery truck approached the U.S. Marines headquarters in southern Beirut, near the Beirut International Airport. The truck stormed through a barbed-wire fence and crashed through the gate and into the lobby of the building. The suicide bomber detonated his explosives as the truck was already inside the building's entryway, destroying the four-story building. About 20 seconds later, an almost identical-style attack was launched against the barracks of the French Parachutist unit, part of a multinational peacekeeping force in Lebanon. A suicide bomber drove a truck down a ramp into the building's underground parking garage, detonating the explosives and destroying the six-story building. The attacks used an unprecedented 12,000 pounds of explosives.

Consequence: 242 Americans were killed and there were multiple injuries in the first attack. In the second attack, 58 French were killed and 15 injured. Aside from the heavy loss of life and damage to buildings, these attacks served as perhaps the most symbolic in modern history in terms of international impact, having triggered the U.S. withdrawal from Lebanon.

Analysis: The attacks served as a catalyst for not only the American but also the British, French, and Italian peacekeepers' withdrawal from Lebanon. The attack illustrated the vulnerability of military peace missions to terrorism. The attackers showed ingenuity in gathering information on the target and in coordinating the attack between all the relevant participants. This was one of the first simultaneous terrorist attacks in general and the first simultaneous and double suicide terrorism attack, serving as an example to outside groups.

Figure 4.3. Case Study: U.S. Marine headquarters and French barracks.

United States as a terrorist entity.[19] The station has also been banned by France, Spain, Australia, and the Netherlands.

Sheikh Muhammed Hussein Fadlallah was the formal spiritual leader of Hezbollah in Lebanon; however, the teachings of Iran's supreme spiritual leader, Ali Khamene'i, have the leading standing and influence.[20] The political leader and most influential figure in Hezbollah is secretary general Sheikh Hassan Nasrallah.[21]

Hezbollah is a hierarchical organization with three main operational levels. The first level, comprised of seven members called the Majilis al-Shura, acts as the supreme decision-making authority in regard to all ideological, political, and military matters. The second is the executive council, responsible for daily operations and decisions. The third is the supervisory committee, composed of 15 clergymen responsible for propaganda and support for Hezbollah activities.[22]

The attacks against the U.S. Embassy and Marine Barracks in Beirut were masterminded by Imad Mughniya,[23] the founder of Hezbollah's operational branch, which has grown to become one of the most powerful terrorist organizations in the world.[24] Mugniyah was associated with the Beirut barracks and United States Embassy bombings in 1983 as well as the kidnapping of dozens of foreigners in Lebanon in the 1980s. He was indicted in Argentina for his role in the 1992 Israeli Embassy attack in Buenos Aires. He is thought to have killed more Americans than any other militant before the 9/11 attacks, and the bombings and kidnappings he is alleged to have organized are credited with all but eliminating American presence in Lebanon in the 1980s.[25] In light of these attacks and his activities, Mughniya was placed on top of the FBI's most wanted list.

Mughniya was killed by a car bomb in Damascus on February 13, 2008. Speculations have since grown that Hezbollah or its Iranian patron will carry out a major attack in retaliation (see Figure 4.4).[26]

Hezbollah's International Reach

Hezbollah is generally considered a regional terrorist organization that concentrates its efforts in Lebanon and the vicinity. However, Hezbollah's involvement in international terrorism has actually been vast and far-reaching. After the 1983 suicide attacks and subsequent withdrawal of Western troops, Hezbollah began kidnapping Westerners from various countries for political and economic gain. Among the countries affected were France, Germany, Great Britain, and Switzerland; also U.S. Citizens of Arab countries were not immune as hostages were taken from Saudi Arabia, Kuwait, and other countries.[27] The organization also hijacked aircraft belonging to foreign airlines, including Air-France, TWA, and the Kuwait national airline.[28] Hezbollah carried out bombings in France, Cyprus, and other countries in Europe and South America.[29]

Hezbollah established logistical infrastructure for its operations in locations around the world by smuggling weapons and explosives. Some of the locations included France, Spain, and Cyprus.[30]

In the late 1980s, the organization's direct involvement in terrorist attacks abroad diminished significantly, especially in regard to its operations on the European continent. Its decision to refrain from conducting international attacks can probably be attributed to instructions received from Iran.[31] That changed in September 1992, however, when Lebanese members of Hezbollah killed four senior opposition members of the Kurdish KDP party in a restaurant called Mikonos in Berlin. The Hezbollah members involved were subsequently captured and sentenced to extended terms in prison for their role in the attack. They revealed that they had been recruited in Germany and had reported to a senior Iranian intelligence official.[32]

ANALYSIS: POSSIBLE RAMIFICATIONS OF THE MUGHNIYAH TARGETING

Hezbollah is a presumed proxy of Iran and is also believed to be influenced to a certain degree by Syria. There are mixed opinions as to the level of Syrian influence on Hezbollah. Some view Syria as a key influencing factor, whereas others see it as a facilitating factor. Few even perceive Syria as a non-factor. One thing is almost certain: Iran has major influence on the organization.

This dynamic leads to the question of what is in the best interest of Iran and Hezbollah in terms of responding to the Mughniyah targeting on February 13, 2008. It is clear that Hezbollah and Iran would like to carry out a significant attack, both in commemoration of Mughniyah and to prove that they are not deterred. At the same time, however, they would like to prevent attacks on similar Hezbollah/Iranian targets in the future. Hezbollah therefore does not seem to be immediately interested in initiating another round of the Second Lebanese War, and it would prefer to either choose a time when it is better prepared to confront Israel or carry out a major attack that cannot be unequivocally and immediately connected to Hezbollah by the international community.

As for Iran, it too would like to retaliate, but it is not in its interest to give further justification for a preemptive strike on its reactors. Iran would thus probably prefer to postpone its retaliation until after November 2008, at which time a new administration in the United States will be taking office and might be more hesitant to attack Iran before it "learns" the issues in-depth. On the other hand, it should be a working assumption that both Hezbollah and Iran are planning and plotting a major attack that will be carried out at a time and place of their convenience. Such an attack may be planned within an earlier timeframe than that mentioned above, if other circumstances fall into place, such as the vulnerability of the preferred target, local, regional, and international developments, and internal pressure to carry out an attack.

Hezbollah will most likely prefer to leave a major mark by carrying out an attack that has never or seldom been seen before. For example, after the killing of Abbas al-Musawi in 1992, Hezbollah responded with subsequent attacks in Argentina. Although Hezbollah is considered a regional terrorism organization, it certainly has international capabilities that it will seek to prove once again with an attack in retaliation for the killing of Mughniyah.

Figure 4.4. Analysis: Possible ramifications of the Mughniyah targeting.

Hezbollah members committed two particularly destructive suicide bombings in Buenos Aires. In an attack against the Israeli embassy building in March 1992, 29 people were killed and dozens were injured. Two years later, the bombing of the AMIA Jewish community building on July 18, 1994 left 86 people dead and dozens more injured.[33] Hezbollah plotted another suicide attack against a target in Bangkok in March 1994, but it ultimately failed due to an operational mishap.[34] Hezbollah has also trained activists and a number of other terrorist organizations, including Iraqi militia in Iran, that pose a threat to U.S. and coalition forces.[35]

In spite of the attempts by Lebanon, Syria, and Iran to present Hezbollah as a legitimate resistance movement fighting for the liberation of Israeli occupied territories, Nasrallah himself declared in an interview on December 18, 2001 that "to earn victory we have to fight on all fronts. We have to be global and integral."[36]

Traces of Hezbollah's global strategy can even be found in the United States. In July 2000, U.S. federal agents arrested 18 alleged supporters of Hezbollah, suspected of participating in a ring that raised and sent funds and military equipment, including night vision equipment, global positioning devices, mine detection equipment, cellular phones, and blasting equipment, to Hezbollah in Lebanon.[37] Sweden's Sapo intelligence agency identified 15 people in Sweden who had direct links with al-Qaeda and Hezbollah terrorist organizations. They were suspected of assisting the terrorist organizations with information, communications, and financing.[38] In June 2002, Singapore accused Hezbollah of recruiting Singaporeans in a failed 1990s plot to attack U.S. and Israeli ships in the Singapore Straits.[39]

Various Islamic extremist organizations have firmly established themselves in what is known as the "tri-border area," where Argentina, Brazil, and Paraguay converge, with Hezbollah holding a strong presence.[40] Experts believe that Hezbollah earns substantial income from various illicit activities in the tri-border area, in addition to financial support from the government of Iran and income derived from narcotics trafficking in Lebanon's Al Beqa'a Valley.[41]

Apart from its international activities, Hezbollah's stronghold is in Lebanon, where it serves as an Iranian proxy and trains thousands of activists in villages located in the Beqa'a valley, in Beirut, and in southern Lebanon.

In the summer of 2006, Hezbollah was able to endure a 34-day war with Israel. Hezbollah did not anticipate Israel's retaliation for the murder and abduction of IDF soldiers on July 12, 2006. In fact, the organization regretted that its actions led to an escalation in violence and what has been termed the Second Lebanon War.[42] Still, despite these issues and the heavy losses it suffered in the confrontation, Hezbollah managed to survive, preserve its terrorist and guerilla attack capabilities, and subsequently rebuild. Israel was able to target Hezbollah, damaging infrastructure and killing hundreds of Hezbollah terrorists, but was not able to eradicate the Shiite terrorist organization. Hezbollah managed to rocket Israeli cities throughout the 34-day conflict and, despite its losses, has been able to restore its militant and political standing in Lebanon.

The Iran–Hezbollah–al-Qaeda Connection

Today, Iran is the most active state sponsor of terrorism in the world.[43] Shiite cooperation between Iran and its terrorist proxy Hezbollah dates as far back as 1983, when Hezbollah suicide bombers attacked American and French peacekeeping forces in Lebanon.[44] The cooperation has yet to cease.

The joint Shiite–Sunni venture between Iran, Hezbollah, and al-Qaeda against the West has also been evident on a number of occasions:

- In an indictment against an al-Qaeda operative responsible for the attack on U.S. embassies in Tanzania and Nairobi that resulted in hundreds of fatalities, the U.S. Justice Department alleged that bin Laden had "stated privately ... that al-Qaeda should put aside its differences with Shiite Muslim terrorist organizations, including the Government of Iran and its affiliated terrorist group Hezbollah, to cooperate against the perceived common enemy, the United States and its allies. ..." Thus, the indictment explained: "Al-Qaeda also forged alliances ... with the government of Iran and its associated terrorist group Hezbollah for the purpose of working together against their perceived common enemies in the West, particularly the United States."[45]
- The 9/11 Commission states that "senior al-Qaeda operatives and trainers traveled to Iran to receive training in explosives. In the fall of 1993, another such delegation went to the Beka'a Valley in Lebanon for further training in explosives as well as intelligence and security."[46]
- In a statement published by the U.S. Attorney General, following the al-Qaeda bombing of the Khobar Towers, it was charged that "elements of the Iranian government inspired, supported, and supervised members of the Saudi Hezbollah. In particular ... the charged defendants reported their surveillance activities to Iranian officials and were supported and directed in those activities by Iranian officials."[47]
- There are even reports that on July 26, 2002, bin Laden and his family received safe refuge from Iran as American forces began closing in on him in Afghanistan.[48]

At Iran's annual "World without Zionism" conference held in October 2005, Iranian President Mahmud Ahmadinejad told his audience, "We are in the process of a historical war between the World of Arrogance and the Islamic world, and this war has been going on for hundreds of years." He elaborated by emphasizing: "The annihilation of the Zionist regime will come. ... Israel must be wiped off the map. ... And God willing, with the force of God behind it, we shall soon experience a world without the United States and Zionism" ... "Very soon," he proclaimed, "this stain of disgrace will vanish from the center of the Islamic world—and this is attainable."[49]

Putting the risk in context, Benjamin Netenyahu has commented that:

> The risk is imminent. Unchallenged, this ideology will grow with time. If Iran develops nuclear weapons it will be the first time in history that an extreme ideology bent on destruction of 'infidels' and world domination will be armed with nuclear weapons.[50]

Hezbollah's Direct Threat to U.S. Homeland

Although the majority of Hezbollah activity has taken place in Lebanon and the organization has yet to carry out a direct attack on the U.S. homeland, there are signs indicating a change. Evidence, allegations, and related cases suggest that Hezbollah has established a sophisticated intelligence apparatus that reaches into the United States.[51] Traces of such capabilities are evident in the case of Spinelli vs. U.S. in December 2007, in which three Lebanese-born American civil servants were convicted, among other things, of fraud and illegal accessing of a federal computer system in order to obtain information about Hezbollah.[52]

Along with her associate Nada Nadim Prouty, a former FBI agent and CIA case officer, the primary suspect in this case was Samar Spinelli, a U.S. Marine captain that commited citizen and passport fraud by intending to fraudulently obtain U.S. citizenship.[53] Another associate of Spinelli's is Prouty's sister, Lebanese born Elfat El Aouar, who was sentenced to 18 months in prison. El Aouar is married to fugitive Tala Chahine, an alleged Hezbollah financial operative who is believed to have fled to Lebanon before being charged in 2006 with tax evasion in connection with a scheme to illicitaly transfer $20 million to Lebanon.[54]

This case does not prove that Hezbollah has access to sensitive U.S. intelligence information, nor does it verify Hezbollah's intentions to target the United States—but it does illustrate that Hezbollah may be increasingly active in the United States or interested in U.S. targets.

Al-Qaeda and its proxies, as well as Iran and its proxies, believe in the goal of Jihad. Although U.S. agencies have stressed the threat from al-Qaeda and associated Sunni extremist groups since the attacks of 9/11, rather than from Hezbollah and other Shiite Muslim groups, it is clear that Hezbollah is a growing danger.[55] The Hezbollah that has targeted Israel and its interests on numerous occasions may very well do the same to the United States.[56]

SRI LANKA

Sri Lanka has suffered from decades of ethnic conflict, fueled by a separatist struggle between its Sinhalese majority and Tamil minority. Ethnic tensions have erupted into violence and civil war, with militant groups employing terrorist tactics in their fight for autonomy. Among more than 30 militant nationalist groups, the Liberation Tigers of Tamil Eelam (LTTE) has fought at the forefront of the Tamil separatist movement, a terrorist organization notorious for its extensive use of suicide bombings, which once numbered more than 200 and accounted for more than half the world's suicide attacks.[57] As of 2000, the group was considered by experts to be "unequivocally the most effective and brutal terrorist organization ever to utilize suicide terrorism,"[58] boasting an "unrivaled" track record for suicide attacks.[59] In fact, the LTTE was once responsible for more suicide attacks than any other terrorist organization in the world, and until 2003 it had perpetrated more attacks than all the Palestinian terrorist groups combined.[60]

A History of Conflict

Home today to more than 20 million people, the island of Sri Lanka, formerly known as Ceylon, gained independence from British colonial rule in 1948 and has since been a sovereign nation with a democratic multiparty government. The population is represented primarily by two ethnic groups: the Sinhalese, composing about 70% of the population, and the Tamils, making up 18% and numbering about 200,000. The Sinhalese are traditionally Buddhist and live primarily in the southwestern portion of the island while the Tamils are primarily Hindu, with a small percentage of Christians. The Tamils, who arguably received preferential treatment under the British, live primarily in southern India and in the northern and eastern areas of Sri Lanka, which they refer to as "Tamil Eelam." Their religion and language, Tamil, set the minority group apart from the Sinhalese majority, who currently dominate the country and the ruling government. The left-wing Sri Lanka Freedom Party took control of the country in 1956; this sparked discontent and rioting among the Tamil minority, who viewed the party as favoring Sinhalese culture and language. In the 1970s, a new constitution directed the state to grant Buddhism a "foremost place" in the country, calling for the country to "protect and foster" it. Soon after, government agricultural projects asserted rights over traditionally Tamil lands;[61] and in 1972, legislation was enacted that confirmed Sinhalese as the national language, effectively made Buddhism the state religion, and eliminated protections for minorities, effectively shutting the Tamil population out of higher education institutions through a system of quotas.[62] Sinhalese have since dominated political, economic, and cultural life in Sri Lanka, increasing their control of the government.[63]

In response, Tamil nationalist movements became active in a campaign to fight perceived discrimination and pursue greater Tamil autonomy, through both political and militant channels. Tens of thousands of Tamil insurgents have since spent decades protesting Sinhalese rule, claiming that the government is an oppressive regime under which they suffer numerous forms of discrimination.[64] War between the two parties began in 1972, following a brutal anti-government revolution led by military mastermind Velupillai Prabhakaran, who later went on to establish a military youth movement called the Tamil New Tigers (TNT) that soon after transformed into the LTTE.[65]

The Tamil Tigers and Suicide Terrorism

Officially founded in 1976 by Velupillai Prabhakaran and other members of TNT, the LTTE, also known as the Tamil Tigers, is composed primarily of Hindus and Christian Marxists committed to the creation of an independent Tamil nation in northern and eastern Sri Lanka. The organization has been engaged in a decades-long guerilla and terrorist campaign against the Sri Lankan government, targeting Sinhalese, and, at times, Muslim civilians and security forces. Their first terrorist attack took place in September 1978 with the mid-air bombing of a passenger plane.[66] Targets have since included (a) key government and military personnel, (b) ships, and (c) economic and public infrastructure such as buses, trains, oil tanks, and power stations. By 1983, tension and isolated violence escalated into a civil war between the Sri Lankan government and the LTTE and other groups, claiming more than 60,000 lives.

SUICIDE SQUAD: THE BLACK TIGERS BRIGADE

Unique among terrorist organizations, the Black Tigers Brigade has strict requirements for its members. All members—male or female—are required to wear a vial of sodium or potassium cyanide around their neck. If they fail in their mission and are about to be captured, a member is obligated to bite down on the vial, sending the deadly poison directly through the body via the lacerations on the gums from the broken glass. Over 20 years, at least 600 Tamil Tigers have committed suicide in this manner, and only a handful have ever been captured alive (Pape, 143).

Figure 4.5. Suicide squad: The Black Tigers Brigade.

The Tamil Tigers, based in the Wanni region of Sri Lanka, have about 10,000 armed members, under half of which are trained fighters. Many are recruited primarily from the lower Hindu castes, but committed members and high-level leaders generally come from the higher-status castes and tend to be well-educated. The organization has two primary branches: a professional military wing and a political branch, with a central governing committee overseeing all activities and subdivisions. Velupillai Prabhakaran, widely depicted as a military genius, is the founder and current leader of the organization.

A unique feature of the organization is its notorious "Black Tigers" brigade, a distinct subdivision of the organization's governing committee that serves as an "elite suicide commando unit" (Figure 4.5).[67] The brigade was designed specifically to carry out suicide attacks and is itself responsible for the deaths of over 901 people and two world leaders.[68] Sri Lanka's first exposure to suicide terrorism came in July 1987, when the Black Tigers launched LTTE's first suicide attack, a truck bombing attack against a Sinhalese military camp in Vadamarachi, which killed 40 security force personnel.[69] The attack was modeled after the 1983 suicide bombing of the U.S. Marine barracks in Lebanon, from which LTTE leaders drew significant inspiration.

Suicide terrorism has since become one of the LTTE's primary methods of attack, having at one point[70] far surpassed the suicide campaigns of groups like Hezbollah or Hamas in terms of frequency and lethality.[71]

Since the first series of attacks in 1987, the organization has carried out approximately 200 suicide attacks against civilian and political targets, with political assassinations being a primary objective. Targets of LTTE attacks are military and civilian, symbolic and strategic—including civilians in mass transit facilities, Buddhist shrines, and office buildings. The group has also used suicide attacks to target its Tamil opposition.[72] The tactic of deploying suicide bombers was used in a number of assassinations, including the murder of former Indian Prime Minister Rajiv Gandhi in 1991 in which a female suicide bomber detonated a prototype suicide vest (see Table 4.1). Sri Lankan Defense Minister Ranjan Wijeratne was also a victim that same year. Sri Lankan President Ranasinghe Premadasa was the most senior Sri Lankan official assassinated through this tactic, killed in 1993.

TABLE 4.1. Causalities of Terrorist Violence in Sri Lanka 2002–2008

Year	Civilians	Security Force Personnel	Terrorists	Total
2002	14	1	0	15
2003	31	2	26	59
2004	33	7	69	109
2005	153	90	87	330
2005	153	90	87	330
2006	981	826	2319	4126
2007	525	499	3345	4369
2008[a]	307	578	5891	6776
Total	**2044**	**2003**	**11,737**	**15,784**

[a]Data until August 4, 2008.

Source: South Asia Terrorism Portal, http://satp.org/satporgtp/countries/shrilanka/database/annual_casualties.htm

One of the group's most devastating attacks occurred in October 1997, when a truck bomb exploded at the new World Trade Center in Sri Lanka's capital city of Colombo. The attack killed 18 people and injured over 100 others, causing serious damage to nearby structures.[73] While physical destruction is a clear characteristic of the attacks, the Tamil Tigers have traditionally favored symbolically significant targets. The group is the only terrorist organization to have assassinated three government leaders using suicide terrorism.[74]

Employing suicide attacks as an assassination method is not the only way in which the LTTE has proved to be a pioneer in the field. The organization pioneered the use of concealed suicide bomb vests, today utilized by numerous organizations worldwide. They have also proven to have the most effective maritime suicide attack capabilities among all their terrorist counterparts. Women also play a large role in the LTTE organization, composing about one-third of its member and participating in 30–40% of the group's suicide activities.[75] The LTTE is also known for its use of children in its terrorist campaigns, relying on brigades of boys and girls between the ages of 10 and 16.[76]

In fact, these specialties have actually come to represent several subdivisions of the Tamil Tigers overall organizational structure. In addition to the "Black Tigers," the organization is further subdivided between the "Sea Tigers" (a naval attack unit), the "Baby Tigers" (child soldier division), and the "Women's Military Units of the LTTE" (Figures 4.6 and 4.7).[77]

Consequences and Counterterrorism Efforts

Amidst sporadic informal peace talks and waves of heightened violence, effective policies to thwart the threat posed by the Tamil Tigers and resolve the ongoing conflict have yet to be fully implemented. The government had been reluctant to declare an ongoing all-out-war on the organization, but have attempted to design

VOICES OF TERROR

"As Black Tigers, they are the physical embodiment of self-determination and liberation. They employ their lives as missiles armed with the kind of determination and purpose that is unmatched by any conventional weapon that the Sinhala forces may deploy. There lies the strength and honor of our Black Tigers."
—Nandini, Female LTTE member, statement made in 1995 (Pape, 143).

Figure 4.6. Voices of terror.

both preventative and operational methods to deter the Tamil Tigers—using sometimes "brutal and arbitrary" force to squash them, with Tamil civilians also affected by counter-terrorist operations.[78] Counterterrorism tactics, especially during periods of heavy resistance against the LTTE, have come under heavy criticism from foreign organizations, the subject of several damaging human rights reports. Police action has been characterized as being brutally violent and ultimately ineffective against the Tigers suicide attacks, often escalating the violence or creating humanitarian crises.

The government established the Prevention of Terrorism Act (PTA) in 1979, which gave the army and police the right to hold prisoners incommunicado for up to 18 months without trial.[79] In effect, the police and army used the law to arrest without warrants, search and seize, and detain suspects for long periods of time without trial or communication with the outside world. The PTA appeared only to escalate the level of violence and attacks by the Tamil Tigers.

In the political and diplomatic realm, several national organizations and foreign governments have encouraged the Sri Lankan government to participate in negotiations with Tamil separatists. Peace talks were initiated after the 1987 riots, with India serving as a facilitator. The Indo-Sri Lankan peace accord of 1987 granted some autonomy to the Tamils in northern and eastern Sri Lanka, and it established the Indian Peace-Keeping Force (IPKF) to replace Sri Lankan troops. The peace accord ultimately failed, however, when radical Sinhalese groups launched a guerilla insurgency against the Sri Lankan government; a youth movement called Janatha Vimukhti Peramuna claimed that the accords destroyed Sri Lanka's territorial integrity because its actions undermined the efforts of both sides. Peacekeeping forces withdrew several years later, and violent confrontations between Sri Lankan forces and the LTTE continued.[80]

Talks recommenced in 2002, when the Tamil Tigers agreed to a cease-fire and informal peace talks with the Sri Lankan government, brokered largely by Norway. The Sri Lankan president agreed on a plan for interim self-rule, aimed at ending the longstanding conflict; but LTTE rejected the approach, leaving talks in a deadlock.[81] From 2002, however, the group significantly reduced its reliance on suicide terrorism tactic, maintaining the cease-fire outside of a few isolated attacks. The devastation of the 2004 tsunami, which left 30,000 dead, actually helped to maintain what experts referred to as a very "fragile truce."[82]

CASE STUDY

LTTE SUICIDE ASSASSINATION

Date of Incident: March 21, 1991
Location: Sriperumbudur, southern Indian state of Tamil Nadu, 30 miles from the capital of Madras
Type of Incident: Assassination by suicide attack
No. of Attackers: 1 female suicide bomber
Organization: LTTE, Black Tigers Brigade

Motivation for Attack: According to the Supreme Court judgment, the attack was carried out specifically in order to assassinate Prime Minister Rajiv Gandhi, due to personal animosity toward him by LTTE leader Prabhakaran.

Description: Indian Prime minister Rajiv Gandhi was killed by a female suicide bomber on March 21, 1991 at a campaign rally in the town of Sriperumbudur, located in the southern Indian state of Tamil Nadu, 30 miles from the capital of Madras. Forty-six-year-old Gandhi was greeting the large crowd in a reception area when the young female suicide bomber approached Gandhi, wearing a custom-made explosives belt hidden under her sari.[1] The bomber, Thenmuli Rajaratnam, also known as Gayatri Dhanu, detonated the explosives strapped to her body as she bowed down to greet him in the reception area. The blast killed Gandhi, the bomber, and 14 other people.[1]

Consequence: 16 total deaths, including Prime Minister Rajiv Gandhi—the intended target of the suicide assassination attack.

Analysis: The incident demonstrated a number of unique characteristics of the LTTE, including their use of female suicide bombers, their introduction and reliance on the suicide explosive vest, and the use of suicide tactic for the purpose of assassination. It was also their first and only attack outside of Sri Lanka, showing the potential international reach of their activities. The attack was one of the first examples of the use of suicide terrorism for assassination purposes, setting a new precedent that continued to characterize LTTE activities in the years that followed.

Incident Timeline

March 21, 1991

8:00 P.M.: Prime Minister Rajiv Gandhi arrives at Chennai and is driven by motorcade to Sriperumbudur, stopping at a number of other campaign venues along the way. He is interviewed by a foreign journalist while riding in the car.

10:00 P.M.: Prime Minister Gandhi arrives at the reception area and steps onto the red carpet, greeting a waiting crowd.

10:10 P.M.: The female suicide bomber approaches Gandhi to welcome him, bowing before him, a sign of respect. With her right hand, she detonates the explosive charge for the bomb she is carrying, hidden under her sari.

Figure 4.7. Case Study—LTTE suicide assassination.

The reduction in conflicts between LTTE members and government forces lasted only until riots broke out in April 2006, with explosions killing at least 16 people. The LTTE attacked a Naval convoy shortly after, regarded as the "most blatant" violation of the 2002 peace agreement. Peace talks ceased entirely when an LTTE suicide attack on a Sri Lankan naval convoy killed 93 and wounded 150 in October 2006. Attempted talks continued to arise, and continued to fail, leading to more clashes between the LTTE and Sri Lankan forces. Anti-terror measures were enacted the following December, giving security forces more power to search, arrest, and question suspects.[83] As of mid-2008, there were no signs of a cease-fire, with a renewed civil war seeming almost inevitable.[84]

The Tamil Tigers' capabilities are almost unrivaled and must be studied. However, the 1991 assassination of Gandhi is considered the only terrorist attack carried out by LTTE outside the boarders of Sri Lanka which may infer a low risk in terms of the threat to the U.S. homeland.

The LTTE is an example of an organization that has been a pioneer in suicide terrorism tactics. The organization used suicide terrorism because of its operational effectiveness as opposed to adhering to a religious call. The members of the organization were motivated by national intents and are primarily Hindu in faith, in contrast to most Islamic terrorist organizations that have adopted suicide terrorism.

Perhaps most importantly, from an American law enforcement perspective, the Tamil Tigers are predominantly a local organization that rarely carries out attacks outside Sri Lankan borders.

ISRAEL

Since its foundation in 1948, and even before, the people of Israel have suffered from years of war, conflict, and terrorism. Rooted in a difficult dispute over self-determination and territory, the Arab–Israeli conflict has been part of ongoing tension, conflict, and war in the region. In the context of the conflict and with the growing involvement of militant Islamic groups, terrorism has become a common tactic of organizations such as Hamas, Fatah, Hezbollah, and the Palestinian Islamic Jihad (PIJ), who have employed bombings, hijackings, armed attacks, and suicide terrorism against Israeli civilian, military, and government targets.

Palestinians have used the term *Intifada* when referring to their violent rebellions against what they perceive as Israeli "occupation" of Palestinian territory. It was during the first Intifada, from 1987 to 1993, that a Palestinian organization first initiated a suicide terrorist attack against an Israeli target, launched in April 1993.[85] Between 1993 and 2008, 188 successful attacks were launched against Israel targets—at shopping malls, on buses, at street corners, and in other densely populated places. In that time period, more than 200 suicide attacks were thwarted. Six decades after Israel's establishment, the country continues to be plagued by terror in general and, for the past 15 years, from suicide terrorism in particular.

Historical Background

After years of a Zionist struggle to establish an independent Jewish state in Israel, the UN accepted a resolution on November 29, 1947 to divide the region into two states: one Arab and one Jewish. While Israel compromised and accepted the partition plan, the Palestinian leadership and the Arab League, comprised of the surrounding Arab countries, rejected it. After Israel declared its independence on May 14, 1948, neighboring Arab nations invaded, leading to Israel's War of Independence or the "Nakba" (Disaster) as many Palestinians refer to it. After a hard fought war, Israel gained its independence.

The Palestinian Liberation Organization (PLO) was founded in 1964 with the goal of "waging the battle of liberation" for the Palestinian people.[86] The organization began its campaign of terror against Israel in 1965, launching its first terrorist attack against Israel's national water carrier.

After the 1967 Six-Day War, Yasser Arafat was appointed chairman of the PLO. Following his appointment in 1969, Israel faced a multitude of hijackings and hostage situations throughout the 1970s. The 1980s were characterized by Katusha rockets launched from Lebanon and the first Intifada in 1987. In 1993, suicide terrorism was introduced to Israeli soil and has since been the most lethal form of terrorist attack, accounting for more than 1500 killed civilians in one of the smallest countries in the world.[87]

Israeli refusal to recognize the legitimacy of the PLO since its founding in 1964 was rooted in the organization's terrorist activities and its decades-long obstinacy in recognizing Israel and its right to exist. Israel's denunciation of the Palestinians' national rights was summed up by Israel's late Prime Minister Golda Meir who remarked in 1974 that "there is no such thing as a Palestinian People."[88]

More war and conflict followed the October 1973 Yom Kippur War, but diplomatic attempts to settle the conflict reached an historic peak in the early 1990s. On Monday, Sept. 13, 1993, the Declaration of Principles (DOP) was signed on the south lawn of the White House, sealed with a historic handshake between the late Israeli Prime Minister Yitzhak Rabin, and the Palestine Liberation Organization (PLO) Chairman Yasser Arafat. Rabin took the opportunity to note that the agreement was the first between the Palestinians and Israel since the founding of the State of Israel, and marked "a historic moment which hopefully will bring about an end to 100 years of bloodshed and misery between the Palestinians and Jews, between Palestinians and Israel."[89]

Reconciliation was based on a historic compromise. The principles of partition and mutual acceptance were officially acknowledged as the basis for a settlement in a long and bitter conflict.[90] In the ostensible acceptance of the principle of partition of land and mutual legitimacy, an attempt was made to find a solution to the problem of sharing the cramped living space between the Jordan River and the Mediterranean Sea. On that day in September 1993, many believed mutual denial had made way for mutual recognition and that by implementing practical negotiations, an end to the conflict and senseless bloodshed would be achieved. Unfortunately, however, the conflict has continued.

In the context of the Oslo agreement, the Israeli military Israel Defense Forces (IDF) pulled its forces out of populated Palestinian centers and other parts of the territories. In 1994, the Palestinian Authority (PA) was created to control the area instead, set to govern the Palestinian population and secure the area from terrorism.

Following the signing of the Oslo Accords on September 13, 1993, however, Palestinian terrorist groups—particularly Hamas, Fatah and the Palestinian Islamic Jihad (PIJ)—unleashed an unprecedented wave of suicide attacks against Israeli civilians. The suicide terrorism campaign had two primary peaks: from February to March 1996, and in March 2002.[91]

A Decade and a Half of Suicide Terrorism

Between the years 1993 and 2008, almost 400 suicide attacks were launched against Israeli targets, with 188 of them achieving success. More than 80% these attacks occurred between the years 2000 and 2005. Had Israeli security forces failed to thwart at least 200 similar attacks during that time span, the resulting the causality figures would have been even more dramatic.[92]

Suicide attacks represented only about 1% of all terrorist attacks against Israel during that time frame. Yet, in terms of effectiveness, suicide attacks were by far the most lethal tactic, responsible for about 49% of fatalities and 59% of injuries resulting from terrorism.[93]

As demonstrated by Figure 4.8, Israel has experienced a number of suicide terror waves, with the peak reaching 60 fatal attacks in 2002. In terms of counterterrorism, Israel was relatively successful in mitigating the suicide terrorism phenomenon after transforming an ineffective policy of deterrence and appeasement to one of prevention and preemption. This rejuvenated counterterrorism policy was highlighted by a number of specific measures, including the

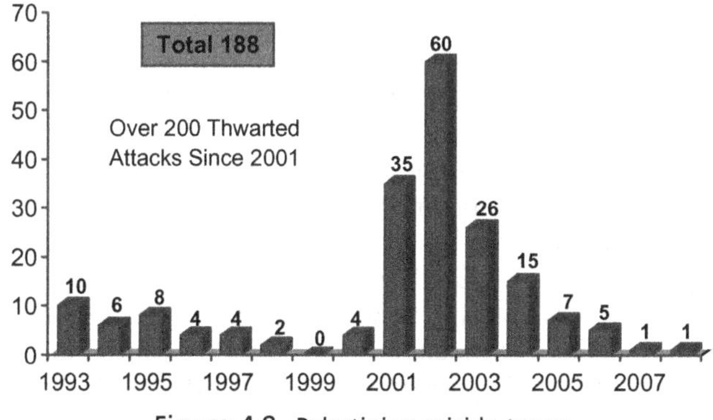

Figure 4.8. Palestinian suicide terror.

establishment of a security fence that serves as a physical barrier, or at least hindrance, to potential perpetrators of attacks. More importantly, the new policy included a concentrated campaign of targeted killing, focusing on terrorist leaders and operatives prior to their strike.[94]

Misconceptions

The premise of Israel's willingness to sign the Oslo Accords was that it would enhance Israel's security. The working assumption was that (a) the political process and land concessions would decrease the motivation for terror; and (b) if terrorism would not vanish, the PA would be more effective than Israel in combating Palestinian terrorism because their confrontation would not be "constrained by Bagatz or B'tselem."[95]

The premise may have turned out to be a misconception. Land concessions arguably furthered the quest for additional yields and may have enhanced the motivation for acts of terror. Terrorist actions were perceived by Palestinian terrorist organizations as a catalyst for concessions rather than an obstacle. Concurrently, not only did the PA usually refrain from combating terrorism, they often covertly instigated it.[96] See Figures 4.9 and 4.10.

From Deterrence to Prevention and Preemption

Israel's 60-year-old security doctrine, originally developed by Israel's first Prime Minister, and Defense Minister, David Ben-Gurion, did not considerably change in principle until the suicide terrorism phenomenon hit Israel. At that point, the doctrine shifted its focus from deterrence (and at times appeasement) to prevention and preemption. Since it is very difficult to deter suicide terrorism, and because Israel's early warning capabilities were significantly undermined after the IDF withdrew from large cities in the territories, Israel needed to put more weight on prevention (the security fence and other physical measures) and preemption (targeted killing).

There are those that argue that Israel's counterterrorism tactics have bred more Palestinian terrorism. For example, Mia Bloom argues that some Israeli counterterrorism tactics may have motivated Palestinian factions to increase terrorism and boosted public support for acts of terror, encouraging rather than deterring future attacks and leading to a "dead-locked battle of assassination-suicide bombing, assassination—suicide bombing ..."[97]

However, most security and counterterrorism experts believe that Israel's counterterrorism measures were most effective when they applied significant force against terrorist strongholds. A good example of such a counter terrorism campaign began on March 29, 2002 in response to an unprecedented wave of Palestinian suicide attacks, referred to as *Operation Defensive Shield.*

The operation included an incursion into the PA's capital of Ramallah, followed by raids into six of the largest cities in the West Bank and their surrounding towns, villages, and refugee camps. The Israeli Defense Forces imposed strict

CASE STUDY

BEIT LID JUNCTURE

Date of Incident: January 22, 1995
Location: Beit Lid Juncture (25 kilometers North of Tel Aviv)
Type of Incident: Modular Suicide Attacks
No. of Attackers: 2
Organization: Islamic Jihad

Capabilities: This incident illustrated the Islamic Jihad's capability to pick a prime target and gather information on it. More significantly, it showed ingenuity of carrying out a modular attack in which respondents are targeted after they come to provide assistance to the attacked target.

Description: On January 22, 1995, at the Beit Lid juncture, 25 kilometers north of Tel Aviv, a double suicide attack was carried out by Islamic Jihad terrorists, serving as a turning point in Israel's perspective of suicide terrorism. Although this was not the first suicide terror attack in Israel, it was the first to use two bombers in the same incident, exploding their charges one after the other. The first bomber targeted a crowded area at the juncture while the second explosion was meant to maximize damage once response and rescue personnel arrived on the scene. As a result, 21 people were killed and 69 wounded.

Consequence: 21 killed paratroopers and 69 injured

Analysis: The significance of this incident was that it was the first incident in Israel to use two bombers in the same incident, exploding their charges almost simultaneously. The first bomber targeted a crowded area at the juncture while the second explosion was meant to maximize damage once rescue personnel arrived on the scene. This incident was copied by other organizations in subsequent attacks and affected first respondent procedures. Following the attack, Israel's Prime Minister and Defense Minister at the time, Yitzhak Rabin, said, in closed quarters, that terror had become a *strategic* threat to Israel and threatens its existence.[1] In the mid-1980s, Benjamin Netanyahu, then a diplomat, dubbed terrorism as a strategic threat not only to Israel but also to the entire free world.[2]

Sources:

1. An interview carried out by the author with former head of Israel's Security Agency, Mr. Karmi Gilon, April 28, 1996.
2. Benjamin Netanyahu, *Terrorism—How the West Can Win*, The Jonathan Institute, 1986.

Figure 4.9. Case Study—Beit Lid Juncture.

curfews on civilian populations and restricted movement in an effort to isolate and uproot terrorist cells. The underlining objective of the operation was to regain security control, for a limited period of time, over regions that were transferred to PA control and to "cause them losses, victims, so that they feel a heavy

CASE STUDY

PASSOVER MASSACRE—PARK HOTEL IN NETANYA

Date of Incident: March 27, 2002
Location: Park Hotel in Netanya
Type of Incident: Suicide attacks
No. of Attackers: 1
Organization: Hamas

Capabilities: This incident illustrated Hamas's capability to pick a prime target, the most fitting time to attack it, and the ability to carry out a covert attack by having the attacker disguised as a woman.

Description: Disguised as a women, Abdel-Basset Odeh, a member of the Hamas Iz a Din al-Kassam Brigades from the West Bank city of Tulkarem, walked into the dining room of the Park Hotel in Netanya and detonated an explosive device in the midst of 250 civilians celebrating the Passover Seder. Thirty people were killed and 140 injured—20 seriously—in that March 2002 suicide bombing, commonly referred to as the "Passover Massacre."

Consequence: 30 killed civilians and 140 injured

Analysis: After a series of Palestinian suicide attacks, this attack served as the one that "broke the camel's back." It marked the peak of an unprecedented wave of Palestinian suicide terrorism. In the wake of the attack, Israeli Prime Minister Ariel Sharon and his cabinet ordered the immediate recruitment of 20,000 reservists in an emergency call-up and the following day launched Operation Defensive Shield, which managed to mitigate suicide terrorism in Israel considerably. Israel launched an attack on all terrorist strongholds in the West Bank in an effort to uproot Palestinian terrorist infrastructure. Prior to the Passover Massacre and based on the Oslo Accords between Israel and the Palestinian Authority signed on September 13, 1993, a significant part of Israel's counterterrorism security was entrusted in the hands of the Palestinian Authority's willingness and capability to confront terrorists in the West Bank and Gaza strip. The Passover Massacre changed that notion and compelled Israel to rejuvenate its counterterrorism policy.

Figure 4.10. Case Study—Passover Massacre.

price."[98] The operation resulted in more than 4250 suspects detained, hundreds of suspected terrorists killed, thousands wounded, and a sharp decline in Palestinian suicide terrorism.[99]

When the Prime Minister of Israel, Mr. Benjamin Netanyahu, was asked what the guiding policy for counterterrorism should be, he replied as follows:

> The policy should be based on an un-proportional response to terrorism. For example, in 1999, after Hezbollah launched Katusha rockets on Northern Israel, we responded with a massive bombing of key infrastructure in Lebanon, causing millions of dollars in damage. The result was a long and quit period for Northern Israel.

> The key in deterrence and prevention is that the response or perceived response to a terrorist attack will be un-proportional to the attack itself.[100]

According to the numbers, suicide terrorism did decline in Israel from its peak in 2002 (60 successful attacks) to one successful attack in 2007 and again in 2008, illustrating that Israeli counterterrorism efforts were successful in preventing suicide attacks.

Relevance to U.S. Homeland Security

Israel's counterterrorism policy and its political process with its neighbors are also of interest to the United States and its homeland security department. In addition to the fact that the United States and Israel share similar morals and doctrines concerning democracy and freedom, Israel faces counterterrorism challenges that often serve as previews to those of the United States domestically and in regard to U.S. targets abroad.

Aside from the conventional wars fought by Israel in 1948, 1956, 1967, 1973, 1982 and in the summer of 2006, Israel has continuously confronted the challenges of terrorism. Throughout the early 1950s, Israel was attacked by insurgents from Jordan, Syria, and Gaza, who had infiltrated porous borders, referred to as "Fedayun" ("suicide squads").[101] Some of these infiltrators suffered from economic hardship and were actually often looking for a job rather than an attack. Such infiltrations may bear a resemblance to the illegal border crossings between Mexico and the United States. They are usually motivated by economic considerations but the problem also highlights the void in homeland security.

The main lessons that can be learned from the Israeli experience with terrorism evolve around securing homeland borders and taking the fight to the streets and homes of terrorist operatives before they reach the home front. Although the United States has taken many precautions and invested a lot of funding into counterterrorism measures since 9/11, United States borders remain vulnerable. The long and peaceful borders between the United States and Canada in the north and the United States and Mexico in the south can be infiltrated, as can the hundreds of marine ports that do not inspect thousands of containers that arrive at their facilities daily. Hundreds of flights can still be boarded by passengers with malicious intent.

While this section will not detail recommended mitigation measures, U.S. law enforcement should assume that although no major attack has been carried out on the U.S. homeland since 9/11, al-Qaeda and organizations of its ilk are still capable of carrying out attacks. A clear basis for such an assumption was the plot to down ten airliners in August 2007, scheduled to make their way from Europe to the United States.

Operational Lessons

- Counterterrorism actions can be most effective when carried out by the threatened entity rather then by subcontracting national security responsibilities to proxies that may have different national interests.

- Control over areas that are known to be possible terrorist strongholds can facilitate intelligence and counterterrorism capabilities.
- Counterterrorism needs to be a constant and consistent effort that does not allow terrorist organizations to regroup, recruit and rearm.
- Deterrence and prevention may be achieved if the perceived response to a terrorist attack is unproportional to the attack itself.

INDIA

Since its independence in 1947, India has faced different waves of insurgency and terrorism throughout the country, largely rooted to tensions with Pakistani-militant groups over control of Kashmir, a disputed area bordering Pakistan and India. Great Britain dismantled its Indian mandate in 1947, and since partition of the subcontinent, India and Pakistan have faced ongoing tensions rooted in their differing religious beliefs, history, and long-running dispute. Tensions have escalated into violence, wars, and terrorism.

One of the first modern suicide attacks in India was carried out by the LTTE in 1991, with the suicide-attack assassination of former Prime Minister Rajiv Gandhi. Another reported incident of a suicide attack took place in April 2000, when a 14-year-old suicide bomber targeted India's central army base in downtown Srinager, in Indian-Administered Kashmir (IAK).[102] Soon after, in December 2001, India experienced an extremely devastating attack, when a suicide bomber targeted the Indian parliament, leading to a tense standoff between the two nuclear powers.[103] In both cases, Pakistani/Kashmiri militant groups seeking Kashmir's integration with Pakistan were presumed responsible.

In addition to facing terrorist activities based out of Punjab, Jammu, Kashmir, and the interior provinces, as well as movements from the northeast bordering Myanmar and Bangladesh, India has also faced religious-based terrorism from sectarian groups. Despite the different militant and terrorist group's active in India, the tactic of suicide terrorism has largely been adopted by Pakistan's Islamic Jihad groups operating in Jammu and Kashmir and in other parts in India. While terrorist attacks and bombings are all too common in India compared to other regions of the world, the specific tactic of suicide terrorism has been used in a limited number of attacks launched in the country.

History of Conflict

In 1947, with the departure of British rule, the Indian subcontinent was partitioned into Hindu-dominated but largely secular India and into the newly created Muslim dominated state of Pakistan. Princely states previously loyal to Britain—including the state of Jammu and Kashmir—were given the option of accession to either India or Pakistan, taking into consideration the geographical position and religion of inhabitants.

Following partition, many Muslims, Sikhs, and Hindus found themselves on the wrong side of partitioned provinces, leading to rioting and population movement. An estimated 500,000 people died in the violence that ensued following independence, and millions more became homeless. India and Pakistan went to war in October 1947 after Pakistan supported a Muslim insurgency in Kashmir. India supported Kashmir's maharaja under the condition that the state would accede to India. Kashmir signed an instrument of accession on Oct. 26, 1947 with India and cease-fire was called in early 1949. While Jammu and Kashmir acceded to India under the direction of its Hindu leader, the region is still under dispute, as Pakistanis believe it should have become part of Pakistan because the majority of the state's population is Muslim. Although diplomatic relations were established between the two countries, years to follow were marked by conflict, and the two countries have gone to war with each other four times since independence.

After the 1947 war, the two countries went to war again in 1965 when a clash between border patrols erupted into major fighting in the Rann of Kutch, a sparsely inhabited region along the southwestern India–Pakistan borders. Pakistan claimed victory when India withdrew, and about four months later, in August, hostilities broke out again and turned into the second Indo–Pakistani war.[104] Both countries agreed to a UN-sponsored ceasefire in September 1965 and later signed an agreement to solve their disputes through peaceful means.

Pakistan fell into civil war, however, when East Pakistan demanded autonomy and independence, leading to the creation of Bangladesh in 1971. Millions fled to India with the outbreak of war. In 1974, Kashmir reached an accord with the Indian government making it "a constitute unit of the union of India," which Pakistan immediately rejected.[104]

In 1989, armed resistance to Indian rule broke out in Kashmir, with groups calling for both independence and union with Pakistan. Militant Muslim groups began emerging in the 1990s, and the movement largely shifted from being nationalistic and secular in nature to adopting Islamic ideology. The arrival in Kashmir of large numbers of Jihadi fighters—"Afghan Alumni"—partially inspired the shift toward a more Islamic ideology.

Conflict between India and Pakistan erupted again in 1999, and Pakistani-based Kashmiri militants were held responsible for several attacks along the ceasefire line, leading to the build up of troops along the Indo–Pakistan border. In the mid-2000s there were improvements in relations, but tensions arouse once more after the 2007 Samjhauta Express bombings, in which bombs were set off in two carriages filled with passengers on a train connecting Delhi, India and Lahore, Pakistan. Sixty-eight people were killed and dozens more were injured.[105]

The region has suffered from violence and war, alternating between periods marked by diplomacy and ceasefire, to violence, war, a nuclear arms race, and ongoing accusations of encouraging terrorist activity. Recent terrorist attacks in Mumbai have pushed tensions forward once again.

India's Experience with Suicide Terrorism

The most active Kashmiri and Pakistan militant groups, which have been linked to the Pakistan Inter-Services Intelligence (ISI), are Islamic terror organizations Jaish-e-Muhammed (JeM) and Lashkar-e-Taiba (LeT). The Jaish-e-Mohammed (Army of Mohamed) is an Islamic extremist group based in Pakistan and was formed by Masood Azhar upon his release from prison in India in early 2000. The group seeks to overthrow Indian rule in Kashmir and unite Kashmir with Pakistan.[106]

LeT (Army of the Pure) operates out of Pakistan and Kashmir and seeks to unify Indian-Administered Kashmir (IAK) with Pakistan. It has been active since 1993, formed as the military wing of the Pakistani Islamist organization, Markaz-ad-Dawa-wal-Irshad, which was founded in 1989 and recruited volunteers to fight alongside the Taliban.[107] It was officially banned by Pakistan in 2002, having claimed responsibility for a number of attacks, including a January 2001 attack on Srinagar airport that killed five Indians and an attack in April 2002 against Indian border security forces that left at least four dead. After having been outlawed, the group did not claim responsibility for their attacks, although the group was held allegedly responsible for many. Both JeM and LeT have alleged links to al-Qaeda.[108]

Although suicide terrorism committed by local Pakistani groups emerged only in 2000, several trends have since emerged, including the reliance on Vehicle-Borne Improvised Explosive Devices (VBIED) in launching suicide attacks against government officials and institutions. There also appears to be a trend of coordinating VBIED-based suicide attacks with small-arms attacks, such as using explosive-laden cars to penetrate security barriers so that armed terrorists can launch further attacks. Such attacks are believed to be linked to Pakistan Inter-Services Intelligence (ISI)-backed Islamic terror organizations Jaish-e-Muhammed (JeM), or Lashkar-e-Taiba (LeT).[109]

On October 1, 2001, JeM claimed responsibility for a suicide attack on the state assembly building in Kashmir. The attackers drove a VBIED toward the gate of the assembly building. After the explosion, other JeM members entered the compound on foot and tried to storm the building to attack their primary targets—state legislators.

A similar attack occurred on July 5, 2005 in Ayodhya, a city in northeast India, against a makeshift Ram temple. A suicide bomber driving a jeep rammed into the gate of the temple, while terrorists with small arms tried to storm in after.

Most recently, on November 26-28, 2008, LeT was believed to be responsible for a large-scale terrorist attack against 13 different targets across Mumbai, launching more than 10 coordinated shooting and bombing attacks across the city, as well as taking hundreds of hostage (see Mumbai Case Study, Figure 4.11).

Many JeM and LeT attacks are not strictly suicide attacks, but rather are terrorist attacks that entail a high risk of death to the perpetrators. The above-mentioned attacks are some of the major attacks that could be defined as suicide terrorism against Indian targets in India or IAK.

CASE STUDY

MUMBAI ATTACKS

Date of Incident: November 26–28, 2008

Location: Mumbai, India

Target: Thirteen different targets across Mumbai, including the Taj Mahal Palace and Tower hotel, Mazagaon docks, a taxi at Vile Parle, the Leopold Café, Cama Hospital, a Jewish Chabad Center, the Metro Cinema, and targets near the *Times of India* Building and St. Xavier's College.

Type of Incident: More than 10 coordinated shooting and bombing attacks across Mumbai, hostage taking, sea-based assault

No. of Attackers: 10 (Nine were killed and one was captured by security forces)

Organization: Allegedly Lashkar-e-Taiba (LeT), Pakistan-based Islamic Terrorist organization

Motivation for the attack: It is possible that LeT sought to draw attention to the issue of Kashmir, create tensions between India and Pakistan, and make a statement of strength through a spectacular attack.

Description: The attackers traveled by sea to India from Karachi, Pakistan, entering Mumbai in speedboats on board trawlers. Around 8:00 p.m. on November 26, the ten Pakistani attackers came ashore in two different locations in Colaba, traveling in inflatable speedboats. The attackers split into two groups and headed in two different directions. Local fisherman made a report to police, but it received little attention.

Two gunmen attacked the Chhatrapati Shivaji Terminus around 9:20 p.m., entering the passenger hall and opening fire, throwing grenades, and killing at least 50 people.

A Chabad Lubavitch Jewish Center, the Nariman House, in Colaba, also known as the Mumbai Chabad house, was taken over by two attackers who held several residents hostage. After a long gun battle with NSG commandos who stormed the building, the two attackers were killed. However, the hostages were killed inside the house by the attackers. By the morning of November 27, the Chabad House was secured.

The Taj Mahal Palace hotel as well as the Oberoi Trident hotel, were both attacked on November 26. Six explosions were reported at the Taj Hotel and one at the Oberoi Trident, causing heavy damage to both buildings. Hostages were taken during the attacks, and the siege over the Taj Mahal hotel lasted over 60 hours. While it was believed that the hostage situation had ended the morning of November 27th, there were actually still two attackers holding hostages at the Taj Mahal hotel. During the attacks, both hotels were surrounded by Indian Security forces, including the Rapid Action Force and Marine Commandos, as well as National Security Guards (NSG) Commandos. Many foreign diplomats were in the hotel at the time of the attack, and about 450 people were staying at the Taj at the time, with another 380 at the Oberoi. Foreign nationals and tourists were specifically targeted.

Figure 4.11.

By the early morning of November 28, all sites were secured by Mumbai Police and Security forces except for the Taj Mahal Palace. India's National Security Guards launched a counter attack on November 29, leading to the death of the last remaining attackers at the Taj Mahal Palace and ending the 60-hour campaign.

Consequences: More than 160 people were killed, including 28 foreign nationals from 10 countries, and 300 wounded. A wave of resignations from the Indian political community followed the attacks.

Analysis: Although the attacks against Mumbai were not suicide attacks in their classic definition, the magnitude of the attacks and the use of several different tactics demonstrates the growing capabilities of local terrorist groups. The attacks were highly-advanced and very well coordinated, apparently took at least a year to plan. Some experts believe LeT was supported by al-Qaeda or the Pakistani Intelligence Services.[1] The group launched simultaneous, coordinated attacks focusing on foreign nationals, western targets, and tourists—a trademark al-Qaeda style. The attackers entered India through the port, using speedboats to reach their destination in a sea-based assault. They employed arms, bombs, and hostage taking as their modus operandi, and launched 10 coordinated attacks in order to maximize the damage and chaos created by the attacks.

The attackers allegedly spent several months planning the attacks, and were extremely familiar with the target sites and city. The attackers allegedly used Google Earth in order to familiarize themselves with the locations of the target buildings, and additionally may have received assistance from Mumbai residents or local operatives, who scouted the Mumbai targets prior to the attacks. In order to stay awake for more than 50 hours, the attackers allegedly took cocaine, LSD, and steroids during the attacks to sustain their energy. The attackers allegedly had constant communication with handlers from Pakistan, and were provided training and supplies. The attacks allegedly were planned in Bangladesh and refined in India, with support provided from Indian-based militant groups and criminal organizations.

Source:
[1]Eric Stakelbeck, The Mumbai Bombings: Who, How and Why? *CBN News*. Dec, 1, 2008. http://www.cbn.com/CBNnews/492562.aspx

Figure 4.11. *Continued.*

Within Pakistan, however, suicide terrorism has become a powerful force over the past several years. Although it was used before 2006, its frequency dramatically increased in the years that followed, and the targets shifted from being local Shiite targets to more government, foreign, police and military targets. According to data from the South Asia Terrorism Portal, Pakistan experienced 114 suicide bombings in the last two years. Prior to 2006, up to seven attacks a year had been reported. In 2007 alone, 56 suicide bombings in Pakistan left 729 people dead, including 552 civilians and 177 security personnel. Of these attacks, 27 occurred in Pakistan's Northwest Frontier Province (NWFP) with another 13 in the Federally Administered Tribal Areas (FATA).[110] Global Jihad groups are becoming increasingly active in Pakistan as well, having been responsible for major terrorist attacks such as those against the Danish Embassy and the Marriot hotel in Pakistan.

As long as territorial or religious/cultural issues continue to be disputed, it is likely that suicide attacks will continue to some extent in India, the IAK, Pakistan, and perhaps elsewhere.

TURKEY

Often characterized by its dichotomies—east versus west, modernity versus antiquity, Islam versus secularism—Turkey has struggled not only in its role between two continents but also in its position in countering two ongoing threats. Since the republic's establishment in 1923, internal civil strife among the Kurdish minority, and—more recently—the demands of Islamist groups and the Global Jihad movement have significantly shaped Turkish identity and policy. In the context of an ongoing internal insurgency and a brutal civil war, Turkey experienced a three-year suicide terrorism campaign in the 1990s launched by the Kurdistan Workers Party, or PKK, a nationalist/separatist movement committed to achieving equal rights and autonomy for the Kurdish minority.

While this short-lived campaign ultimately failed to gain popular support, Turkey's experience with suicide terrorism has not been limited to those formative years, nor is it framed by the local civil conflict alone. The country has also been the target of attacks by local and international Islamist groups, including international terrorist organization al-Qaeda, which was connected to a series of brutal suicide bombings in Istanbul in 2003. Since the PKK's terrorist campaign largely dissipated in the late 1990s, other Kurdish rebel groups have perpetrated attacks in its place, targeting civilian and tourist sites throughout Turkey.

Civil War and Civil Strife: The PKK's Answer to the "Kurdish Question"

Turkey, founded as a secular state, is home to a number of ethnic and religious groups who have, throughout the history of the republic, challenged or rebelled against the state, demanding autonomy and increased civil rights. The Kurds, who number between 25 and 30 million across the Middle East, make up about 20% of Turkey's total population (around 12 million people).[111] Like their ethnically Turkish counterparts, the Kurds are a Sunni Muslim people, but are divided from the majority along linguistic, tribal, and ethnic lines. Their traditional ethnic homeland encompasses portions of southeastern Turkey, northeastern Iraq, and parts of Syria, Iran, and Armenia. Prior to World War I, the Kurds were traditionally a nomadic people. The possibility of an independent Kurdistan was ruled out after the abandonment of the Treaty of Sèvres in 1923 and uprisings and revolts in the 1920s and 1930s failed to achieve nationality rights for Kurds. Across the Middle East, the Kurds represent a minority group that has since historically struggled to gain autonomy, often clashing with ruling government forces.[112] In

Turkey, this struggle is known as the "Kurdish question" and has been a constant source of instability for Turkey.

Kurds are not an officially recognized minority group in Turkey. The rise of Kurdish nationalism in the 1970s stemmed from both the lack of minority rights and the economic deprivation in the southeastern regions of Turkey, where most of the Turkish Kurdish population lives. Under Turkish authority, Kurds have suffered various forms of legal discrimination. Major points of contention include language rights, cultural autonomy, and the failure to receive official minority status by the state.

Before the 1980 military coup, the Kurdish language was completely banned—forbidden in all government institutions, including schools and courts. In the early 1980s the government outlawed the possession of written or audio materials in Kurdish, and it forbid traditional Kurdish dress in the cities.[113] Kurdish language newspapers and journals—established despite the ban—were quickly closed down and editors were arrested.[114] From 1983 to 1991, it was illegal to speak Kurdish in public, and only in 2003 were parents allowed to give their children Kurdish names. Efforts of assimilating the Kurdish population have spurred frustration and the development of a militant Kurdish culture.[115]

Kurdish rebellions, prompted by Turkish independence in 1923, erupted throughout the 1920s and 1930s, but were harshly suppressed by Turkish president and founder Mustapha Kemal Attaturk. The Kurdish struggle remained relatively dormant following the rebellions, but general discontent and strife among the Kurdish population continued to mount until 1984, when a civil war and guerilla campaign for Kurdish independence and rights was launched in southeastern Turkey, led by Kurdish leader Abdullah Öcalan of the PKK and supported partially by the Syrian government.[116]

Turkey's Experience with Suicide Terror

The Kurdistan Workers Party (Partiya Karkeran Kurdistan, PKK), led by Abdullah Öcalan, began as a Marxist-leftist group in 1978 seeking to achieve an independent Kurdish state or full Kurdish autonomy (Figure 4.12). With between 5000 and 10,000 armed fighters, in the 1980s the PKK began directing attacks against the Turkish military, government property, government officials, Turks living in Kurdish populated regions, Kurds accused of collaborating with the government, foreigners, and Turkish diplomats abroad.

In its more than 20 years of operation, the PKK has generally relied on guerrilla activity in an attempt to achieve its goals, primarily targeting Turkish police and military targets in southeastern Turkey. Under Turkish military occupation and upon facing major military setbacks in its rebellion campaign, however, the organization turned to suicide terrorism in 1996—a tactic initiated by its leader, Abdullah Öcalan, and that was seemingly inspired by the activities of groups like Hamas and the Tamil Tigers (LTTE) of Sri Lanka.

In fact, many of the PKK attackers were women, very much reflecting the precedent set by the LTTE.[117] The first three female suicide bombers in Turkey

IN BRIEF: THE KURDISTAN WORKERS PARTY (PKK)

- Alternate names: Kongra-Gel (KGK), Kurdistan Freedom and Democracy Congress (KADEK), Kurdistan People's Congress, Freedom and Democracy congress of Kurdistan
- Founded in 1984 by PKK leader Abdullah Öcalan
- Goals: Establishment of an Independent Kurdish state, equal rights for Kurdish minority in Turkey
- Ideology: Separatist/Nationalist, Secular, Leftist; founded as a Marxist-Leninist
- Modus Operandi/Tactics: guerilla war launched in 1984, suicide campaign from 1996–1999; car bombings, kidnappings, attacks against foreign diplomats in Europe
- Headquarters/Areas of Operation: Southeastern Turkey, Turkey, Northern Iraq, bases in Syria and Lebanon; active cells and funding support in Europe; United States
- Activists/Members: Turkish Kurds, Sunni Muslim, 5,000–10,000 armed fighters
- Structure: Hierarchical

Figure 4.12. In Brief: The Kurdistan Workers Party (PKK).

actually disguised themselves as pregnant women, strapping explosives to their stomachs. These attacks, carried out on June 30, October 25, and October 29 of 1996, resulted in the deaths of 14 people.[118] With these attacks, the PKK's previous plan of armed struggle shifted to include suicide terrorism—and female suicide terrorism in particular.

Perpetrating their suicide campaigns largely in waves, the PKK carried out its first 14 suicide attacks against government and military targets between June 1996 and July 1999, described by Robert A. Pape as one of the world's "least aggressive modern suicide terrorist campaign(s)."[119] The attacks left up to 22 people dead in addition to the attackers themselves, resulting in an average of fewer than 2 causalities per attack—a relatively small figure compared to the international average of 12 since 1980.[120] The limited nature of these attacks, and their voluntary halt, can be attributed to the group's leadership, group dynamics, and ultimately a failure in achieving internal and public support.

The PKK voluntarily halted its first suicide terrorism campaign in 1999 with the arrest of its leader, Abdullah Öcalan. Directly following his arrest, Öcalan initially made appeals to his followers to commit suicide attacks in protest of government plans to execute him (see Figure 4.13). A series of bombings did indeed follow, and the frequency of suicide attacks periodically increased. These attacks, paired with pressure from the European community, were successful in compelling the government to reverse the decision to execute Öcalan. Instead he was sentenced to life imprisonment.

However, following Öcalan's initial appeals to increase the violence, he apologized for the violence perpetuated by his party and called for a cease-fire. In

CASE STUDY

CAPTURE OF ABDULLAH ÖCALAN

Date of Incident: February 16, 1999
Location: Nariobi, Kenya
Type of Incident: Counterterrorism Operation: Capture and Arrest of PKK Leader Abdullah Öcalan
Organization: The Kurdistan Workers Party (PKK)
Involved Forces: U.S., Kenyan and Turkish Security Forces

Manner of Confrontation: Turkish operatives, masked and armed, ambushed the convoy of PKK leader Abdullah Öcalan as he left the premises of the Greek Embassy in Nairobi, Kenya to board his private jet at the airport. American intelligence and law enforcement officers, in addition to Kenyan security officials, were already in Nairobi investigating the bombing of the U.S. Embassy there in August. When they discovered Öcalan had arrived in Nairobi, they placed the Greek Embassy under surveillance and monitored Öcalan's cell phone conversations through satellites. A Turkish commando team flew to Nairobi once receiving a tip from Kenyan and American authorities as to Öcalan's whereabouts. He was captured after agreeing to be driven to the airport by a Kenyan cooperating with Turkish security forces.

Description and Background: Prior to arriving in Kenya, Öcalan had spent five months traveling between different locations seeking refuge. After Turkey and America stepped up pressure on the Syrian government, threatening the country, Öcalan was forced in October 1998 to leave Syria, his long-time sanctuary. American diplomatic pressure, supported by intelligence-gathering, helped put Öcalan on a plane from Syria and put pressure on European nations to refuse him refuge—forcing him to a desperate search to different countries.

Öcalan arrived in Moscow on October 9. After seeking a base in Europe, on November 2, he flew to Rome, where he was held on a German warrant charging him with terrorism. Germany dropped the charges, and after two months in Italy and a diplomatic breakdown in relations between Turkey and Italy, Öcalan secretly left for St. Petersburg on January 16; he flew to Athens on January 30 on a private plane, owned by a retired military officer sympathetic with the Kurdish cause. Two days later, Greek officials tried to fly Öcalan to the Netherlands, but his plane was barred, and he returned to Greece. Finally, the next day he flew with a Greek official to Nairobi, where the Greek government agreed to provide him temporary shelter at the embassy.

After two weeks in Nairobi, Öcalan was told he could fly to Amsterdam. He was captured on the way to the airport and subsequently held for interrogation and trial on a Turkish island in the Sea of Marmara.

Consequence: After his capture, Abdullah Öcalan was transferred to Turkey to face trial for acts of terrorism.

Figure 4.13. Case Study: Capture of Abdullah Öcalan.

Mainstream Turkish media portrayed Öcalan—handcuffed and blindfolded and being led by Turkish security forces—as weak and broken, ready to betray his cause, telling them that he was willing to cooperate and loved the Turks. It was a scene, captured on a Turkish intelligence-service video, that caused an upsurge in nationalism among Turks and caused rage among Kurds and PKK supporters. Kurdish protest demonstrations raged throughout Europe in the days following the capture, with supporters attacking Greek Consulates and Embassies and attempting to storm the Israeli Consulate in Berlin on account of rumored Israeli involvement in the capture. Three were killed and 16 wounded when these supporters opened fire at the consulate.

Turks praised the arrest as a breakthrough for Turkey's ongoing fight against the PKK. U.S. support in the capture was also applauded by Turks.

Directly following his arrest, Öcalan initially made appeals to his followers to commit suicide attacks in protest of government plans to execute him. A series of bombings followed. Pressure from the European community in terms of human rights also mounted, and Öcalan was sentenced to life imprisonment instead of execution. Öcalan eventually called for his followers to give up violence, and the group soon after declared a cease-fire.

Analysis: The successful arrest and capture of Abdullah Öcalan effectively paralyzed the PKK. The organization declared a cease-fire soon after his arrest, heeding the leader's call and changing its name and tactics to reflect its move toward more peaceful politics. There were isolated incidents of violence in the years following his arrest, but PKK activities were largely limited with Öcalan's capture—a dramatic and significant victory for Turkish forces and the counterterrorism community. Intelligence sharing and international cooperation were crucial in the arrest and capture of the leader.

The trial of Öcalan was carefully scrutinized by the international community, seen as a "test" to examine Turkey's progress in the arena of Human Rights—ensuring Öcalan received a fair and transparent trial. Others noted it as the opportunity to seek a solution to the "Kurdish problem." Öcalan was sentenced to death in June 1999, but the sentence was commuted to life imprisonment with no chance of parole in October 2002, after the death penalty was abolished in Turkey in August 2002.

Sources: http://news.bbc.co.uk/2/hi/europe/380861.stm; http://www.globalsecurity.org/military/library/news/1999/02/wwwh9f18.htm; http://query.nytimes.com/gst/fullpage.html?res=9E03E3D8143DF933A15751C0A96F958260&sec=&spon=&pagewanted=2; http://www.time.com/time/daily/special/ocalan/bitterend.html; http://edition.cnn.com/WORLD/meast/9906/29/ocalan.verdict.03/

Figure 4.13. *Continued.*

exchange for Kurdish cultural rights, he proposed an agreement to preserve the integrity of the state and called for an end to the armed struggle.[121]

There were a few isolated incidents of conflict from 2000 to 2004, but for the most part, following Öcalan's arrest, the PKK dissipated and activities were largely limited to the guerilla warfare arena, focusing on military targets and not civilian targets. The unilateral cease-fire ended, however, in August 2004, coinciding with an increase on illegal PKK activity that included a number of attacks on military, police, and governmental targets near the Iraqi border.

In the years that followed, suicide attacks on tourist targets along the western coast and in the capital of Turkey were largely blamed on the PKK or PKK sympathizers, but the group itself generally denied involvement. Several Kurdish separatist groups or individuals with alleged links to the PKK were held responsible for a majority of these attacks, including a suicide attack in Ankara in 2007 which left six people dead and injured dozens more.[122] The PKK was implicated in the attack but denied responsibility.[123] Sporadic violence and short-term ceasefires have characterized the situation since. The threat of the Kurdish separatist movement, however, took a back seat to a new form of terrorism that emerged in the spotlight in 2003.

Islamic Groups Step Up to the Plate

In what was dubbed Turkey's 9/11,[124] on November 15, 2003, two suicide bombers drove explosive-laden trucks into the Neve Shalom and Beit Israel synagogues in Istanbul, killing 24 and injuring more than 300. Just five days later, the British consulate and the HSBC bank headquarters were targeted, throwing the city into chaos as two suicide bombers killed another 27 people when they drove explosive-laden trucks into the consulate and bank. These devastating attacks were perpetrated by local Turkish activists from southeastern Turkey, allegedly linked to the international terrorist organization, al-Qaeda, and local Islamist terror organizations acting as al-Qaeda affiliates.[125] Turkish (Kurdish) Hezbollah was linked to the attack, as was the Turkish Islamic militant group IBDA-C, or the Great Eastern Raiders' Front, which took responsibility for the attack in a phone call to the semi-official Anatolia news agency (see incident case study, Figure 4.15).[126]

The attack was a rude awakening for Turkey, whose previous experience with suicide terrorism had been largely limited to smaller-scale attacks by Kurdish separatists. The large-scale al-Qaeda-style attack highlighted the homegrown Global Jihadi and Islamic terrorism trend taking hold in Turkey.[127]

Al-Qaeda has since been linked to several active cells in Turkey, with police raids exposing Jihadi materials, training pamphlets, and bomb-making materials. Accompanied by over 40 arrests in April 2008, police even discovered a "parallel Jihad society"—a well-developed educational, legal, and religious underground network nurturing future Jihadists.[128] Yet it was not al-Qaeda that introduced such Islamist ideology and modus operandi into the Turkish terrorism experience.

At the height of the PKK's separatist campaign in the 1980s, a number of Islamic terrorist organizations emerged in Turkey, reaping havoc not only on civilians, "anti-Islamic" establishments, and official government forces, but also on the PKK itself—which was considered a rival group destroying Islamic values. Turkish Hezbollah, unrelated to the Lebanese Shiite base organization of the same name, was established in the early 1980s with the goal of overthrowing the ruling secular regime and founding an Islamic state in Turkey. Its activities only later targeted Turkish secularists and government forces; however, in its first 10 years it focused instead on PKK activists and supporters in southeastern Turkey, who they viewed as being "Islam's enemy"[129] and accused of committing atrocities against Muslims in southeastern Turkey. The Kurdish Sunni Islamic terrorist organization arose in response to the PKK's secularist approach in establishing an independent Kurdistan.

As such, various human rights activists have gone so far as to claim that Turkish security forces supported and sponsored the group, or at least encouraged their activities by awarding them immunity from security forces.[130]

Turkish Hezbollah primarily adopted kidnapping, torture, and shootings among its favored tactics, with a number of small-scale suicide bombings sprinkled in,[131] claiming more than a thousand lives between 1992 and 1995.[132] A government crackdown in 2000 largely suppressed the group's operational capabilities, arresting hundreds of activists and killing one of the foremost leaders of a deadly faction of the group, Huseyin Velioglu, in a police raid on his house in Istanbul.

Other fundamentalist Islamic organizations active in Turkey include the Islamic Jihad, the Islamic Movement Organization, and the Islamic Great Eastern Raiders Front (IBDA-C). These loosely organized groups advocate for an Islamic state in Turkey and also have an anti-Western agenda. They have claimed a number of attacks under a variety of names, including a suicide attack against a Free Masons Lodge in Ankara in 1994 and against another Masonic lodge in Istanbul in May 2004.[133]

The IBDA-C, which took responsibility for the November 2003 bombings, is a Sunni Salafist group that supports replacing the current "corrupt" and pro-Western government with Islamic rule in Turkey. Since their establishment in the mid-1970s, their attacks have primarily focused on civilian or perceived anti-Islamic targets, such as churches, charities, pro-secular establishments and journalists, banks, and clubs. According to several sources, Turkish police believe that the group may falsely claim responsibility for attacks in order to "elevate its image."[134]

Turkish Counterterrorism Tactics

In responding to PKK guerilla attacks and activities, military clashes spiraled into a large-scale and long-term civil war in Turkey, resulting in thousands of civilian and military deaths. Based in northern Iraq, in 1984 the PKK began its militant campaign by launching a series of cross-border raids, seeking to control a number of rural Kurdish towns. The Turkish military responded with extensive counterinsurgency tactics, including large military operations employing up to 35,000 troops. In the 1980s and continuing into the 1990s, Turkish military forces occupied and dominated large areas of traditionally Kurdish populated lands, using extensive measures to suppress any rebellion movement, including mass arrests, extra-judicial killings, torture, and harassment by security forces. An estimated 35,000 Kurdish civilians were killed by Turkish military operations during this time, with still thousands more killed by the PKK themselves.[135] By 1993 about 200,000 security forces were involved in the struggle.[136]

Turkey's military initiated an aggressive campaign aimed at completely destroying the PKK's military infrastructure. The government gave civilian Kurds the option of either fighting on the side of the Turkish government as proxies, or abandoning their homes. According to human rights groups, Turkish forces evacuated or destroyed 2664 villages and hamlets in southeastern Turkey, creating almost 3 million internal refugees and various internal economic and social problems.[137] At least 362,000 Kurds were "forcibly resettled" during that time, dramatically reducing the rural population of the southeast by 12%.[138]

Turkish military forces withdrew from much of the region in the early 1990s, but continued to conduct counter-PKK operations throughout late 1998.[139]

These policies may have actually also had another unintended consequence—inadvertently fueling the PKK's shift to suicide terrorism in 1996. Under military occupation and facing heavy losses, PKK member morale was low. Some experts argue that suicide missions were chosen as a means of demonstrating the group's threat level and provoking the state into responding harshly, because "support for the PKK increased as Turkish repression intensified."[140] In effect, PKK suicide attacks were designed to incite "an overreaction on the part of government forces," outraging and radicalizing the civilian Kurdish population and ultimately convincing them to support the PKK.[141] In a seemingly never-ending cycle of violence, support for the PKK only further provoked harsh counterterrorism policies. Human rights groups have argued that such tactics were counterproductive because they drove "more and more civilians into the arms of the PKK."[142] Turkey's approach focused on incapacitating PKK militant activity, but in additionally affecting Kurdish civilians, it may have broadened PKK's support base and simply exacerbated tensions.

Also as a result of the heavy-handed policies in southeastern Turkey, Kurds increasingly migrated to Western Europe, where activists were able to build broad bases of support and financing. The threat of terrorist activity was simultaneously exported as well. In an interview in March 1996, Öcalan warned Germany that they could be the next target of suicide bombings, threatening them for repressing Kurdish dissidents in Germany: "Up until now, my guerillas knew how to avoid death. ... Now they will learn how to die. Every Kurd will become a living bomb."[143]

Turkey eventually changed its counterterrorism strategy, focusing on breaking down public support for the PKK—"winning their hearts and minds"—by easing their military operations and granting more rights to the Kurdish population, essentially "add[ing] carrots to its arsenal of sticks," as expert Mia Bloom writes:

> It appeared that they finally understood that the counter-terror strategies of the past exacerbated rather than mitigated conflict and had created very counterproductive consequences. Previously, the government had rarely distinguished between real PKK operatives and the larger Kurdish civilian population. By lumping together the insurgents with the larger population and punishing them both equally, they had alienated the Kurds from the government and driven them in to the arms of the PKK. The shift in policy effectively meant that the government was willing to grant greater cultural rights. The government added carrots to its arsenal of sticks.[144]

The government replaced aggressive ground operations with air–ground coordination, bulked up police and security forces in the region, and promised to provide economic and cultural aid packages to Kurds.

Public support for PKK activities was therefore short-lived. The PKK itself targeted Kurdish civilians that it viewed as "traitors," aiming to spread fear among the rural population and stop Turkish government influence and attempts of assimilation. The series of attacks that followed Öcalan's arrest largely failed to resonate positively with the Kurdish public. They did not provoke hard-line counter-insurgency tactics by government forces as they had in the past. In addition, conditions for the Kurdish population were improving, so support for the PKK had more of a price. Instead of public outrage of government violence and increased nationalist support for the PKK, the Kurdish population grew increasingly weary of the violence perpetrated by the organization.[145]

The PKK began to disintegrate with the loss of its leader, shifting instead into a political party. While the PKK changed its name to KADEK to reflect its move toward peaceful politics, many still associated the party—and especially its military branch—with terrorist activities and organizations. Yet Turkey's successful capture of Abdullah Öcalan in 1999 largely led to the end of PKK terrorism.

In seeking EU admission, Turkey took significant steps to improve its human rights record and grant increased rights to its Kurdish population. Reforms have been implemented granting Kurds more cultural rights, such as a lifting of the language ban and limited Kurdish broadcasting. Despite improvements, however, longtime tensions with Kurdish militants boiled over once more in late 2007 and early 2008, when cross-border attacks launched by the PKK against military targets in southeastern Turkey killed 12 Turkish soldiers in October 2007 and prompted a week-long Turkish military incursion and air strikes against PKK camps in northern Iraq.[146] Fears of Kurdish separatism increased in the wake of the U.S. invasion of Iraq, as a strong Kurdish entity allied with the United States formed in Northern Iraq.

In its efforts to prevent a backlash from the moderate Kurdish population, the Turkish government continues to develop economic aid and cultural reform packages to improve the humanitarian situation in southeastern Turkey and recognize Kurdish cultural autonomy. Fear of separatist activity, however, remains high.

Lessons From Turkey's Experience with Suicide Terrorism

- Winning the hearts and minds of the population may be just as effective as or more effective than breaking down military capabilities. A terrorist group needs both (a) public support for its cause and (b) military capabilities to conduct suicide attacks. For this reason, among others, counterterrorism strategies should not disregard the effects of military operations and policy on the local civilian population.

- International legitimacy and respect for human rights in counterterrorism policy may also be important for long-term success in fighting terrorism.
- Alienating the civilian population can lead to broader-based support for terrorist groups, including funding, radicalization, and recruitment both locally and abroad.

AL-QAEDA

An architect of international terror, al-Qaeda has played an extremely influential role in the global dissemination of the suicide attack tactic. In spreading its ideology of Global Jihad, the group has sought to inspire and motivate activists all over the world. These activists have adopted not only the cause behind the Global Jihad movement, but also what has become an al-Qaeda trademark tactic—suicide terror. Supporting this process of internationalization is the group's changing organizational structure, characterized by a shift from hierarchical control and command to an international network of diffuse and loosely connected cells.

While al-Qaeda began carrying out such attacks about 15 years after the modus operandi was first introduced in Lebanon, the organization is considered to be (a) one of the most dominant groups using the tactic today and (b) "the main force behind the internationalization of suicide terrorism."[147] The notion of self-sacrifice, or Istishhad, has become its guiding value, and suicide attacks have become its preferred method. According to Yoram Schweitzer, by transforming the concept of Istishhad into a core symbol and practice for the jihad movement, al-Qaeda is largely responsible for its globalization.[148] It has furthermore made it part of its mission to multiply and spread such attacks around the world, itself striving to serve as a "role model" to affiliate organizations and independent activists.

For al-Qaeda, suicide attacks are preferred not only because of their logistical effectiveness and lethality, but also for their communicative value. The notion of martyrdom and self-sacrifice conveys a specific message and confers added propaganda value. It garners media attention and serves as a highly effective form of psychological warfare, communicating not only the skills and determination of the attackers, but also subtle messages that are intended to demoralize the enemy. Such an act helps to portray the conflict as a David and Goliath type struggle and encourages Islamists to become inspired by the cause.[149]

With al-Qaeda behind but not always tightly at the reigns, the international terrorist organization and its affiliates have been responsible for hundreds of suicide attacks 30. The porous nature of international borders is demonstrated all too well by the organization, whose attacks have plagued locations as geographically diverse as Western Europe, North America, the Middle East, and Southeast Asia. Al-Qaeda has been held directly responsible for at least *seven* major terrorist attacks since 1998, all of which were suicide attacks with relatively high casualty figures.[150] Organizations affiliated with al-Qaeda have been responsible for numerous more attacks.

Establishing the "Base"

Answering a call to defensive Jihad, young Muslims from all over the world flocked to Afghanistan between 1979 and 1989 to join in the fight against invading Soviet forces. Communist efforts to secure control over the country spurred a fierce national resistance movement led by local Afghans and supported by "billions of dollars' worth of secret assistance" from Saudi Arabia and the United States.[151] Osama bin Ladin, 23 years old when he arrived from Saudi Arabia in 1979, was one of thousands of volunteers to join the "holy war." Bin Laden quickly rose to power within the movement, along with Palestinian theologian and scholar Abdallah Azzam. When the movement emerged victorious against the Soviets in 1988, bin Ladin and Assam recognized the importance of preventing the organization's dissipation, instead taking advantage of the trained and eager volunteers to maintain and extend the Jihad movement.[152]

In 1988, they established al-Qaeda ("the base") out of these activists. The organization included several different branches, such as operations and intelligence, and military, financial, and political committees, as well as a committee that dealt with propaganda and media affairs. Azzam's death in 1989 left bin Laden as the principal leader of the organization, committed to establishing Islamic regimes all over the world and spreading militant Islam beyond the country's borders.[153]

Once the organization's activists were trained in al-Qaeda camps in Sudan and Afghanistan in the 1990s, they returned to their home countries, establishing local cells or spreading the message of Global Jihad to preexisting local movements and organizations. About 20,000 people from 47 different countries passed through the camps from the mid-1990s until October 2001.[154] Leaders of small organizations that trained in al-Qaeda camps were "imbued with radical ideology and the means to create or revitalize local terrorist groups."[155] As Schweitzer describes, this transnational mode of operation was based on a large number of committed fighters from diverse backgrounds and nationalities but with shared ideologies. As such, they could be distributed around the world to advance and strengthen Global Jihad.[156]

Al-Qaeda operations introduced a new organizational paradigm, described by Schweitzer as a "cross-nation Muslim community dispersed all over the globe" employing extreme violence to all opposed to its ideology.[157] In its early years, al-Qaeda functioned primarily as a support system to outside groups and networks committing attacks, providing financial and logistical assistance. Its first direct attack was in August 1998 against the U.S embassies in Kenya and Tanzania, intended, *inter alia*, to set an example in terms of tactics and targets to outside groups and affiliates inspired by its ideology and operations.[158]

Al-Qaeda's primary objective of establishing regimes operating according to Islamic law began by targeting Muslim countries such as Saudi Arabia, Egypt, Pakistan, and Indonesia. Soon countries with large Muslim minorities were targeted, and then the organization's strategy shifted to focus on the "far enemy"—the United States and the Jews—as a means of expelling their presence from the Middle East and enabling the overthrow of local regimes.[159]

By the summer of 1996, bin Laden had declared war against the United States, and in 1998 he extended the "Jihad against Jews and Crusaders" even further, declaring: "to kill the Americans and their allies—civilians and military—is an individual duty for every Muslim who can do it in any country in which it is possible to do it" (bin Laden, 1998).[160]

Before 2001, al-Qaeda was supported by Afghanistan's Taliban government, which provided a safe haven for its operations and training camps as well as offered financial assistance. According to the 9/11 Commission report, between 10,000 and 20,000 fighters were trained in al-Qaeda camps from 1996 to 2001. Al-Qaeda planned several attacks during this time period. In addition, the organization provided financial assistance and fighters to conflicts in Chechnya, the Balkans, and Kashmir.[161]

Al-Qaeda has since been responsible for some of the most devastating and deadly suicide attacks of modern time, including the September 11, 2001 attacks on the Pentagon and the World Trade Center. The 9/11 attacks prompted the United States to invade Afghanistan, al-Qaeda's base, and may have led to the war in Iraq in 2003. Al-Qaeda's infrastructure was significantly weakened by the 2001 invasion, but—despite claims by George Bush and other U.S. officials that al-Qaeda was largely defeated and "on the run"[162]—the organization has been connected to deadly bombings around the world since 9/11, including attacks in Madrid in April 2003, Istanbul in November 2003, London in July 2005, and a number of 9/11-style plots that were thwarted. From 1996 to 2005, the organization opted for suicide bombing plots more than 20 times in its attacks against the United States and its allies.[163]

Al-Qaeda Organization: Affiliates and Networks

A distinguishing characteristic of the al-Qaeda phenomenon is that the organization's central core has relied heavily on small and local cells, as well as larger Islamist terrorist groups, to serve as proxy or affiliate "subcontractors" furthering its objectives and carrying out major terrorist attacks around the world on its behalf.[164] The following organizations have been linked to al-Qaeda, serving as affiliate groups:[165]

- Egyptian Islamic Jihad
- The Libyan Islamic Fighting Group
- Islamic Army of Aden (Yemen)
- Jama'at al-Tawhid wal Jihad (Iraq)—late Abu Musab Zarqawi's al-Qaeda in Mesopotamia
- Lashkar-e-Taiba and Jaish-e-Muhammad (Kashmir)
- Islamic Movement of Uzbekistan
- Al-Qaeda in the Islamic Maghreb (Algeria) (formerly Salafist Group for Call and Combat)

- Armed Islamic Group (Algeria)
- Abu Sayyaf Group (Malaysia, Philippines)
- Jemaah Islamiya (Southeast Asia)
- Al-Ittihad al-Islami (AIAI)
- Asbat al-Ansar
- Salafist Group for Call and Combat (GSPC)
- Ansar al-Isram
- Moro Islamic Liberation Front (MILF)

These groups operate as independent terrorist organizations, but are all considered to largely share al-Qaeda's basic Sunni Muslim fundamentalist beliefs. They recruit and operate locally, spreading al-Qaeda's global jihad message and often benefiting from its financial or operational support.[166] It is not clear the extent to which al-Qaeda has directly participated in the activities of these organizations, but numerous leaders and members have been linked to al-Qaeda and have cited the international terrorist group as a source of inspiration or support. Many of their leaders were trained in al-Qaeda camps in Afghanistan following the war.

The heavy blows suffered by al-Qaeda in the post-9/11 American invasion of Afghanistan damaged the organization's infrastructure and capabilities, arguably forcing it to rely more heavily on affiliate organizations, independent cells, and a generally more dispersed organizational strategy. Financial assets were confiscated, training camps were destroyed, and leadership, while not eliminated, was weakened and dispersed, with communication among members greatly reduced.[167] As a result, many experts argue that al-Qaeda has transformed to operate according to a "franchise" system, with existing groups becoming offshoots of the organization as they transform their local grievances to become part of a global cause.[168]

Some experts argue that "command and control" still lies in the hands of Osama bin Laden and other top-level al-Qaeda operatives. Yet several of al-Qaeda's major attacks appear to have been initiated locally or independently, with external support. Several leading scholars argue in fact that, as described by David Ronfeldt, "Al-Qaeda is now more important as an ideology than an organization, a network than a hierarchy, and a movement than a group."[169]

According to terrorism expert Professor Bruce Hoffman, however, al-Qaeda's organization is not entirely dispersed, but has maintained a degree of centralized control, combining both a "bottom-up" approach and a "top-down" approach in directing its attacks.

> The al-Qaeda of today combines, as it always has, both a "bottom up" approach—encouraging independent thought and action from low (or lower)-level operatives—and a "top-down" one—issuing orders and still coordinating a far-flung terrorist enterprise with both highly synchronized and autonomous moving parts.[170]

Within this framework, Hoffman breaks down the al-Qaeda movement into four distinct dimensions. The first is "al-Qaeda central," compromising the core leadership of the organization. These commanders exert coordination—and direct command and control—in terms of commissioning and directing attacks, planning operations, and approving their execution.

The second category is al-Qaeda affiliates and associates, formal terrorist or insurgent groups that benefit from al-Qaeda's guidance, training, financial, or other assistance. On the third level, Hoffman identifies "Al-Qaeda locals," dispersed cells of al-Qaeda activists who have some direct connection to the organization. The fourth category is the "al-Qaeda Network," home-grown Islamic radicals or converts without any direct connection to al-Qaeda or any other formal terrorist organization. These activists are prepared to carry out attacks in support of al-Qaeda's mission regardless of the fact that their connection to the organization is "more inspirational than actual."[171]

Al-Qaeda Attacks

With its myriad of affiliates and associates, determining which attacks al-Qaeda has been directly responsible for is a challenge for experts, law enforcement, and security personnel. Until the attacks of September 11, al-Qaeda did not directly claim responsibility for its attacks, as opposed to its partner groups, which often or always claim responsibility under their name or another name—even if they had no involvement in an attack.[172] In general, in order to be considered a "direct" attack by al-Qaeda, there must be some level of command and control by a senior member of al-Qaeda from the central core.

Among its first suicide operations, al-Qaeda was responsible for the August 7, 1998 suicide truck bombings of the U.S. embassies in Nairobi, Kenya, and Dar es Salaam, Tanzania. In nearly two simultaneous attacks, 291 people were killed and 5000 wounded in the downtown area of Nairobi near the U.S. Embassy. In Dar es Salaam, 10 were killed and another 77 wounded when a bomb-laden truck was driven into the embassy.[173] In the years that followed, several planned attacks were thwarted by security forces, including an attack targeting the baggage claims of the Los Angeles International Airport. On October 12, 2000, the U.S.S. *Cole* was attacked when a boat-bomb that appeared to be a service vessel exploded along the American destroyer, killing 17 American sailors.

By far the most devastating and significant suicide attack of the century is attributable to al-Qaeda—the infamous September 11, 2001 attacks on the World Trade Center and the Pentagon, which resulted in the deaths of more than 3000 people, marking a turning point both in terms of the lethality of suicide attacks and in the way governments respond to terrorism. Twelve of the 13 suicide attackers-hijackers were Saudi nationals, and one was from the UAE.[174] They were recruited and trained by al-Qaeda and sent to the United States to perpetrate the attack. Aside from leading to two wars—in Afghanistan in 2001 and Iraq in 2003—the 9/11 attacks changed the United States and the world. The attacks also catapulted al-Qaeda into a household name. According to Robert

Pape, using data collected from the Chicago Project on Suicide Terrorism, from 1995 to 2003, 71 al-Qaeda terrorists successfully committed suicide missions.[175]

Additional al-Qaeda attacks include:

- April 11, 2002: Suicide attacker drives a bomb-laden truck into the ancient synagogue in Djerba, Tunisia, killing 21 people, mostly German tourists. The attacker, a French-Tunisian, was recruited by al-Qaeda after attending a training camp in Afghanistan and expressing his willingness to commit suicide. He was reportedly handled "from afar" by senior al-Qaeda operative Khaled Sheikh Muhammad.[176]
- October 2002 attack on a French tanker off the coast of Yemen.
- Several bombings in Pakistan in the spring of 2002.
- November 28, 2002: A suicide attacker driving a car-bomb detonated at a hotel, killing 10 Kenyans and three Israelis in Mombassa, Kenya. Twenty minutes earlier, two missiles were launched at an Israeli jet liner over Mombassa, missing the target.
- May 2003: Car bomb attacks on three housing compounds in Riyadh, Saudi Arabia, killing 34 people.
- July 2005 bombings of London public transportation system.
- February 2006 attack on petroleum processing facility in Saudi Arabia.

North Africa and the Middle East have been common targets for al-Qaeda attacks. Nationals from Saudi Arabia have both been a large part of the al-Qaeda movement and provided financial backing for the organization. The country is also one of al-Qaeda's primary targets, with local cells made of Afghan alumni having carried out various suicide attacks, including attacks on housing compounds in Riydah in May and November 2003 and attacks against oil companies and petroleum plants. Egypt too has been victim to such attacks. On April 2005, two separate attacks targeted Cairo and a devastating suicide attack in July 2005 on the resort of Sharm al Sheikh left 90 dead.

Iraq has experienced the most suicide attacks in the world, resulting in the deaths of hundreds of local Iraqis and foreigners (Figure 4.14). Since March 2003, in the context of the American-led war in Iraq, a suicide terrorism campaign led by al-Qaeda's representative in Iraq, the late Abu Musab al Zarqawi, has plagued the country.[177]

Four days after the attacks on a Riydah compound, on May 15, 2003, 13 suicide bombers attacked five different targets in Casablanca, Morocco, leaving a total of 45 people dead. The targets included a hotel, two buildings housing Jewish organizations, the Belgian consulate, and the "Casa de Espana," a meeting place for Spanish expatriates.

In a "franchise"-style operation, a group called al Salafia al Jihadia (or al-Sirat al-Mustaqim), an offshoot of the al-Qaeda-affiliated Moroccan Islamic Combatant group, took responsibility for the attack. Iraq al-Qaeda leader at the time, Abu Musab al-Zarqawi, reportedly may have been linked to the attack as well.

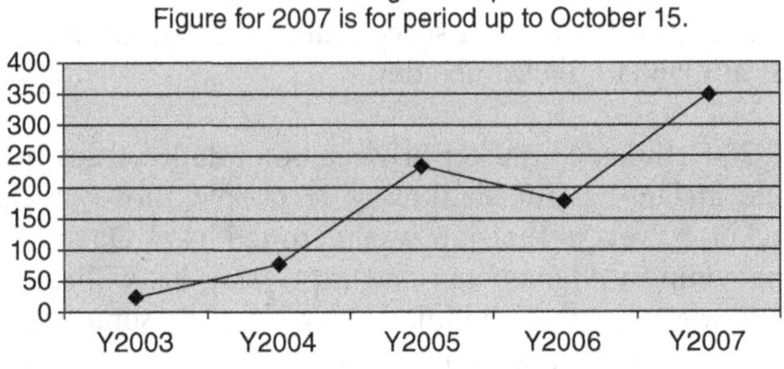

Figure 4.14. Chart of Iraq suicide bombings.

> [Al-Salafiya al-Jihadiya] is not in a hierarchical relationship with Al-Qaeda. Instead, the appropriate analogy here is that they were its "fellow travelers," akin to the sputniks used by the Bolsheviks. In some ways the terrorist attack on Casablanca resembled the fake "Rolex" watches and "LVMH" luggage sold in Moroccan markets. The Al-Qaeda brand has been applied to what is in fact a local product, which initially appears impressive.

In his investigation of the Casablanca attacks, Jack Kalpakian further describes that there are major differences in the tools, scale, training, and targets between central al-Qaeda operations and those of the al-Salafiya al-Jihadiya, as is the case with many al-Qaeda "franchise" organizations.[178]

According to Schweitzer, direct al-Qaeda attacks generally have longer planning periods, extending a number of years as opposed to short 9- to 12-month periods. Al-Qaeda attacks usually have two or more attackers and simultaneous explosions, focusing on symbolic targets and then "soft" targets. Partner groups have a range in number of attackers and tend to combine suicide attacks with other forms of attack.[179]

A traditional hallmark al-Qaeda-style attack involves simultaneous bombings to maximize damage as well as the use of large truck bombs containing explosive material such as ammonium nitrate and diesel. Such attacks are generally costly; Jemaah Islamiya spent up to $30,000 on each truck bomb in their Bali bombings, funded by al-Qaeda.[180] Al-Qaeda attacks often combine explosives with the use of vehicles—be them boats, buses, trucks, or aircraft—as delivery systems in order to increase the number of casualties.

In addition to symbolic targets, al-Qaeda is also very interested in launching its suicide attacks against economic targets, causing significant damage to local and world economies and disrupting daily life. In fact, in a videotape aired on Al Jazeera in late 2001, Osama bin Laden bragged about the damage to the American economy as a result of physical damage to New York and drops in the stock market. In a later video, bin Laden said: "The youths of God are preparing for you things that would fill your hearts with terror and target your economic lifeline." Al Zawahiri has also called for the destruction of the American economy in his video recordings.[181]

TABLE 4.2. Al-Qaeda and Affiliate Characteristics

	Al-Qaeda	Affiliates
Planning:	Precise and extended, sometimes lasting a number of years.	Short-term, usually no more than a year.
Management and Command:	Overall command by a senior member of al-Qaeda's central command and supervised by an operational commander in the target country.	Supervised by the commander of the local terrorist network.
Number of Attackers:	Usually two or more.	From individual attackers to groups.
Nature of Attack:	Usually simultaneous and parallel mode.	Integrated with other modes of attack.
Targets:	Focus on symbolic, hard and soft targets.	Usually "soft" targets.
Claiming Responsibility:	No direct claim of responsibility until September 11, 2001. After 9/11, a move to direct claim of responsibility.	Always claim responsibility, usually under different names. Clearly associated with al-Qaeda and the idea of Global Jihad

Source: Schweitzer (2006, p. 144).

The working assumption, as detailed in Table 4.2, that al-Qaeda requires a long duration to plan and carry out attacks, may be less accurate today. By using technological and media outlets to disseminate its message, al-Qaeda has been able to more easily facilitate its recruiting, planning, and implementation processes. In the past, operatives would have to wait weeks or perhaps even months to receive instructions by means of smuggled tapes through porous borders; today, however, information on possible targets can be easily gathered by open sources, accurate maps can be found online, and instructions can be relayed immediately by satellite or other means. Furthermore, with the increasing reliance on "homegrown" terrorists, law enforcement cannot assume that attacks require an extensive planning period.

Al-Qaeda in Europe

In the European context, bombings in Madrid, London, and Istanbul have had devastating local consequences (Figures 4.15 and 4.16). While not actually a suicide attack, on March 11, 2004, 10 bombs were detonated in four different locations on the Madrid commuter trains, killing 191, injuring another 2050, and resulting in dramatic political ramifications.[182]

CASE STUDY

ISTANBUL, TURKEY DOUBLE SUICIDE ATTACKS

Date of Incident: November 15 and 20, 2005
Location: Istanbul, Turkey
Type of Incident: Double Suicide Attacks
No. of Attackers: 4 suicide bombers, Turkish Nationals
Organization: Local affiliate network associated with Turkish Hezbollah or the IBDA-C

Planning and Preparation before Attack: The two major attacks were locally planned and perpetrated, using resources and infrastructure of an already established Islamist terrorist cell associated with Turkish Hezbollah, or the Turkish Islamic militant group IBDA-C, the Great Eastern Raiders' Front, which took responsibility for the attack. The leaders of the cell were trained in al-Qaeda camps in Afghanistan, and they had met with Bin Laden prior to the attacks to get his "blessing." They maintained communication with the organization through the Internet as they prepared the attacks. Bin Laden reportedly gave authorization for the attack and provided funding.

Motivation for Attack: When groups claimed responsibility for the attacks, they warned that new attacks against the United States and its allies were being planned. Turkey is a supporter of the U.S.-led war on terrorism and has strong relations with Israel. Turkish President and Foreign Minister at the time, Abduallah Gul, has remarked that terrorists in Turkey take issue with the country as a modern, democratic Muslim state. The attacks may have been intended to undermine Turkey's pro-Western and secular policies. In one report, a suspect in the bombings said it was the aim of the organization to "take action against American and Israeli targets and to break their dominance over Islamic countries." Another noted that the global Muslim community had been oppressed and that Jihad could be conducted locally.

Description: On November 15, two almost simultaneous attacks were launched against Istanbul's two main synagogues, the Beit Israel and Neve Shalom. Two suicide terrorists from eastern Turkey, ages 22 and 29, detonated two pickup trucks filled with hundreds of kilograms of explosives and chemical fertilizers, concealed beneath boxes of cleaning substances. Twenty-three people were killed and 300 wounded in the attacks. Only five days later, on November 20, 2005, another cell of the same al-Qaeda supported Turkish network carried out a double suicide attack, targeting the British consulate and a branch of HSBC Bank. The two car bombs exploded 12 minutes apart from each other, killing 34 people and injuring 500.

Consequence: The bombings left a total 57 dead and injured more than 800. They also caused damage to nearby buildings. In the November 20th attack, bombers may have waited for traffic lights in front of the HSBC bank to turn red in order to maximize damage. A top British official, Consul General Roger Short, was killed in the attack targeting the British consulate. Authorities also note that the attack may have been timed to coincide with a visit to the United Kingdom by President George W. Bush.

Figure 4.15. Case Study: Istanbul, Turkey double suicide attacks.

Analysis: The incident was one of the worst suicide terrorist attacks in Turkish history. The country had previously suffered smaller-scale suicide attacks by separatist groups, but the November 15 and 20 attacks were perpetrated on a larger scale in terms of the lethality and the central location of the targets. They further highlighted that the Global Jihad trend had spread to Turkey. Al-Qaeda appears to have harnessed the capabilities of already trained and established terrorists in Turkey to conduct an attack on its behalf, offering financial support but allowing the planners to make the primary decisions and prepare locally. Al-Qaeda leadership discussed preferred targets but gave the local network the freedom to choose them. These activists once belonged to local terrorist organizations within Turkey and decided to travel and train in al-Qaeda camps in Afghanistan. According to transcripts from one report, the planners had organized themselves into a cell before they approached al-Qaeda in Afghanistan, looking for its assistance and blessing. The attack serves as an indication of al-Qaeda's ability to harness local infrastructure, especially within Muslim communities, to find willing activists for their global cause, a trend that has spread.

Sources:

BBC News reports; Fox News, MSNBC, CNN, USA Today, http://www.washingtonpost.com/wp-dyn/content/article/2007/02/12/AR2007021201715_pf.html

Figure 4.15. *Continued.*

Immediately following the attack, Spanish authorities identified the Basque separatist group ETA as the primary suspect. The ETA had a history of carrying out attacks during elections, and the remote-control detonation used was hallmark ETA style as opposed to al-Qaeda's suicide attacks.[183] However, this time it was not ETA.

The bombings took place three days before the national election, in which the presiding party led by Prime Minister Aznar was leading the polls. Analysts say the attacks may have been timed in order to influence Spanish voters, provoking a backlash against the ruling party, unpopular for having sent troops to Iraq and considered a strong ally of the United States. As it became clear in the days following the attack that a local cell of Islamic radicals was involved, and not ETA, voters reacted at the polls. The party lost the election on March 14, and a new socialist party government of Prime Minister Jose Luis Rodriguez Zapatero kept the campaign pledge of withdrawing Spain's 1300-troops from Iraq.[184] This served as a clear and immediate example of the effectiveness of terrorism in realizing a terrorist organization's goals.

The Abu Hafs al-Masri Brigade claimed responsibility for the attack on behalf of al-Qaeda. A videotape of al-Qaeda's supposed "European military spokesman" was released two days after the attack, claiming responsibility and stating that it had been intended as punishment for Spain's support of the U.S.-led war in Iraq.[185] The translation of the direct statement was as follows:

> The attack was in retaliation for Spain's involvement in the war in Iraq. It was a response to the crimes you have committed around the world, chiefly in Iraq and Afghanistan, and there will be more, God willing.[186]

CASE STUDY

LONDON ATTACKS

Date of Incident: July 7, 2005
Location: London, England
Type of Incident: Multiple suicide bombings against public transport system
No. of Attackers: 4 suicide bombers
Organization: Local cell, al-Qaeda affiliate

Motivation for Attack: According to a recorded video by the attack's ringleader, Mohammad Siddique Khan, aired on Al Jazeera TV and discovered in his will, the bombers were seeking to avenge perceived injustices carried out by the West against Muslims. Khan's will places greater emphasis on the importance of martyrdom as supreme evidence of his religious commitment. According to the government-commissioned report, the attacks were largely motivated by concerns over British foreign policy, seen as deliberately anti-Muslim, as well as the religious benefits of martyrdom.

Description: In a coordinated attack, four suicide bombers detonated themselves along London's public transport system during the morning rush hour. Three bombs went off nearly simultaneously—within 50 seconds of each other—around 8:50 A.M. in the underground train system, two along the underground trains outside Liverpool Street and Edgware Road stations and another between King's Cross and Russel Square. One hour later, a fourth bomb was detonated on a double-decker bus in Tavistock Square, near King's Cross.

Consequence: The bombings left 52 commuters and the four suicide bombers dead, and they injured more than 770. The attacks also caused a severe day-long disruption of the city's transport and the mobile telecommunications infrastructure countrywide.

Analysis: The incident was the worst terrorist attack in British mainland history. Just one month prior to the incident, Britain's Joint Terrorism Assessment Center concluded that there were no groups with the capabilities or intent to launch a suicide attack against the United Kingdom, subsequently downgrading the United Kingdom's threat level. As a result of the attacks, calls to withdraw British troops from Iraq gained momentum, and British involvement in the Iraq War and its relationship with the United States was undermined.

The attack was planned with a small budget by four men reportedly using the Internet for weapons preparation. Three of these men were British with Pakistani decent, and one was Jamaican. While authorities initially thought the attacks were the result of disenfranchised British Muslims, self-radicalized and self-selected and only operating locally, security officials have learned that the ringleader and a bomber visited Pakistani terrorist camps previous to the attack, where they likely received al-Qaeda training. Still, there is no evidence of direct support or planning by al-Qaeda. Damage caused by this relatively inexpensive attack was maximized by targeting crowed public transportation hubs. In addition, by launching simultaneous attacks, bombers strengthened the attacks' psychological impact on the public.

Sources:

BBC News reports; http://news.bbc.co.uk/2/hi/uk_news/4749649.stm http://www.iwar.org.uk/homesec/resources/7-7/report.pdf; Bruce Hoffman, http://www.rand.org/pubs/testimonies/2006/RAND_CT263.pdf; Al Jazeera, Al Qa'ida Claims London Bombing, September 3, 2005, http://english.aljazeera.net/English/archive/archive?ArchiveId=14586

Figure 4.16. Case Study: London attacks.

However, police authorities have not found evidence of direct al-Qaeda involvement—not in the planning, financing, or initiation of the attack. Authorities have identified loose links to the organization and to the radical Moroccan Islamic Combat Group and the Salafia Jihadia group, responsible for the attacks in Casablanca.

The Madrid attack was considered by some analysts to be an act of "home-grown terrorism," in which a local radical group sympathetic to al-Qaeda plans and perpetrates an attack independently.[187] Al-Qaeda often attempts to exploit the alienation felt by Muslim immigrants in the west in order to rally them behind a common cause of Jihad, inspiring them to form local cells or networks to conduct attacks.[188]

Another such example was the devastating bombings in Istanbul, Turkey on the 15th and 20th of November 2003. In this case, a local Turkish network was directly linked to al-Qaeda, although planning and perpetration of the attack was conducted locally. On November 15, two suicide bombers drove explosive-laden trucks into two synagogues in Istanbul, killing 24 and injuring more than 300. Just five days later, the British consulate and the HSBC bank headquarters were targeted, as two suicide bombers killed another 27 people with explosive-laden trucks. Local Turkish activists from southeastern Turkey, allegedly linked to Islamist terror organizations acting as al-Qaeda affiliates, were responsible for the attacks.[189] Turkish (Kurdish) Hezbollah was linked to the attack—as was the Turkish Islamic militant group IBDA-C, or the Great Eastern Raiders' Front.[190]

Al-Qaeda in Southeast Asia

Previous cooperation in Afghanistan linked the leaders of al-Qaeda with local affiliates throughout Southeast Asia, in particular to one of al-Qaeda's most active affiliate groups, Jemaah Islamiyah (JI) (Figure 4.17). JI is a militant Islamist group based in the world's largest Muslim country, Indonesia, and active across all of Southeast Asia. The organization's primary objective is to establish a pan-Islamic state across the region, from the Philippines to Indonesia.[191]

JI has cells operating in Indonesia, Malaysia, Singapore, and possibly the Philippines, Cambodia, and Thailand. Its most high profile attacks have included a series of suicide bombings in Bali in October 2002 and 2005. The October 12, 2002 attack was the worst incident of terrorism in Indonesia's history, killing a total of 202 people—mostly young Australians—and injuring hundreds more at two nightclubs. On October 1, 2005, three nearly simultaneous suicide bombings killed at least 19 people and wounded more than 100 in three restaurants in Bali's Jimbaran and Kutu districts. Jemaah Islamiah is also responsible for carrying out the 2003 JW Marriot hotel bombing in Kuningan, Jakarta and the 2004 Australian embassy bombing in Jakarta. JI has been directly and indirectly involved in dozens of bombings across the southern Philippines, usually in cooperation with the Abu Sayyaf Group.[192]

While the extent of cooperation between al-Qaeda and JI is unclear, some experts refer to the group as al-Qaeda's Southeast Asian wing, because it has

IN BRIEF: JEMAAH ISLAMIYAH

- Founded as JI in 1997, existed in ideological form since the late 1940s, offshoot of the Darul Islam movement, an organization active since the 1950s
- Goals: Establishment of a pan-Islamic state across the region, from the Philippines to Indonesia
- Ideology: Militant Islam, Global Jihad
- Modus Operandi/Tactics: Guerilla war launched in 1984, suicide campaign from 1996 to 1999; car bombings, kidnappings, attacks against foreign diplomats in Europe
- Headquarters/Areas of Operation: Indonesia; JI has cells operating in Indonesia, Malaysia, Singapore, and possibly the Philippines, Cambodia, and Thailand
- Activists/Members: Sunni Muslim, fighters
- Leaders: Abu Bakar Bashi *and* Abdullah Sungkar (founders), Hambali
- Structure: Organizational guidelines are outlined in the PUPJI. At the top of the organization is the *amir*, who appoints the leaders and controls different councils: governing council, a *fatwa* council, and a disciplinary council. The governing council is led by a central command.
- Connections to other groups: Al-Qaeda, the Abu Sayyaf Group, the Moro Islamic Liberation Front (MILF), the Misuari Renegade/Breakaway Group (MRG/MBG), Darul Islam, Fazlul Rahman's (HUJI), the Rohinga Solidarity Organization, and others.

Figure 4.17. In Brief: Jemaah Islamiyah.

merged its local operations with al-Qaeda's agenda. JI has benefited from al-Qaeda funding and, although officially active for almost 10 years before developing strong ties with al-Qaeda, only adopted the suicide terrorism tactic once cooperation between the groups intensified and financial assistance was provided.[193]

Consequences and Counterterrorism

The attacks of September 11, 2001 provoked an immediate and multi-tiered response from the United States and the international community. Although early al-Qaeda attacks heightened awareness of the threat, the expanding nature of al-Qaeda was arguably underestimated and U.S. capabilities and preparedness to manage the threat remained limited prior to 9/11.[194] Terrorist threats were primarily approached as law enforcement challenges. According to the 9/11 Commission Report, many U.S. officials regarded terrorists primarily as agents of states or as domestic criminals, as opposed to the "new type of terrorism" posed by bin Laden and his expansive global network. While terrorist plots were effectively thwarted and operatives were captured, there was no one cohesive strategic policy aimed at countering al-Qaeda.[195]

Immediately following 9/11, the U.S. government reorganized to initiate a multi-level response to the devastating attacks. In order to minimize the immediate impact, the administration addressed the issues of federal emergency assistance, victim compensation, reopening the financial markets, and restoring civil aviation. Because there was no one government organization dedicated or prepared to dealing with the aftermath of the attack, Vice President Dick Cheney recommended that a new entity be established to coordinate all the relevant departments. By September 20, 2001, a Homeland Security Advisor and Homeland Security Council were announced.[196]

In efforts to diminish al-Qaeda operations in the United States and locate suspects, individuals that had violated immigration laws were arrested in the context of the FBI's investigation of the 9/11 attacks. They were held until cleared of any connections to terrorist activities, including 9/11 plotter Zacarias Moussaoui. In the first weeks following the attacks, the U.S. administration also worked quickly to improve its ability to collect intelligence inside the country and to share information between intelligence and law enforcement agencies, an issue that, according to the 9/11 Commission Report, was not a high priority before 9/11. The U.S. Patriot Act was the outcome of these initial efforts, formerly signed into law on October 26, 2001.[197]

After the September 11th terrorist attacks on the United States, President George W. Bush declared a "war on terrorism," committing the country to the destruction of al-Qaeda and Osama bin Laden's network. The war was not limited to al-Qaeda itself, but to terrorist groups functioning all over the world—a U.S.-led campaign that completely changed the political landscape. A pre-9/11 draft presidential directive on al-Qaeda expanded to become a global war on terrorism, making no distinction between terrorists and those who harbor them.[198] The U.S. goal was to eliminate all terrorist networks, eliminate Taliban support for the organization, dry up their financial support, and prevent them from acquiring weapons of mass destruction. The directive was formally signed on October 25, 2001, after fighting in Afghanistan began.[199] The United States sought a wide coalition in its war in Afghanistan, dubbed Operation Enduring Freedom, gaining the support of the United Kingdom and Canada and other allied nations, whether directly through the provision of troops or through other logistical or material support. The international effort to confront al-Qaeda after 9/11 has been primarily led by the United States, although assistance by allied coalition forces and local groups in Afghanistan and Iraq have aided U.S. efforts.

Since the war began, several senior al-Qaeda operatives have been captured or killed, including 9/11 mastermind Khalid Sheik Mohammed, considered third in command of al-Qaeda and captured near the Pakistani capital of Islamabad on March 1, 2003. Khalid Sheik Mohammed has been allegedly involved in almost every major al-Qaeda attack, including the 1998 attacks against the U.S embassies in Kenya and Tanzania, the USS *Cole* attack, and the attacks against the Djerba Synagogue.[200] Some other al-Qaeda operatives captured or killed include the following:[201]

- Waleed bin Attash (arrested April 29, 2003), alleged al-Qaeda operational commander
- Mohammed Atef (a.k.a. Abu Hafs) (reported killed in mid-November 2001 in a U.S. airstrike near Kabul), bin Laden's most senior deputy and a family member, co-founder of the al-Qaeda terror network, a member of the group's ruling council, and a top military commander
- Abu Zubaydah (captured by Pakistani authorities, March 2002), senior planner of al-Qaeda terrorist operations
- Mustafa Ahmed al-Hawsawi (a.k.a. Shaihk Saiid; Sa'd al-Sharif), financial support for 9/11
- Mohammed Omar Abdel-Rahman (arrested February 2003 in Pakistan), allegedly a senior al-Qaeda operative

While much of al-Qaeda's leadership has been captured or killed and the organization's infrastructure in Afghanistan was largely dismantled by U.S.-led operations, recent intelligence reports indicate that Osama bin Laden has relocated and rebuilt its base in ungoverned Pakistani tribal areas (Pakistan's Federally Administrated Tribal Areas—FATA). Bin Laden has established training camps and recaped ability to attack and reach out to activists and affiliates all over the world.[202] Camps in the Pakistani tribal areas are smaller than those operating in Afghanistan before the 2001 invasion, but are estimated to host as many as 2000 local and foreign fighters, according to news reports.[203]

Al-Qaeda is still considered to be one of the greatest threats to the United States and its allies, because it has rebuilt its operational capabilities, restored some central control, and replaced captured or killed leaders. According to State Department reports, Osama bin Laden has remained the "ideological figurehead," and Ayman al-Zawahiri has become the organization's primary operational planner.

The organization has also continued to expand its network by reaching out to new and established affiliate groups, "exploiting local grievances for their own local and global purposes."[204] For example, Algerian insurgents have reportedly joined the movement, in a form of outsourcing that indicates that al-Qaeda's capabilities and organizational structure will allow it to continue launching attacks. The Algerian Salafits Group for Preaching and Combat (GSPC), led by Abdelmalek Droukdal, made contact with Abu Musab al-Zarqawi, the late leader of al-Qaeda in Mesopotamia, paving the way for what has been dubbed a "corporate merger."[205] The group has since grown into one of al-Qaeda's strongest affiliate forces. Renamed al-Qaeda in the Islamic Maghreb, these militants have since accepted al-Qaeda support and adopted signature al-Qaeda tactics and targets, as demonstrated by the truck bombings of UN and court offices in Algiers in December 2007, which killed 41 people and injured 170. The group launched more than eight suicide bombings using vehicles in 2007. As this nationalist insurgency transforms into part of the Global Jihad movement, al-Qaeda has benefited through hundreds of experienced fighters and connections to activists living in Europe (see Figure 4.18).[206]

VOICES FROM THE FIELD

"These groups, as best we can tell, have a fair amount of independence. They get inspiration, they get sometimes guidance, probably some training, probably some money from the al-Qaeda leadership ... it's not as centralized a movement as it was, say, in 2001. But in some ways, the fact that it has spread in the way that it has, in my view, makes it perhaps more dangerous."

Defense Secretary Robert M. Gates on the merger between al-Qaeda and affiliate groups (June 2008)

Figure 4.18. Voices from the field.

This North African branch of al-Qaeda has been able to establish small training camps for militants from Tunisia, Morocco, and Nigeria, identified as hot spots in the State Department's annual report on global terrorism.[207]

In a now famous speech, in May 2003 President George W. Bush declared after military victories in Iraq that "al-Qaeda is on the run. The group of terrorists who attacked our country is slowly but surely being decimated."[208] Yet, however loosely coordinated attacks in recent years have been, al-Qaeda has proved that it continues to seek out targets around the world and is capable, whether directly or through its network of affiliates and independent cells, to commit attacks. As Professor Hoffman aptly notes, al-Qaeda today is rather, "on the march."

> Although al-Qaeda is often spoken of as if it is in retreat—a broken and beaten organization incapable of mounting attacks, its leadership cut off, living in caves somewhere in remotest Waziristan—the truth is that the organization is not on the run but on the march. It has regrouped and reorganized from the setbacks it suffered during the initial phases of the global war on terrorism and is marshaling its forces to continue the epic struggle begun more than 10 years ago. ... Rather than "al-Qaeda R.I.P.," we face an al-Qaeda that has risen from the grave.[209]

The threat of al-Qaeda attacks is currently most acute to Europe in general and France in particular, but there are experts that assess that the threat level to the United States is similar to the level it was prior to 9/11. Analysts note that the majority of attacks in recent years have been carried out by local groups or regional affiliates, as opposed to the core and central leadership of al-Qaeda.

However, as it becomes the new headquarters for al-Qaeda leadership, Pakistan has become a breeding ground for radicalization and may come to serve not only as a "stronghold" on the ideological battlefield, but on the physical front as well. The United States and Europe remain vulnerable to attacks by the Global Jihad movement.[210]

The Future of al-Qaeda

Although the al-Qaeda phenomenon may change form, most experts agree that it is not likely to dissipate in the near future. Different organizations and local networks will most likely continue to harness the al-Qaeda umbrella to support their activities, with or without direct al-Qaeda control or direction.

In order to confront the threat, counterterrorism measures, in terms of the preemptive and preventive approaches currently being applied, must continue. Concerning the battlefront, as long as the fight is taken to areas of al-Qaeda presence, it will be more difficult, although feasible, for al-Qaeda and its affiliates to carry out major attacks against the U.S. homeland. As for the homefront, key factors in successfully confronting the threat include "public awareness," which is significantly higher than it was prior to 9/11, in addition to "public resilience," which has fortunately not been tested directly in the United States since 9/11.

In line with the current trends in international terrorism, the home front has become a primary battlefront. The awareness and resilience of civilians, soldiers at home, and law enforcement is crucial. Suspicious activities should be immediately reported to the authorities, and efficient investigations should be carried out. Citizens should be alert of suspicious activities, especially when individuals appear to be conducting intelligence gathering operations on potential targets. Signs may include people wandering in the vicinity of high-valued facilities or public buildings, taking photos or videotaping. In addition, individuals who purchase an excess amount of materials that could be used to prepare bombs—such as pipes, canisters, pesticides and ammonia nitrate fertilizers—should alarm store owners. Without undermining the grief of victims, the aftermath and consequences of attacks should be mitigated in such a manner so that civilian and commercial life can be restored as soon as possible.

The most important component of confronting al-Qaeda and similar organizations is intelligence. Efficient collection, dissemination, and operational use of intelligence is the most crucial factor in successfully countering the threat. Achieving such intelligence is only possible with wide international cooperation.

CHECHNYA

The use of suicide terrorism is a relatively new tactic in the Chechen conflict, with the first strike having occurred in just 2000. The strategy has nevertheless been adopted as a favored mode of attack among its perpetrators, primarily because of its effectiveness and lethality. Between the first attack in June 2000 and the last to date in mid-2005, approximately 940 people have been killed in 28 different suicide attacks carried out by 112 Chechen separatists.[211] One of the most interesting aspects of this trend, on a local level, is the fact that more than one-third of these attacks—with some estimates as high as 43%—were carried out by female suicide bombers, known as "black widows."

History of Conflict

The phenomenon of Chechen suicide terrorism emerged during a second wave of conflict in a more than century-long struggle. With roots dating back to the early 19th century, the Chechen conflict can be characterized primarily as a separatist struggle, inspired by the movement to establish a Chechen state independent from mother Russia. Chechen rebel Iman Shamil and his fighters attempted to establish an independent Islamic state as early as 1858, eventually defeated by the Russians after decades of violent resistance. The movement experienced a significant resurgence with the collapse of the Soviet Union in 1991, when Chechen president Dzhokhar Dudayev declared independence from Russia—a move largely ignored by Soviet leaders.[212]

The conflict has since escalated into two regional wars: The first one occurred between 1994 and 1996; and the second one, which has yet to subside, started in 1999. During the years between 1996 and 1999, when Chechnya experienced quasi-independence after Russian forces failed to defeat the Chechens, the region fell into a period of disorder and violence, marked by hostage-taking, lawlessness, and warlords.[213] As the conflict has continued into the 21st century, Chechen separatist groups and resistance movements have increasingly adopted the tactics of guerilla warfare and terrorism, conducting attacks against targets both inside and outside mainland Russia on behalf of their cause—calling for an independent Chechnya and the withdrawal of Russian troops.[214]

Chechen Separatists and Suicide Terrorism

In the first two years that suicide terrorism was employed by Chechen groups, attacks were conducted on a relatively small scale and targeted only military bases within Chechnya. The first suicide attack occurred on June 7, 2000, when two female terrorists drove an explosives-laden truck into the temporary headquarters of a Russian Special Forces detachment in the village of Alkhan Yurt in Chechnya, killing two and injuring five others.[215] Targets of suicide attacks have since expanded to include civilian transportation, military bases, and government facilities in Moscow, Chechnya, and the surrounding North Caucus and Southern Russian region. Estimates vary, but about 46% of these attacks were directed against strictly civilian targets, with military bases accounting for about 39% of attacks. More than half the attacks that have occurred within Chechnya were targeted against military compounds (see Table 4.3).[216]

However, suicide attacks carried out by Chechens since 2002 have become increasingly grandiose, focused on civilian targets often within Russia that involve large amounts of weaponry, wide use of suicide terrorists, and, consequently, heavy media coverage. One of the most high-profile incidents to mark the increasing reliance on suicide terrorism was the takeover of the Dubrovka theater in Moscow in October 2002, in which 40 attackers, armed with belt-bombs and other weapons, took approximately 800 people hostage over a three-day period, threatening to blow up themselves and the theater if their demands were not meet. An unsuccessful rescue attempt by the Russian Special forces ended the siege but left 129 hostages dead (see Figure 4.19).[217]

A similarly devastating incident took place on September 1–3, 2004 in northern Ossetia in a town called Beslan (Figure 4.20). Thirty-two terrorists carrying

TABLE 4.3. Suicide Attacks by Year, Type, and Location[6]

Year	Military Bases	Government Places	Civilian Places	Total	Chechnya	Southern Russian Region	Moscow	Total
2000	4	0	0	4	4	0	0	4
2001	1	0	0	1	1	0	0	1
2002	1	1	1	3	2	0	1	3
2003	4	2	5	11	4	4	3	11
2004	0	1	5	6	0	2	4	6
2005	1	0	2	3	3	0	0	3
Total	11	4	13	28	14	6	8	28
	(39%)	(14%)	(47%)		(50%)	(21%)	(29%)	

explosives and a variety of weapons, some wearing belt-bombs, took hundreds of school children and adults hostage. The incident, which began as a negotiating hostage situation and ended as a suicide attack, took the lives of over 300 people, half of whom were children.[218] Russian media reports speculated that foreign, non-Chechen operatives—including Arabs—took part in the attack.[219]

These unprecedented attacks are notable not only for their lethality, but also for their distinct similarities to al-Qaeda-style operations. The theater attack, in addition to a September 2004 attack on a Beslan school, required diligent planning, training, and dedication to self-sacrifice. The incidents involved multiple perpetrators and resulted in numerous casualties.

Such attacks point to a larger trend in the Chechen conflict. During the second war, the Chechen conflict evolved from a strictly nationalistic struggle into, at times, a religiously motivated campaign. Religious rhetoric and ideology increasingly entered the otherwise nationalistic and separatist motivations of these Chechen groups, many of which have gradually adopted the goals, tactics, and ideology of the Global Jihad movement.

After the withdrawal of Russian troops in 1996, the post-war atmosphere in Chechnya helped foster the spread of radical Islam. Saudi preachers flooded into the North Caucus region, finding new recruits and spreading a message associated with the Global Jihad movement. Chechens were trained in al-Qaeda camps in Afghanistan, receiving support from al-Qaeda and serving as agents of influence upon their return. They helped disseminate the concept of Global Jihad in general and the idea of self-sacrifice in the name of Allah in particular. Other reports cite the establishment of training camps within Chechnya and the subsequent recruitment of local Jihadists.[220]

Several foreign Jihad fighters shifted their focus from Afghanistan to the conflict in Chechnya—taking up arms against the Russians and adopting the al-Qaeda Jihad philosophy of assisting Muslims in conflicts all over the world against Christian or Jewish infidels. In the late 1990s, several foreign Jihadists joined the Chechen campaign, including Ayman al-Zawahiri, who later became

IN-DEPTH CASE STUDY

TERROR IN THE THEATER

On Wednesday, October 23, 2002 a group identified as members of the 29th Division of the Chechen Army stormed the Dubrovka theater building, a cultural center on Melnikov Street in the south of Moscow, and took it over just as the second half of the "Nord Ost" performance was about to begin. After some of the audience managed to escape during the initial confusion and about 75 foreigners were released, Movsar Barayev, the leader of the group, announced that he and 40 of his comrades (19 of whom were women) were "suicide attackers" prepared to die for their cause. The group held 650 civilians hostage in the heart of Moscow.

The attackers' leader, Movsar Barayev, demanded the complete withdrawal of Russian troops from Chechnya, threatening to kill the hostages if his demands weren't met within seven days. In a videotape released to the Arabic news channel "Al Jazeera," Barayev announced: "Each one of us is willing to sacrifice himself for the sake of God and the independence of Chechnya. I swear by God that we are keener on dying than you are on living."[1]

At the time the theater was attacked it was holding a performance of a Russian musical. "People in camouflage uniform ran onto the stage when the show was already in progress. They started shooting into the air from assault rifles,"[2] a woman who managed to escape from the theater told Russian media.

Heavily armed elite Russian troops surrounded the building. Russia's Federal Security Service (FSB) and Interior Ministry Special Forces arrived at the scene and an emergency headquarters was established. The Russian authorities began negotiations with the Chechen rebels, and a Russian official reported that the attackers were putting forward an "unacceptable demand to end to military action in Chechnya."

Cell phone conversations with hostages trapped in the building revealed that there were over 40 Chechen terrorists, many of them women. The terrorists, armed with grenades and other explosives strapped to their bodies, deployed a range of explosives throughout the theater and indicated they would blow up the entire building if government security personnel attempted to attack.

While security forces were taking up positions, the rebels released 180 hostages, mainly Muslims, including 24 children and two pregnant women, after killing one woman who tried to escape as the rebels took over the building.[3] The crisis continued onto its second day when elite troops trotted past clusters of locals in the doorways of surrounding blocks of flats, taking up positions within sight of the besieged theater.

During the siege, several influential officials, including a former Russian prime minister, entered the theater complex in an attempt to negotiate with the hostage-takers. A Russian spokesman said that the Chechen rebels would be allowed to leave freely if they released the hostages unharmed. Seven other Russian men and women were released earlier that day; and although the rebels said they were prepared to release all 75 foreign nationals into the hands of diplomats, the move was repeatedly postponed. The rebels continued to threaten to kill hostages if their demands were not taken seriously. The situation was tense inside the theater, and conditions were growing worse as the captives did not receive food or water.

Russian President Vladimir Putin offered a guarantee of safety to the heavily armed terrorists if they released all the hostages. After a night of heavy explosions and repeated bursts of gunfire, Russian Special Forces stormed the building early on Saturday morning. The dramatic end to the three-day standoff came just before the dawn deadline set by the terrorists, who threatened to begin killing the remaining hostages

Figure 4.19. In-depth case study: Terror in the theater.

if the Russian government did not meet their demands. Russian Special Forces, from the Federal Security Service (FSB), spread a sedative gas before storming the theater. The gas appeared to have instantly subdued the terrorists but also caused 120 hostage deaths. Only one hostage died of gunshot wounds. About 600 of the 646 rescued people where hospitalized because of effects of the gas, while 150 were placed in intensive care. Almost all the Chechen hostage-takers were reported killed, with claims some were shot at close range while incapacitated.

Russia's Deputy Interior Minister, Vladimir Vasilyev, said a sleeping gas had been used to "allow us to neutralize, among others, those women kamikazes who were literally encased in explosives with their fingers on the detonators." He added, *"We have proved that Russia cannot be brought down to its knees."*

Russian President Vladimir Putin apologized for the captives' deaths in a television address, saying: "Please forgive us. The memory of the victims must unite all of us." Russian television footage from inside the theater showed the camouflage-clad body of the gunmen's leader Movsar Barayev, lying on his back amid blood and broken glass. In the theater hall, the corpses of several of the female captors, dressed in black robes and head coverings, were found sprawled among the theater's seats. The Russian government declared the outcome a victory in the continuing war against Chechnya, but many Russians—including politicians, journalists, and intelligence agents—were not convinced.

[1]Al Jazeera satellite TV channel October 24, 2002.

[2]http://news.bbc.co.uk/1/low/world/europe/2354753.stm

[3]Associated Press http://www.islamundressed.com/C-Current_News.htm

Case Study Details

Date of Incident: October 23, 2002 until October 26, 2002
Location: Dubrovka Theater, Moscow Russia
Type of Incident: Hostage taking and intent to commit suicide attack
No. of Hostages: 650
No. of Attackers: 41, 19 of which were women
Organization: 29th Division of the Chechen Army, (SPIR)

Declared Demands: Withdrawal of Russian troops in Chechnya and independence of Chechnya.

Motivation for Attack: The attackers intended to present their cause before a global audience and to carry out a major attack of mass casualties in order leave a mark to remember.

Manner of Confrontation: Tactical negotiation process that facilitated the release of a number of hostages and eventually a military effort to free hostages.

Consequence: 120 killed hostages and 41 killed captors

Analysis: This incident illustrated striking capabilities and extreme motivations of terrorist organizations to take hundreds of civilians captive in the heart of a major capital. In their effort to publicize their cause, the attack was meant to leave a mark by means of mass casualties. On the other hand, for the first time, security forces made use of a sedative and incapacitating gas based on the drug fentanyl in an effort to free the captives. The use of this rescue method resulted in many civilian casualties but may lead to additional modes of tactical countermeasures in the future.

Figure 4.19. *Continued.*

Operational Details

- Immediately after receiving information of the hostage incident at approximately 9:00 P.M. on October 23, 2002, special forces began training and conducting simulations on similar buildings.
- Training continued throughout the crisis.
- After detailed planning, the attempted rescue raid commenced at approximately 5:00 A.M. on October 26, 2002.
- Special forces simultaneously broke into the building in a number of different entrances.
- 530 hostages were freed. Many of them were hospitalized due to inhaling gas.
- Approximately 120 hostages were killed on site due to gas inhaling.
- 41 terrorists killed
- 2 security force personnel injured

Key Tactical Notes

- The force consisted of 13 teams of between 10 and 20 operatives. In total, 160 operatives entered the complex.
- 12 sharpshooter teams were placed in observation positions.
- Members of the force were reported to have been able to infiltrate the complex before the main force entered.
- The forces began advancing two hours before storming the complex.
- Thirty minutes before storming the complex, inaccurate shoots were forfired 4–5 minutes by the terrorists after they thought they spotted troops.
- The troops did not fire back and after the terrorists' alert level climaxed. Shooting subsided once they believed the threat had passed.
- At approximately 5:00 A.M., once the terrorists alert level had decreased, forces stormed the theater.
- The overtaking of the theater included:
  - Sharp shooting—killed 8 terrorists
  - Dissemination of gas through the air-conditioning system
  - Simultaneous entry of forces in a number of different entrances
  - Dismantling of booby traps and explosives
  - Evacuation of hostages
- During entry, the forces used gun silencers.
- Special forces first targeted Movsar Barayev, the attackers' leader.
- One of the forces entered from behind the stage, others entered through various windows and other entrances.
- The use of the sedative gas was very effective in neutralizing the terrorists but also had an adverse effect on the hostages, killing scores. The hostages were especially vulnerable to the gas because they were physically weak after being denied food and water prior to the rescue operation.

Figure 4.19. *Continued.*

IN-DEPTH CASE STUDY

BESLAN SCHOOL SIEGE[1]

Description: The Beslan school hostage crisis began when a group of armed Chechen terrorists, identified as the Riyadus-Salikhin Martyr Battalion led by Shamil Basayev, took approximately 1200 children and adults hostage on September 1, 2004, at School Number One (SNO) in the town of Beslan, in the North Caucasus region of Russia. After a three-day standoff, Russian security forces stormed the building using tanks and heavy weaponry. The siege led to a series of explosions and a fire that engulfed the school. Hostage takers and the Russian security forces engaged in a chaotic gun battle before the situation was stabilized. More than 350 hostages were killed, including 186 children and hundreds more were wounded.

Case Study Details

Date of incident: September 1–September 3, 2004
Location: School Number One (SNO) in the town of Beslan
Type of Incident: Hostage-taking and willingness to commit suicide attack
No. of Hostages: Approximately 1200

No. of Attackers: There are discrepancies concerning the number of attackers, but most reports state that there were 32 including 2 women equipped with suicide vests.

Perpetrators: Chechen group called Riyadus-Salikhin "martyr battalion"

Planning and Preparation before Attack: In-depth information gathering was carried out by the attackers to facilitate the attack. There are also reports that men disguised as repairmen concealed weapons and explosives in the school, months before the actual attack.

Declared Demands: Withdrawal of Russian troops in Chechnya and independence of Chechnya.

Motivation for Attack: The attackers intended to present their cause before a global audience and carry out a major attack of mass casualties in order leave a mark to remember.

Manner of Confrontation: Tactical negotiation process that facilitated the release of a number of hostages and eventually a military effort to free hostages.

Consequence: At least 385 fatalities and over 870 injured.

Analysis: This incident illustrated significant terrorist capabilities. Faulty Russian intelligence, combined with detailed information gathering and planning by the terrorists, facilitated the takeover of more than 1000 hostages. Putting mines in the auditorium

Figure 4.20. In-depth case study: Beslan school siege.

where most of the hostages were gathered and placing explosive devices in the rest of the building, in addition to surrounding the hostages with tripwires, made any attempt to free hostages by force virtually impossible. The attackers learned lessons from previous incidents. They broke windows in the building for ventilation, negating the prospects of using gas. In order to hinder sharpshooting efforts, they placed hostages at window entrances to serve as "human shields." This incident, similar to the one in the Moscow Theater almost two years earlier, was carried out by sophisticated attackers that wanted their cause to be heard by the world and were willing to commit suicide in order to further their goals. Unlike the Theater incident, the terrorists in Beslan carried out relatively practical negotiations.

Key Operational Lesson: Response and rescue efforts should take into consideration the possibility that operational plans do not proceed as planned and that rescue forces might be compelled to operate at a time that is not optimal.

Incident Timeline

Wednesday, September 1, 2004

9:00 A.M.: On the first day of the school year, a number of military camouflaged trucks break into the school grounds and a number of heavily armed men and two women wearing suicide bomb belts take control of the school. During the initial confusion, about 50 people manage to escape. About 1200 are taken hostage in the school gym. Eight people are killed and about 12 injured during the takeover.

10:00–11:00 A.M.: The attackers begin placing mines in the gym and other school buildings using improvised explosive devices packed with nuts, bolts, and nails. Inside the gym, the hostages are surrounded with tripwires, and explosives are stringed up over the hostages for maximum human damage. Response and rescue units arrive to the scene and take up positions.

11:15 A.M.: Hostage-takers force hostage children to stand exposed in the windows, acting as human shields.

Noon: Reports of explosions and shootings in the school.

2:00 P.M.: Reports of 10 hostages being killed by attackers as a sign of "seriousness."

3:00 P.M.: The two female hostage-takers detonate themselves or are detonated by their leader.

4:40 P.M.: Twelve children and one adult escape after hiding in nearby boiler room.

7:30 P.M.: Authorities establish first contact with hostage-takers.

8:00 P.M.: Twelve hostages are killed as another sign of "seriousness."

Figure 4.20. *Continued.*

Thursday September 2, 2004

2:00 A.M.**:** After negotiations throughout the night have failed, a number of the terrorists attempted an escape and were gunned down by sharpshooters. The attempted escape served as the initial sign of a split among the terrorists. Subsequently, the terrorists' leader killed three members of the group.

10:00 A.M.**–4:45** P.M.**:** Negotiations with hostage-takers were carried out while Russian security forces trained and conducted pre-entry simulations. The negotiations succeeded in releasing 26 mothers and nursing babies.

9:00 P.M.**:** Ongoing negotiation with hostage-takers continued throughout the night and succeeded in delivering food, water, and medicine to the hostages.

Friday September 3, 2004

12:00 P.M.**:** Attackers agree to let medical workers retrieve the bodies of 22 killed hostages dumped outside the school buildings.

12:10 P.M.**:** After agreement was reached with hostage-takers, two trucks approach the school to pick up the bodies.

12:40 P.M.**:** Medical workers approaching the building are fired at by the terrorists.

1:06 P.M.**:** In the course of clearing the bodies, two large almost simultaneous explosives are unexpectedly detonated, causing damage and confusion. A number of hostages are killed immediately, many others are wounded. About 30 hostages manage to flee as the terrorists fire on women and children.

The reasons for the explosions are not clear, but the three possible explanations are as follows:

1. One of the bodies was booby-trapped and exploded as it was being cleared.
2. One of the female terrorists mistakenly detonated her suicide vest.
3. One of the mines on the structure's roof was accidentally triggered.

1:10 P.M.**:** At any rate, the explosions caused a cascade of events that led to a forced breach of the Russian forces into the compound and resulted in a tragic ending.

1:30 P.M.**–11:00** P.M.**:** A series of explosions shook the school, followed by a fire that engulfed the building and a chaotic gun battle between the hostage-takers and Russian security forces. A number of hostages were rescued or managed to escape, and others were either killed or injured. Ultimately, over 350 hostages were killed, including 186 children. Hundreds more were wounded.

Sources include: (1) *Russia: Recounting the Beslan Hostage Siege—A Chronology*, by Jeremy Bransten; (2) The Calculus of Chechya's Suicide Bombers, *Chechnya Weekly*, **6**(2), January 13, 2005, by John Reuter; (3) undisclosed sources.

Figure 4.20. *Continued.*

deputy head of al-Qaeda.[221] Mohammed al-Atta expressed desire to join in the campaign, but was eventually redirected by al-Qaeda leader Osama bin Laden to the September 11, 2001 attacks on the United States.[222] Chechens cooperated with al-Qaeda in a propaganda campaign to transform their struggle from an independent, local movement into part of the larger global confrontation between Islam and the Jewish–Crusader alliance, as portrayed by bin Laden and his colleagues.[223]

While the base-motivation behind Chechen terrorist-sponsoring groups is nationalistic in nature, with the flow of foreign funding and the introduction of a more radical and militant brand of Islam into Chechnya, several groups have become more religiously oriented.

For example, the Moscow theater attack was carried out by a group led by Movsar Barayev, referred to as the 29th Division of the Chechen Army or the Special-Purpose Islamic Regiment (SPIR) (see Figure 4.21). The SPIR was loosely formed in 1996 by Arbi Barayev, a leading member of the mainstream Chechen resistance movement under the Chechen President, Aslan Maskhadov. After being stripped of his rank within the republic's armed forces, Barayev turned to his criminal faction, regarded as one of the primary hostage-taking and kidnapping groups operating in Chechnya after the 1994 civil war. He enlisted support from Islamic militants, including Saudi-born commander Amir al-Khattab. After being killed by Russian troops in 2001, Arbi Barayev's nephew, Movsar Suleimanov, took control of the organization and changed his last name

IN BRIEF: SPIR (SPECIAL-PURPOSE ISLAMIC REGIMENT)

- Alternate Names: 29th Division of the Chechen Army, Special-Purpose Islamic Regiment (SPIR), al-Jihad-Fisi-Sabililah Islamic Regiment
- Loosely formed in 1996 by Arbi Barayev, a leading member of the mainstream Chechen resistance movement, as a criminal organization
- Goals: Establishment of a Chechen state independent from mother Russia
- Ideology: Separatist, nationalist, increasingly influenced by Global Jihad movement and radical Sunni Islam, Wahhabi fundamentalism
- Modus Operandi/Tactics: Hostage and kidnapping, criminal activity, oil smuggling, suicide attacks, guerilla operations
- Headquarters/Areas of Operation: Chechnya, Russia, North Caucus region
- Activists/Members: 100 fighters at any given time
- Leaders: Amir Khamzat (Amir Aslan)—true identity unknown, Movsar Barayev, Arbi Barayev
- Connection to other groups: Al-Qaeda; other Chechen groups such as the Islamic International Brigade (IIB), led by Chechen rebel Commander Shamil Basayev; and the Riyadus-Salikhin Reconnaissance and Sabotage Battalion of Chechen Martyrs

Figure 4.21. In brief: SPIR (special-purpose Islamic regiment).

to Barayev to honor his uncle and "to strike horror in many hearts in Russia and to chill the people's veins."[224] Under Movsar Barayev, the organization became increasingly radicalized, adopting the Wahhabi creed of Islam imported to the region through Islamists like Amir al-Khattab, who had connections with al-Qaeda.

Other Chechen terrorist organizations include the International Islamic Brigade (IIB), which is led by one of the most prominent leaders of the movement, Chechen rebel Commander Shamil Basayev—once Russia's most wanted man. Basayev rose to fame after Russian forces invaded Chechnya in 1994; and he was responsible for numerous hostage takings and attacks, including the deadly school hostage siege in Beslan.[225] He was killed by Russian forces in July 2006.

Consequences and Counterterrorism

In response to Chechen suicide terrorism campaigns, the Russian government has taken an aggressive stand against Chechen militant groups. Incursions into Chechnya by Russian forces since the second civil war have been focused on subduing militant actors, with Russian authorities largely presenting the campaign as a crusade against Islamist terrorism. In ongoing clashes between militants and Russian forces, thousands of Chechen fighters, Russian soldiers, and civilians have been killed, with unconfirmed figures being reported by both sides in terms of causalities.

After a series of bombings rocked the country in late 1999, anti-Chechen sentiment reached an all time high in the country, prompting Russian forces to launch an aggressive military campaign in Chechnya—the start of the second civil war. Russian forces sealed off Chechnya's borders and quickly conquered the northern portion of the republic. By the summer of 2001, military and civilian causalities, in addition to reported human rights violations, had far surpassed that of the first civil war. Described by some scholars as a "violent military crackdown,"[226] Russia's campaign against these groups included search raids, large number of arrests, attacks on factories and apartment buildings, and alleged human rights violations.[227] In their initial campaign, Russia's full-scale offensive approach resulted in guerilla resistance, terrorist attacks, and heavy military casualties, challenging Russia's political control over Chechnya.

In general, the government has made significant efforts to portray the conflict as one against Islamic fundamentalists and "bandits" as opposed to an ethnic civil conflict, especially following the September 11 attacks on the United States. The Russian government is increasingly concerned with the radicalization of the Chechen resistance movement and its links to Middle East extremist groups.[228] It is in this capacity that the Russian government has found common ground with the United States, framing its conflict with Chechen militant and terrorist actors as part of the larger "war on terror."

Russian forces were successful in establishing a government in Chechnya considered pro-Moscow with the appointment of Akhmad Kadyrov as the Chechen president in May 2000. However, Kadyrov was killed in a bombing in

2004 by Chechen separatists, a large setback for Russian efforts against the Chechen rebel movement. Another pro-Moscow candidate, Aly Alkhanov, was appointed in 2004. Former Chechen rebel Ramzan Kadyrov gained control of the presidency in April 2007. The rebel turned Moscow-loyalist is the son of Akhmad Kadyrov.[229]

Most of the prominent Chechen separatist leaders, such as radical warlord and terrorist leader Shamil Basayev and Chechen rebel leader Aslan Maskhadov, have been killed by police operations. Basayev was killed in an explosion in Ingushetia while he was escorting a truck carrying explosives. Russian authorities claim the blast was the result of a special police operation. While Russian forces have seriously subdued the rebel movement, with suicide attacks diminished since 2005, violence still erupts throughout the North caucus region in an ongoing civil war.[230]

SUMMARY AND CONCLUSIONS—THE INTERNATIONALIZATION OF SUICIDE TERRORISM

Militant and terrorist organizations around the world have demonstrated their innovation in suicide bombing. A tactic that began in Lebanon when explosive-laden trucks triggered the withdrawal of super powers was soon adopted by others, including in Afghanistan where "kamikaze camels" helped instigate Soviet withdrawal a decade and a half later.

The Israelis, Indians, Sri Lankans, Turks, Russians, Americans, and Iraqis have all had to confront suicide terrorism. As demonstrated throughout this chapter, different tactics have been applied by various organizations around the world. As suicide terrorism has been adopted, it has also been transformed through the creativity and originality of terrorist groups. Chechen terrorists, willing to commit suicide, caused the deaths of hundreds in a Moscow theater and at a school in Beslan. A Tamil terrorist assassinated the Prime Minister of India by wearing a suicide vest, and Palestinian bombers have murdered bus riders in a similar fashion—armed with explosive vests. Trains in Spain and the tube in London experienced comparable bombings and results in terms of carnage. Turkey and the United States were both targeted by al-Qaeda in an effort to change the paradigm. Iraq, the most heavily-hit country in terms of suicide terrorism, has experienced almost every type of attack. For example, in May 2008, a female suicide bomber hid explosives under her shirt as if she was pregnant, detonating herself during a Shiite wedding near Baghdad.

With resources and budget in mind, terrorist organizations have also sought to recruit the mentally disabled. This is a known tactic in Afghanistan, where the Taliban has recruited young men suffering from mental illnesses or drug addictions. Even lingerie has been used to launch attacks, such as the 2005 case in Israel in which a female suicide bomber was caught at a checkpoint with 20 pounds of explosives sewn into her underwear. In November 2007, a Tamil Tigers female suicide bomber wearing an explosive bra, who was mentally disabled, blew herself up unwittingly outside the office of a Tamil minister.[231]

Regardless of ideology, religion, language, or purpose, militant groups worldwide will continue to learn from each other and adopt new tactics as they seek to achieve the upper hand. While suicide bombing may be a decades-old phenomenon, the ingenuity applied by terrorist groups has very much added new life to the tactic.

Terrorist groups have evolved and learned from the operational lessons of their counterparts, which calls for government and security forces to learn from their counterparts and be at least one step ahead. Law enforcement agencies must apply international experience and lessons learned in confronting suicide terrorism and other future threats. Means have varied vastly, however common denominators should enable counterterrorism practitioners to gain from both adopting and adapting successful tactics and learning from failures.

The use or the perceived use of unproportional responses to terrorism may be effective, as long as civilian life is respected. In the long run, however, the use of force alone may not be enough to eradicate terrorism; it should be supplemented by the widespread understanding that terrorism should not be supported by anyone. Education, public awareness, and international law are all tools that should be used to facilitate that goal.

Tactical countermeasures and policy approaches need to be adapted to the idiosyncrasies of each state, but all can certainly learn by looking across borders.

For true success in confronting suicide terrorism campaigns, international cooperation is essential and a broad coalition must be established. As expressed by counterterrorism expert Dr. Boaz Ganor, "It takes a network—to beat a network."

ENDNOTES

1. Yoram Schweitzer and Sari Goldstein Ferber, Al-Qaeda and the Internationalization of Suicide Terrorism, Memorandum No. 78, Jaffe Center for Strategic Studies, Tel Aviv University, November 2005.
2. M. Radu, Terrorism After the Cold War: Trends and Challenges, *Orbis* **46**(2), 275–287, 2002.
3. Ami Pedahzur and Arie Perliger, Introduction: Characteristics of Suicide Attacks, *Root Causes of Suicide Terrorism: The Globalization of Martyrdom*. New York: Routledge, 2006, p. 8.
4. Rohan Gunaratna, Suicide Terrorism; A Global Threat, *Janes*, October 20, 2000; http://www.janes.com/security/international_security/news/usscole/jir001020_1_n.shtml (last retrieved December 19, 2007).
5. In the map, red-colored regions illustrate areas that have been attacked by Red suicide terror. Yellow-colored areas have yet to be attacked by means of suicide terrorism.
6. David Ucko, April 2002 Suicide Attacks—A Tactical Weapon System, *The International Institute for Strategic Studies*. http://www.iiss.org/staffexpertise/list-experts-by-name/colonel-christopher-langton/recent-articles/suicide-attacks–a-tactical-weapon-system/
7. *Ibid*, p. 7.
8. Primarily Sunni, Shi'a, Druze, Isma'ilite, Alawite, or Nusayri.
9. Primarily Maronite Catholic and Greek Orthodox.

10. Lebanon: Exposing the Country's Jihadist Movement, *Stratfor*, May 20, 2008. http://www.stratfor.com/analysis/lebanon_exposing_countrys_jihadist_movement (last retrieved May 21, 2008).
11. On December 15, 1981, the first modern suicide attack was carried out by a Lebanese Shiite affiliated with an organization that called itself "The Army for the Liberation of Kurdistan." The attack targeted the Iraqi Embassy in Lebanon, killing 30 people. On November 11, 1982, a 15-year-old Lebanese Hezbollah suicide bomber demolished an eight-story building housing Israeli forces in Tyre, southern Lebanon, killing 90 people in the attack [the Department of Government at the University of Texas at Austin Database http://dev.laits.utexas.edu/movabletype/blogs/tiger/Suicide_Attacks_world-wide.xls (retrieved January 28, 2008)]. Notwithstanding, the attack that was most responsible for bringing suicide terrorism to the attention of the international community was carried out in Lebanon on April 18, 1983 against the U.S. Embassy in Beirut.
12. Further analysis of Hezbollah and its activities appears later in this chapter.
13. Yoram Schweitzer, Suicide Terrorism: Development & Characteristics, Institute for Counter-Terrorism, April 21, 2000. http://www.ict.org.il/articles/articledet.cfm?articleid=112
14. *Ibid.*
15. Rohan Gunaratna, Suicide Terrorism; A Global Threat, *Janes*, October 20, 2000. http://www.janes.com/security/international_security/news/usscole/jir001020_1_n.shtml (retrieved December 19, 2007).
16. Schweitzer, Note 13.
17. U.S. Department of State, Terrorist Organization List, http://www.basicsproject.org/reports/us_state_dept_terrorist_org_list_45323.pdf
18. Israel Ministry of Foreign Affairs, Hizbullah, (April 11, 2006), http://www.mfa.gov.il/mfa/mfaarchive/1990_1999/1996/4/hizbullah%20-%2011-apr-96
19. U.S. Department of the Treasury, "U.S. Designates Al-Manar as a Specially Designated Global Terrorist Entity Television Station is Arm of Hizballah Terrorist Network," March 23, 2006. http://www.ustreas.gov/press/releases/js4134.htm
20. Hezbollah as a Strategic Arm of Iran, The Intelligence and Terrorism Information Center at the Center for Special Studies, September 8, 2006, http://www.intelligence.org.il/eng/eng_n/html/iran_hezbollah_e1b.htm
21. Profile: Sheikh Hassan Nasrallah, *BBC News*, July 13, 2006, http://news.bbc.co.uk/2/hi/in_depth/5176612.stm
22. http://www.cpt-mi.org/pdf/HezbollahDossierFinal.pdf
23. Top Iranian Defector on Iran's Collaboration with Iraq, North Korea, Al-Qaida, and Hizbullah, *MEMRI Special Dispatch Series*. No. 273, February 20, 2003. http://www.memri.org/bin/articles.cgi?Area=sd&ID=SP47303
24. State Department, Counterterrorism 2008, Imad Fayez Mugniyah Profile, http://www.nctc.gov/site/profiles/mugniyah.html
25. Nicholas Blanford, Mughnieh Murder Could Trigger Retaliation, *The Arab American News*, February 16, 2008, http://www.arabamericannews.com/news/index.php?mod=article&cat=Lebanon&article=662
26. Imad Mughniya Is Dead, *Middle East Strategy at Harvard*, John M. Olin Institute for Strategic Studies at the Weatherhead Center for International Affairs, February 13, 2008, http://blogs.law.harvard.edu/mesh/2008/02/imad_mughniyah_is_dead/

27. Maskit Burgin, Foreign Hostages in Lebanon, in Ariel Merai and Anat Kurz, eds., *International Terrorism in 1987*. Boulder, CO: Westview Press, 1988, p.70.
28. U.S. Department of State, Significant Terrorist Incidents, 1961–2003: A Brief Chronology, Office of the Historian, Bureau of Public Affairs, March 2004, http://www.state.gov/r/pa/ho/pubs/fs/5902.htm
29. Matthew Levitt, Islamic Extremism in Europe: Beyond al-Qaeda, Hamas and Hezbollah in Europe, *Washington Institute* (April 27, 2005), http://www.washingtoninstitute.org/templateC07.php?CID=234
30. *Ibid.*
31. Ely Karmon, Iran–Syria–Hizballah–Hamas: A Coalition Against Nature—Why Does it Work? *The Proteus Monograph Series* 1(5), May 2008, http://www.carlisle.army.mil/proteus/docs/karmon-iran-syria-hizbollah.pdf
32. Roya Hakakian, The End of the Dispensable Iranian, *New York Times*, April 10, 2007.
33. Terrorist Bombings in Argentina (1992–1994), *Jewish Virtual Library* http://www.jewishvirtuallibrary.org/jsource/Terrorism/argentina.html
34. Karmon, Note 31.
35. Michael R. Gordon, Hezbollah Trains Iraqis in Iran, Officials Say, *New York Times*, May 5, 2008.
36. Hassan Nasrallah, interview, *El Mundo* (Madrid), December 18, 2001, cited by Ely Karmon, Fight On All Fronts: Hizballah, the War on Terror, and the War in Iraq, *Washington Institute Research Memorandum* 45, December 2003.
37. David Firestone, 18 are Accused of Plotting to Help Islamic Militants, *New York Times*, July 22, 2000.
38. Fifteen with Links to Terrorist Organizations Identifies in Sweden, *Politiken* (Copenhagen), October 13, 2001, cited by Ely Karmon, Fight On All Fronts: Hizballah, the War on Terror, and the War in Iraq, *Washington Institute Research Memorandum* 45, December 2003.
39. CFR.org Staff, Backgrounder: Hezbollah, *Council on Foreign Relations*, February 14, 2008. http://www.cfr.org/publication/9155/
40. Rex Hudson, *Terrorist and Organized Crime Groups in The Tri-Border Area of South America*, Federal Research Division, Library of Congress, July 2003, http://www.loc.gov/rr/frd/pdf-files/TerrOrgCrime_TBA.pdf
41. See Ref. 40.
42. Hezbollah Leader Nasrallah Regrets War, *CBC News*, August 28, 2006, http://www.cbc.ca/world/story/2006/08/27/nasrallah-abduction.html
43. Peter Wehner, Why They Fight and What It Means for Us, *Wall Street Journal*, January 9, 2007.
44. Andrew McCarthy, Negotiate with Iran? How Many Americans Do They Need to Kill Before We Get the Point? *National Review Online*, December 8, 2006.
45. *Ibid.*
46. *The 9/11 Commission Report: Final Report of the National Commission on Terrorist Attacks Upon the United States*. New York: W.W. Norton & Company, p. 61.
47. U.S. Attorney General announcement, June 21, 2001.
48. Richard Miniter, Bin Laden's Iran alliance, *Washington Times*, October 27, 2004.
49. Remarks by Iranian President Mahmoud Ahmadinejad during a meeting with protesting students at the Iranian Interior Ministry, October 25, 2005.

50. Excerpt from forthcoming book: *The Israeli Tiger*, Benjamin Netanyahu, 2009.
51. Fred Burton and Scott Stewart, Hezbollah: Signs of a Sophisticated Intelligence Apparatus, December 12, 2007, http://www.stratfor.com/weekly/hezbollah_signs_sophisticated_intelligence_apparatus (last retrieved May 28, 2008).
52. Department of Justice Press Release, Marine corps Officer and Two-Time Iraq War Veteran Pleads Guilty to Citizenship Fraud Conspiracy—Conspired with former FBI and CIA employee Nada Prouty in naturalization fraud (December 4, 2007), http://detroit.fbi.gov/dojpressrel/pressrel07/de120407.htm (last retrieved May, 28, 2008).
53. Burtonand Stewart, Note 51.
54. Note 51: Counterintelliegnce Case: Nada Nadim Prouty, The Centre for Counterintelligence and Security Studies, http://cicentre.com/Documents/DOC_Prouty_Case.html
55. Associated Press, Feds: Hezbollah Growing Threat To U.S., January 11, 2007, excerpts from National Intelligence Director John Negroponte's annual intelligence review, http://www.cbsnews.com/stories/2007/01/11/world/printable2352377.shtml
56. Note 54.
57. Yoram Schweitzer, Suicide Terrorism: Development and Characteristics, April 21, 2000, http://www.ict.org.il/articles/articledet.cfm?articleid=112
58. C. Van de Voorde, Sri Lankan Terrorism: Assessing and Responding to the Threat of the Liberation Tigers of Tamil Eelam (LTTE), *Political Practice and Research: An International Journal. Police Practice and Research* **6**(2), 181–199, 2005 (p. 185).
59. Hudson, 2000, p. 50 (cited in C. Van de Voorde, p. 185).
60. Robert A. Pape, *Dying to Win: The Strategic Logic of Suicide Terrorism*. New York: Random House Trade Paperbacks, 2005, p. 139.
61. Pape, p. 141.
62. Diego Gambetta, ed., *Making Sense of Suicide Missions*, New York: Oxford University Press, 2005, p. 47.
63. Backgrounder: The Sri Lankan Conflict, Council on Foreign Relations, September 11, 2006, http://www.cfr.org/publication/11407/sri_lankan_conflict.html
64. C. Van de Voorde, p. 183.
65. C. Van de Voorde, p. 184.
66. C. Van de Voorde, p. 184.
67. C. Van de Voorde, p. 187.
68. Robert Pape, p. 139.
69. Pape, p. 141.
70. Figures from 2004 indicate that the LTTE had committed more suicide attacks than any other organization at the time. See figures from R. Ramasubramanian, *Suicide Terrorism in Sri Lanka: IPCS Research Papers*. New Delhi, India: Institute of Peace and Conflict Studies, August 2004, p. 20. http://www.ipcs.org/IRP05.pdf
71. C. Van de Voorde, p. 187; Jonathan Lyons, Suicide Bombers—Weapon of Choice for Sri Lankan Militants, *Reuters News Agency*, April 20, 2006. http://www.swissinfo.org/eng/international/ticker/detail/Suicide_bombers_weapon_of_choice_for_Sri_Lanka_rebels.html?siteSect=143&sid=6988537&cKey=1156036055000
72. Mia Bloom, *Dying to Kill: The Allure of Suicide Terror*. New York: Columbia University Press, 2005, p. 60.
73. Bloom, pp. 45–75.

74. R. Ramasubramanian, *Suicide Terrorism in Sri Lanka: IPCS Research Papers*, New Delhi, India: Institute of Peace and Conflict Studies, August 2004, http://www.ipcs.org/IRP05.pdf
75. C. Van de Voorde, p. 186.
76. C. Van de Voorde, p. 186.
77. Bloom, p. 60.
78. C. Van de Voorde, p. 187.
79. Bloom, p. 51.
80. C. Van de Voorde, p. 190.
81. C. Van de Voorde, p. 191.
82. C. Van de Voorde, p. 191.
83. Liberation Tigers of Tamil Eelam (LTTE), *GlobalSecurity.org,* http://www.globalsecurity.org/military/world/para/ltte.htm
84. Backgrounder: The Sri Lankan Conflict, Council on Foreign Relations, September 11, 2006. http://www.cfr.org/publication/11407/sri_lankan_conflict.html
85. Revital Sela-Shayovitz, "Suicide Bombers in Israel: Their Motivations, Characteristics, and Prior Activity in Terrorist Organizations," *IJCV: Vol. 1 (2)* 2007, p. 162
86. Statement of Proclamation of the Organization, Permanent Observer Mission of Palestine to the UN, May 28, 1964, Jerusalem. http://www.un.int/palestine/PLO/doc3plo.html
87. For more details regarding the Israeli-Palestinian Conflict and the Israeli experience with terrorism, see Chapter 2 of this book.
88. Z. Flamhaft, *Israel on the Road to Peace: Accepting the Unacceptable*, Boulder, CO: Westview Press, 1996, p. 65.
89. Declaration of Principles on Interim Self-Government Arrangements, Ministry of Foreign Affairs, p.16.
90. A Shlaim, The Oslo Accord, *Journal of Palestine Studies*, **XXIII** (3), p.26.
91. A detailed analysis of the Israeli experience with suicide terrorism, including an overview of the relevant terrorist organizations activities, particularly between the years 2000–2007 is detailed in Chapter 2.
92. Israeli Security Agency Report, 2007
93. Data presented by former Israel Security Agency (ISA) executive Ehud Ilan at 5th International Conference on Terrorism which took place in Herzlyia, Israel, on September 12, 2005.
94. A detailed analysis of Israel's counterterrorism policy is provided in Chapter 2.
95. This was a concept presented by the late Israeli Prime Minister Rabin. He believed that the PA would not be concerned with human rights in its fight against terrorism, and would be free of judicial intervention, such as Bagatz, Israel's High Court of Justice, or of human rights organizations, such as Bezelem, active in Israel.
96. An analysis of PA instigated terrorism is addressed in Chapter 2.
97. Mia Bloom, Palestinian Suicide Bombing: Public Support, Market Share, and Outbidding, *Political Science Quarterly* 119/1 (Spring 2004) 80; quoted in Hillel Frisch, Motivation or Capabilities? Israeli Counterterrorism against Palestinian Suicide Bombings and Violence. *Mideast Security and Policy Studies*, No. 70, The Begin-Sadat Center for Strategic Studies, December 2006. http://www.biu.ac.il/Besa/MSPS70.pdf

98. Israel Prime Minister Ariel Sharon, quoted in Matt Rees' Streets Red with Blood, *Time Magazine*, March 18, 2002.
99. Hirsh Goodman, Defensive Shield: A Post Mortem, *Insight: A Middle East Analysis*, June 2002, Jaffe Center for Strategic Studies, June 5, 2002, United Jewish Communities website. http://www.ujc.org/page.aspx?id=79160; Statistics on Operation Defensive Shield, *Jewish Virtual Library*, http://www.jewishvirtuallibrary.org/jsource/History/defensiveshield.html; Nitsan Alon, Operation Defensive Shield: The Israeli Actions in the West Bank, PeaceWatch #374, *The Washington Institute for Near East Policy*. https://www.washingtoninstitute.org/print.php?template=C05&CID=2065
100. Author's interview with Benjamin Netanyahu, April 23, 2006.
101. Netanel Lorch, The Arab-Israeli Wars, September 2, 2003, *Israel Ministry of Foreign Affairs*. http://www.mfa.gov.il/MFA/History/Modern%20History/Centenary%20of%20Zionism/The%20Arab-Israeli%20Wars
102. M. Burgess, In the Spotlight: Jaish-e-Mohammed (JEM) Terrorism Project. *Center for Defense Information*. April 8, 2002. On line http://www.cdi.org/terrorism/jem-pr.cfm Major Islamist Terrorist Attacks in India in the Post-9/11 Period, *South Asian Terrorism Portal*. On line; Swami, P. 2001. *An Audacious Strike*. Retrieved December 10, 2008, from Frontline: http://www.hindunnet.com/fline/fl1821/18210200.htm
103. Parliament Suicide Attack Stuns India, *BBC News*, December 13, 2001. http://news.bbc.co.uk/2/hi/south_asia/1708853.stm
104. India-Pakistan: Troubled Relations Timeline. *BBC News*. http://news.bbc.co.uk/hi/english/static/in_depth/south_asia/2002/india_pakistan/timeline/1965.stm
105. "Naqvi, Muneeza "66 Die in India-Pakistan Train Attack". Associated Press, *The Washington Post*, February 19, 2007. http://www.washingtonpost.com/wp-dyn/content/article/2007/02/18/AR2007021801136.html; February 19, 2007. "Dozens Dead in India Train blast," *BBC News*, http://news.bbc.co.uk/2/hi/south_asia/6374377.stm
106. M. Burgess, In the Spotlight: Jaish-e-Mohammed (JEM) Terrorism Project. *Center for Defense Information*. April 8, 2002. On line http://www.cdi.org/terrorism/jem-pr.cfm
107. Directorate of Inter-Services Intelligence (ISI)-the largest branch of Pakistan's Intelligence agencies; Jayshree Bajoria, Profile: Lashkar-e-Taiba, *Council on Foreign Relations*. http://www.cfr.org/publication/17882/
108. M. Burgess, In the Spotlight: Jaish-e-Mohammed (JEM) Terrorism Project. *Center for Defense Information*. April 8, 2002. On line http://www.cdi.org/terrorism/jem-pr.cfm
109. Directorate of Inter-Services Intelligence (ISI)-the largest branch of Pakistan's Intelligence agencies
110. Fidayeen (Suicide Squad) Attacks in Pakistan, South Asia Terrorism Portal, http://www.satp.org/satporgtp/countries/pakistan/database/Fidayeenattack.htm
111. Julien Spencer, Turkey Offers Reforms for Kurdish Minority, *The Christian Science Monitor*, March 13, 2008, http://www.csmonitor.com/2008/0312/p99s01-duts.html; and The Kurds in Turkey, *Federation of American Scientists*, www.fas.org/asmp/profiles/turkey_background_kurds.htm
112. Bloom, p. 103.
113. Who are the Kurds? *The Washington Post* (1999) http://www.washingtonpost.com/wp-srv/inatl/daily/feb99/kurdprofile.htmIn
114. Kurds, *Country Studies/Area Handbook Series, Federal Research Division of the Library of Congress* http://countrystudies.us/turkey/28.htm
115. Meline Toumani, Minority Rules, *New York Times*, February 17, 2008, http://www.nytimes.com/2008/02/17/magazine/17turkey-t.html?_r=1&oref=slogin

116. Meline Toumani, Minority Rules, *New York Times*, February 17, 2008, http://www.nytimes.com/2008/02/17/magazine/17turkey-t.html?_r=1&oref=slogin
117. Eleven of the fifteen PKK suicide attacks between 1996 and 1999 were carried out by young female suicide bombers (Bloom, 102).
118. Clara Beyler, Chronology of Suicide Bombings Carried Out by Women, February 12, 2003, http://www.ict.org.il/apage/10726.php
119. Pape, p. 163.
120. Pape, p. 163.
121. Öcalan Urges End to Fighting, *BBC News*, June 1, 1999, http://news.bbc.co.uk/2/hi/europe/357417.stm
122. Associated Press, Turkey Says Suicide Bomber Carried Out Attack, *MSNBC News*, May 23, 2007, http://www.msnbc.msn.com/id/18817425/
123. Reuters and the Associated Press, Suicide Bomber Carried Out Attack in Ankara, Officials Say, *The International Herald Tribune*, May 23, 2007, http://www.iht.com/articles/2007/05/23/news/turkey.php
124. According to a report by Fox news, Sky Turk reporter Mustafa Azizoglu told Fox News "this is not an ordinary attack" and said "this is the eleventh of September for Istanbul" (Istanbul Truck Bomb attack kills 27, http://www.foxnews.com/story/0,2933,103612,00.html).
125. Al Qaeda Fingerprints on Bombings? *CBS News*, December 2, 2003, http://www.cbsnews.com/stories/2003/11/15/terror/main583850.shtml
126. Istanbul Rocked by Double Bombing, *BBC News*, November 20, 2003, http://news.bbc.co.uk/2/hi/europe/3222608.stm
127. Istanbul Rocked by Double Bombing, *BBC News*, November 20, 2003, http://news.bbc.co.uk/2/hi/europe/3222608.stm
128. Thomas Renard, Police Raids Uncover al-Qaeda's Parallel World in Turkey, *Terrorism Focus* **5**(15), April 16, 2008.
129. Bloom, p. 112.
130. Who is Turkey's Hizbollah? *A Human Rights Watch Backgrounder*, February 16, 2000, http://hrw.org/english/docs/2000/02/16/turkey3057.htm
131. Suleyman Ozoren and Cecile Van de Vorde, Turkish Hizballah: A Case Study of Radical Terrorism, *Journal of Turkish Weekly*, p. 8.
132. Who is Turkey's Hizbollah? *A Human Rights Watch Backgrounder*, February 16, 2000, http://hrw.org/english/docs/2000/02/16/turkey3057.htm
133. Islamic Great Eastern Raiders Front, *Global Security.org*, http://www.globalsecurity.org/military/world/para/eastern-raiders.htm
134. Terrorist Organizations and Other Groups of Concern: Islamic Great Eastern Raiders—Front (IBDA-C), From Chapter 8, Other Groups of Concern, Country Reports on Terrorism 2005, U.S. Department of State, April 30, 2006. *The Investigative Project on Terrorism*, http://www.investigativeproject.org/profile/152
135. Estimates vary. The Turkish government claims that 5000 Kurdish civilians were killed during its military operations from 1984 to 1999, while other independent reports cite as many as 35,000 deaths.
136. Mahmood Monshipouri, *Islamism, Secularism, and Human Rights in the Middle East*. Boulder, CO: Lynne Rienner Publishers, 1998, p. 126.

137. Monshipouri, p. 125.
138. Monshipouri, p. 164.
139. Bloom, p. 104.
140. Bloom, p. 105.
141. Bloom, p. 107.
142. Bloom, p. 106.
143. Alan Cowell, Turkish Kurd Rebel Threatens Suicide Bombings in Turkey, *New York Times*, March 30, 1996, http://query.nytimes.com/gst/fullpage.html?res=9507E0D815 39F933A05750C0A960958260
144. Bloom, p. 109.
145. Bloom, p. 109.
146. Turkish Soldiers Killed by Rebels, *BBC News*, October 7, 2007, http://news.bbc.co.uk/2/hi/europe/7033075.stm
147. Note 1, p. 9.
148. Yoram Schweitzer, Al-Qaeda and Epidemic of Suicide Attacks, *Roots of Suicide Terrorism: The Globalization of Martyrdom*, Ami Pedahzur, ed. New York: Routledge, 2006, p. 132.
149. Stephen Holmes, Al Qaeda, September 11, 2001 in Diego Gambetta, ed. *Making Sense of Suicide Missions*, New York: Oxford University Press, 2005.
150. Schweizer, Note 135, p. 137.
151. *The 9/11 Commission Report: Final Report of the National Commission on Terrorist Attacks Upon the United States*, New York: W.W. Norton & Company, p. 56.
152. *Ibid*, pp. 55–56.
153. Marc Sageman, Islam and Al Qaeda, *Root Causes of Suicide Terrorism: The Globalization of Martyrdom*, Ami Pedahzur, ed. New York: Routledge, 2006, p. 23.
154. Mia Bloom, *Dying to Kill: The Allure of Suicide Terror*. New York: Columbia University Press, 2007, p. 138.
155. *Ibid*, Bloom, p. 138 (Note 141).
156. Schweizer, p. 135 (Note 148).
157. Schweitzer and Ferber, p. 16 (Note 1).
158. Schweizer, p. 142 (Note 135).
159. Marc Sageman, *Understanding Terrorist Networks*. Philadelphia: University of Pennsylvania Press, 2004, p. 124.
160. *Ibid*, p. 124.
161. Eben Kaplan, The Rise of Al Qaedaism, *Council on Foreign Relations*, July 18, 2007, http://www.cfr.org/publication/11033/rise_of_alqaedaism.html
162. Press Conference by the President (October 25, 2006), Office of the Press Secretary, The White House, http://www.whitehouse.gov/news/releases/2006/10/20061025.html
163. Dan Eggen and Scott Wilson, Suicide Bombs Potent Tools of Terrorists, *The Washington Post*, July 17, 2005, http://www.washingtonpost.com/wp-dyn/content/article/2005/07/16/AR2005071601363_pf.html
164. Jonathan Schanzer, Al Qaeda's Armies: Middle East Affiliate Groups and the Next Generation of Terror, Book review, *The Washington Institute for Near East Policy*, www.washingtoninstitute.org/templateC04.php?CID=135

165. Jayshree Bajoria, Backgrounder: Al Qaeda, *The Council for Foreign Relations*, April 18, 2008, http://www.cfr.org/publication/9126/
166. Jayshree Bajoria, Backgrounder: Al Qaeda, *The Council for Foreign Relations*, April 18, 2008, http://www.cfr.org/publication/9126/
167. Sageman, p. 53.
168. Reuter, Christopher, *My Life is a Weapon: A Modern History of Suicide Bombing*. Princeton, NJ: Princeton University Press, 2004, p. 146.
169. David Ronfeldt, Al Qaeda and Its Affiliates: A Global Tribe Waging Segmental Warfare? *First Monday* Issue 10(3), http://www.firstmonday.org/issues/issue10_3/ronfeldt/
170. Bruce Hoffman, Challenges for the U.S. Special Operations Command Posed by the Global Terrorist Threat: Al Qaeda on the Run or on the March? Written Testimony submitted to the House Armed Services Subcommittee on Terrorism, Unconventional Threats and Capabilities, February 2006, p. 6.
171. Hoffman, p. 7.
172. Schweitzer, p. 144.
173. Global Security, Attacks on US Embassies in Kenya and Tanzania, http://www.globalsecurity.org/security/ops/98emb.htm
174. Ref. 60, 9/11 Commission Report, p. 231.
175. Pape, Robert A., *Dying to Win: The Strategic Logic of Suicide Terrorism*. New York: Random House, 2005, p. 109.
176. Ref. 1, Yoram Schweitzer and Sari Goldstein Ferber, Al-Qaeda and the Internationalization of Suicide Terrorism, Memorandum No. 78, November 2005, The Jaffee Center for Strategic Studies, Tel Aviv University, pp. 58–59.
177. It is difficult to track the number of suicide attacks in Iraq. According to a book published by the U.S. Institute of Peace (*Suicide Bombers in Iraq*), the amount of suicide attacks in the Iraqi insurgency has surpassed the number of suicide operations by all previous insurgent and terrorist groups combined, including those by Hezbollah in Lebanon, Tamil Tigers in Sri Lanka, and Hamas in Israel. From March 22, 2003 to August 18, 2006, approximately 514 suicide attacks took place in Iraq, according to the book. (http://www.usip.org/newsmedia/hafez_press/). For further details on suicide terrorism in Iraq, see Chapter 3.
178. Jack Kalpakian, Building the Human Bomb: The Case of the 16 May 2003 Attacks in Casablanca, *Studies in Conflict & Terrorism* **28**(2), 120.
179. Schweitzer, p. 144.
180. Zachary Abuza, The Killer's New Tactic: Smaller Bombs, More Often, *The Sidney Morning Herald*, October 3, 2005, http://www.smh.com.au/news/opinion/the-killers-new-tactic-smaller-bombs-more-often/2005/10/02/1128191602907.html
181. Peter Bergen, Al Qaeda's New Tactics, *New York Times*, November 15, 2002, http://query.nytimes.com/gst/fullpage.html?res=9407E6D61730F936A25752C1A9649C8B63
182. Although the attack was not an act of suicide terrorism, some analysts categorize it as such because the primary attackers killed themselves before arrest. When the Spanish police located the attackers, they surrounded their safe house, at which point the fugitives blew themselves up with explosives.

183. Madrid Blasts: Who Is to Blame? *BBC News*, March 18, 2004, http://news.bbc.co.uk/2/hi/europe/3512748.stm
184. Keith B. Richburg, Madrid Attacks May have Targeted Election, *Washington Post*, October 17, 2004, http://www.washingtonpost.com/wp-dyn/articles/A38817-2004Oct16_2.html
185. AP, Madrid Bombing Probe finds no Al Qaeda Link, March 9, 2006, http://www.msnbc.msn.com/id/11753547/
186. Lawrence Wright, The Terror Web, *The New Yorker*, August 2004.
187. AP, Madrid Bombing Probe finds no Al Qaeda Link, March 9, 2006, http://www.msnbc.msn.com/id/11753547
188. Robert S. Leiken, Europe's Angry Muslims, *Foreign Affairs*, July/August, 2005, http://www.foreignaffairs.org/20050701faessay84409-p10/robert-s-leiken/europe-s-angry-muslims.html
189. AP, Al Qaeda Fingerprints on Bombings? December 2, 2003, *CBS News*, http://www.cbsnews.com/stories/2003/11/15/terror/main583850.shtml
190. Istanbul Rocked by Double Bombing, BBC News, http://news.bbc.co.uk/2/hi/europe/3222608.stm
191. Backgrounder: Jemaah Islamiyah, *Council on Foreign Relations*, June 13, 2007, http://www.cfr.org/publication/8948/
192. Backgrounder: Jemaah Islamiyah, *Council on Foreign Relations*, June 13, 2007, http://www.cfr.org/publication/8948/
193. Schweitzer, 143 (Ref. 57).
194. 9/11 Commission Report, pp. 72–73.
195. 9/11 Commission Report, p. 108.
196. 9/11 Commission Report, pp. 325–328.
197. 9/11 Commission Report, p. 328.
198. 9/11 Commission Report, p. 333.
199. 9/11 Commission Report, p. 334.
200. Al Qaeda, Killed or Captured, *MSNBC*, http://msnbc.com/modules/wtc/wtc_global-dragnet/custody_alqaida.htm (retrieved on July 7, 2008).
201. Al Qaeda, Killed or Captured, *MSNBC*, http://msnbc.com/modules/wtc/wtc_global-dragnet/custody_alqaida.htm (retrieved on July 7, 2008).
202. Mark Mazzetti and David Rohde, Amid U.S. Policy Disputes, Qaeda Grows in Pakistan. *New York Times*, June 30, 2008, http://www.nytimes.com/2008/06/30/washington/30tribal.html?adxnnl=1&pagewanted=2&adxnnlx=1214900828-7rcp10L8VocB09F2s1gorw
203. Mark Mazzetti and David Rohde.
204. Country Reports on Terrorism 2007, Released by the Office of the Coordinator for Counterterrorism, U.S. Department of State (April 30, 2008), p. 8, http://www.state.gov/s/ct/rls/crt/2007/103703.htm
205. Souad Mekhennet, Micael Moss, Eric Schmitt, Elaine Sciolino, and Margot Williams, Ragtag Insurgency Gains Lifeling from Al Qaeda, *New York Times*, July 1, 2008, http://www.nytimes.com/2008/07/01/world/africa/01algeria.html?_r=2&hp&oref=slogin&oref=slogin

206. *Ibid* (Note 127).
207. *Ibid* (Note 127).
208. Reuter, 152.
209. Bruce Hoffman, Remember Al Qaeda? They're baaack. *LA Times*, February 20, 2007, http://www.latimes.com/news/opinion/la-oe-hoffman20feb20,0,2283472.story?coll=la-opinion-rightrail
210. Al-Qaeda in 2008: The Struggle for Relevance, *Terrorism Intelligence Report*, Stratfor (December 19, 2007) http://www.stratfor.com/weekly/al_qaeda_2008_struggle_relevance
211. Anne Speckhard and Khapta Ahkmedova, The Making of a Martyr: Chechen Suicide Terrorism, *Journal of Studies in Conflict and Terrorism* **29**(5) (Routledge, Taylor and Francis Group, July/August 2006), p. 3.
212. Speckhard and Ahkmedova, p. 13.
213. Mark Kramer, The Perils of Counter-Insurgency, *International Security* **29**(3), 7.
214. Gail W. Lapidus, Sovereignty: The Tragedy of Chechnya, *International Security* **23**(1), 1998.
215. Speckhard and Ahkmedova, p. 3.
216. Speckhard and Ahkmedova, p. 5.
217. Speckhard and Ahkmedova, p. 5.
218. Matthew Chance, Ryan Chilcote and Jill Dougherty, Russia School Siege Toll Tops 350, *CNN*, September 5, 2004, http://edition.cnn.com/2004/WORLD/europe/09/04/russia.school/
219. Scott Peterson, Al-Qaeda tie to School Hostage Takers Probed, *The Christian Science Monitor*, September 6, 2004, accessed via *USA Today*, http://www.usatoday.com/news/world/2004-09-06-al-qaeda-chechnya_x.htm
220. Information Report/Swift Knight—Usama Ben Laden's Current and Historical Activities, October 1998, released to *Judicial Watch* in 2004; cited in Kramer, 7.
221. Kramer, p. 58.
222. Kramer, p. 58.
223. Reuven Paz, The Impact of the War in Iraq on Islamist Groups and the Culture of Global Jihad, *Prism*, September 2004, www.prism.org
224. In the Spotlight: The Special Purpose Islamic Regiment, CDI Terrorism Project, May 2, 2003, http://www.cdi.org/terrorism/spir-pr.cfm
225. Basayev: Russia's Most Wanted Man, *CNN News*, September 8, 2004, http://edition.cnn.com/2004/WORLD/europe/09/08/russia.basayev/
226. Svante E. Cornell, The War Against Terrorism and the Conflict in Chechnya: A Case for Distinction, *The Fletcher Forum of World Affairs* **27**(2), 167.
227. Cornell, p. 172.
228. Cornell, p. 167.
229. Luke Harding, Former rebel, 30, sworn in as Chechen President, *The Guardian*, April 6, 2007, http://www.guardian.co.uk/world/2007/apr/06/chechnya.lukeharding
230. CBC News In depth: Chechnya, *CBC News*, July 10, 2006, http://www.cbc.ca/news/background/chechnya/
231. Al Qaeda: Creative Recruiting for Suicide Bombers, Stratfor, May 5, 2008.

5

HIGH-RISK SCENARIOS AND FUTURE TRENDS

Ophir Falk

This book has analyzed why and how the suicide terrorism phenomenon emerged in a number of terror hotbeds around the world and has suggested lessons from key case studies. Notwithstanding, one of the largest downfalls of law enforcement and security experts is that they often limit themselves to preparing for the risks of the past rather than complementing preceding experience with training for the risk scenarios of the future.

Worst-case scenario analysis was once an accepted basis for security assessments and planning. However, today due to the vast scale of threats that may range from CBRN to cyber or conventional attacks, along with the limited resources to address these threats, it is widely recognized that *risk-based* decision-making is the best tool to determine appropriate security measures by governments, corporations, and private entities. In order to understand and confront the risk at hand, it is important to correctly identify the threats to such assets, their vulnerabilities, the related consequences of successful attacks and, most importantly, means of risk mitigation.

A risk scenario consists of a potential threat to the target, the target's vulnerability, and the consequence of a successful attack. It is important that the formulated scenario is within the realm of possibility and addresses known capabilities and intents as evidenced by past events, available intelligence, and future trends.

Suicide Terror: Understanding and Confronting the Threat. Edited by Falk and Morgenstern

High-risk scenarios can be only be regarded as such if the probability (construed of the threat and vulnerability levels) and consequence of a successful attack is high. Such risks need to be real and currently relevant.

Having said that, high-consequence scenarios and planning for scenarios of future probability have their merits and importance in preparing for long-term threats.[1]

High-risk scenarios are of abundance. Lior Lotan, the former director of the International Counter-Terrorism Institute, recently alerted the author to domino- or cascading-type attacks whereby daily attacks can be carried out against soft civilian targets (i.e., day 1 attack within shopping mall, day 2 attack on school, day 3 attack on public transportation, day 4 attack on apartments with each attack being carried out in different parts of the country). Such attacks can cause a public perception that "there is nowhere to hide."[2] Nevertheless, we've limited this section to six high-risk scenarios that can serve as stand-alone scenarios.[3] Most of the analyzed scenarios are ones that can be carried out successfully without the attackers committing suicide; however, a suicide attack can make the outcome more lethal or distract law enforcement from the main setting. The level of *risk* that each plausible scenario possesses is based on assessing (a) the *threat*, which is composed of the capability and intentions to carry out attacks of terror in general and its ability to carry out those attacks in nations like the United States in particular, (b) the *vulnerability* of nations like the United States to such an attack scenario, and (c) the *consequence* of a successful attack, in terms of life and treasure. Most importantly, means and methods of mitigating the risk are also suggested.[4]

Subsequent to the analysis of six high-risk scenarios, we will address future trends that may pose high-consequence scenarios.

SCENARIO NO. 1: CYBER AND PHYSICAL ATTACK ON ENERGY DISTRIBUTION SYSTEMS

Description of Scenario

A foreign government, at odds with the United States over sanctions imposed on it, facilitates its terrorist proxy to target U.S. critical infrastructures. Its goal is the disruption of critical infrastructures that drive the American economy.

For that end, the terrorist proxy, notorious for its use of technology, manages to hack into Energy companies' Supervisory Control and Data Acquisition (SCADA) systems that control the distribution of natural gas and oil. The hackers manage to alter the data and operational codes of systems causing immediate disruption. Almost simultaneously, and primarily with an intention to decoy response efforts, a number of low-level terrorist operatives cause physical damage to various key pipeline sections of the 278,000 miles of natural gas pipelines and 160,000 miles of crude oil transport pipelines in the United States[5] by shooting, bombing, and even attacking the pipelines with suicide bombers.

The authorities almost immediately detect the physical damage done to the systems and are able to repair most of the damage within 72 hours. However, significant time elapses until the cyber attack is detected and the system's malfunctions are corrected.

Threat. The level of threat is composed of two primary factors: *intention* and *capability* of potential perpetrators to attack. It is important that the developed threats or scenarios are within the realm of possibility and address known capabilities and intents as evidenced by past events and available intelligence.[6] The analysis below was done after reviewing major incidents, attack trends, and the profile of organizations that were behind those attacks.

INTENTIONS. It is clear that terrorists seek to destroy, incapacitate, or exploit critical infrastructure and key assets in an effort to threaten national security, cause mass casualties, impede the economy, and damage public morale and confidence.

In essence, the targeting of pipelines is primarily an attack on a nation's economic well-being with the direct price in life being of secondary importance. The motivation of terrorist organizations to target America's economy in general and its critical infrastructure in particular is clear. Three statements that support this observation are as follows:

1. In October 2002, the same day that the *Limburg* oil tanker was struck, the Al Jazeera network in Qatar broadcasted an audio tape of bin Laden warning the "crusader community" that Islamic forces were "preparing for you things that would fill your hearts with terror and target your economic lifeline."
2. "In addition to our having experience in using guerrilla warfare and the war of attrition to fight tyrannical superpowers, as we bled Russia for 10 years, until it went bankrupt and was forced to withdraw in defeat ... we are continuing this policy in bleeding America to the point of bankruptcy. ... For example, al-Qaida spent $500,000 on the event, while America, in the incident and its aftermath, lost—according to the lowest estimate—more than $500 billion."[7]
3. "Overthrow of godless regimes [by] gathering information about the enemy, the land, the installations, and the neighbors ... blasting and destroying the places of amusement ... embassies ... vital economic centers ... bridges leading into and out of cities ..." (al-Qaeda Manuel—published by Department of Justice 2004).[8]

CAPABILITIES. The energy sector has been targeted by various organizations throughout the world. The threat is both local and global. In Colombia, for example, the FARC terrorist organization targets that country's energy sector. But the FARC insurgency is primarily localized. The Global Jihadist insurgency, on the other hand, is transnational. Bin Laden, an educated engineer, and his associates possess the resources in terms of knowledge and funding to mount serious attacks against a nation's gas and oil facilities.

The feasibility of physically damaging pipelines is clear and as has been evident on a number of occasions. For example:

1. In October 2001, the Alaska pipeline was shot at from a high-powered rifle, causing more than 6800 barrels of crude oil to spill from the pipeline; the pipeline was shut down for three and a half days. In that incident, state police charged a man living near the pipeline with criminal mischief, saying the shooting was unrelated to any terrorist activity.[9]
2. Pipeline incidents in Colombia: In Colombia, in 1997, there were more than 45 separate attacks on the 500-mile-long Cano Limon-Covenas oil pipeline by the National Liberation Army (ELN) group. In 1998, there were 50 bombing attacks against that pipeline. These incidents included a bombing by the ELN of the OCENSA pipeline, spilling over 30,000 barrels of oil. More than 70 people died when the fire from that explosion spread to nearby villages. In 1999, however, the bombings of Colombia's pipelines by terrorist groups vastly escalated, with 152 bombing incidents recorded. This incident rate further rose to 170 times in 2001, representing $500 million in lost oil production. In 2002, as a result of new government protective efforts, the Cano Limon-Covenas pipeline was bombed 34 times, as of December 12th of that year.[10]
3. On February 24, 2006, Saudi security forces thwarted an attempted suicide attack at the Abqaiq oil processing facility in eastern Saudi Arabia, which is the largest oil processing facility in the world (more than 60% of Saudi production). In that incident, two pick-up trucks carrying two would-be bombers tried to enter the side gate of plant, but the attackers were not successful because they detonated their explosives after security guards fired on them, resulting in minor damage to pipeline and minor injuries to a few plant workers.[11]Although this attack was not successful, it illustrates the capability of targeting high-value oil facilities.
4. This threat reached crisis proportions in Iraq, where the country's economic recovery was damaged by insurgent attacks against major pipelines and the assassination of oil officials.[12]

As for the capability of hacking into SCADA and significantly disrupting the pipeline flow of natural gas and/or oil products, it is feasible and is an imminent threat.[13]

Most Probable Perpetrator. ***Al-Qaeda*** and its global Jihadists satellites represent the primary terrorist threat against the Western world and its allies. Bin Laden is keenly aware of the significance of the energy sectors to the economic well-being of his adversaries, especially the United States and Saudi Arabia.

Therefore, in terms of a terrorist attack the most probable perpetrator is al-Qaeda and organizations of its ilk. However, such a scenario, as suggested in this case, can also be carried out by Iranian sponsored and trained Hezbollah.

Aside from the threat of terror that this book emphasizes, the criminal aspect of sabotage should be highlighted. In fact, it is estimated that 75–80% of all security-related incidents in the energy industry are caused by company "insiders," who may be disgruntled employees or motivated by other extremism.[14]

The level of threat is high.

Vulnerability

The level of vulnerability is composed of three primary factors: the *visibility* and *accessibility* of potential perpetrators to the target and the existing security apparatus' *ability to prevent* the attack. From a law enforcement and security personnel's perspective, the vulnerability is the most important element of the risk. Security personnel have little control over the motivation and capabilities of its enemy, but they do have the ability to be better prepared for attack, to lower the level of vulnerability, and by doing so, mitigate the risk.

The pipeline system is highly vulnerable. It is a geographically vast and complex sector with multiple spread targets that are very *visible* and *accessible* to intruders. For example, the Trans-Alaska Pipeline System is one of the most vulnerable systems in the United States. It carries about one million barrels of oil a day, accounting for 20% of U.S. crude production, and about half of the pipelines' 800 miles is above ground in unpopulated areas.[15] Due to the vast terrain that pipelines cross, and due to other complexities, the ability to prevent attacks is currently limited.

In the realm of information technology, there is vast use of remotely operated SCADA systems that, among other things, control the flow of natural gas in a pipeline system, making them thus vulnerable to cyber-terrorism.[16]

Therefore, the pipeline's **vulnerability to attack is high**.

Consequence

The level of damage or consequence of a successful attack is composed of three primary factors: *causalities, economic ramifications, and public morale.*

Attacks on key oil and gas pipelines could trigger energy outages throughout a state and across regions and could affect global supply.

A successful scenario, such as in this case, could lead to grave economic ramifications. Physical attacks on pipelines in America and around the world have occurred on many occasions, resulting in economic loss.

An example of such a case occurred on October 4, 2001 when the Alaska pipeline was shot at from a high-powered rifle, causing more than 6800 barrels of crude oil to spill from the pipeline and shutting the pipeline down for three and a half days.[17] The estimated direct loss caused by that attack was 17 million.[18] That incident illustrated how resources very limited in sophistication could cause such severe economic damage.

Suicide attacks on pipelines will not necessarily cause more damage than other forms of physical attacks, but they could definately divert attention from the cyber attack.

Cyber-terrorism could shut down the energy industry by causing large-scale disruptions in pipelines, tankers, refineries, and additional ramifications that wreak havoc on the country's economy and way of life. Attacks on key oil and gas pipelines could also trigger power outages across regions since power and energy grids are widely interconnected.[19]

Due to the interdependency of most critical infrastructures and the importance of energy to them all, such an attack could result in (a) food and water supply shortages and (b) malfunction of transportation systems, telecommunications, public utilities, and banking and finance services, leading to stock market crashes with a cascading effect.

Such an attack could cause prices to skyrocket and bring America's economy to a halt. There may even be a domino effect that cascades throughout the world, leading to a deep global recession.

Therefore the consequence of a successful attack is high.

Means of Mitigation

A number of mitigation measures may include the following:

- In order to understand the actual risk at hand, it is important to correctly identify the threats and vulnerabilities of specific facilities and pipeline sections and the related human, economic, and other consequences of an attack. Due to the vast expanse of pipelines, the most vulnerable and risk prone areas need to be prioritized.
- Emergency response teams are usually well-trained, prepared, and equipped to deal with fires, spills, and explosions, but rarely are they trained for such scenarios. Therefore, proper recruitment and training is required.
- After assessing the risk, a comprehensive security plan and implementation of the appropriate security solutions are needed.
- In terms of technological means and equipment, surveillance mechanisms and smart sensors should be deployed to detect suspicious movement, sound, and actions. There are a variety of enhanced technologies available today to address these threats, but there is certainly room for added research and development in this field. Enhanced detection on a national level is needed. At the same time security forces must work to eliminate false alarms.
- Research and development to enhance robustness, redundancy of key components, and reliability of pipelines.
- Enhanced monitoring and control of employee misconduct and malicious intent.

SCENARIO NO. 2: ATTACKS ON AND BY CIVILIAN AIRCRAFT—BACK TO THE FUTURE

For obvious reasons, with the aftermath of the 9/11 attacks, securing the aviation industry has become a major area of focus for governments and airline corporations. However, in spite of improved security measures, similar and other forms of terrorism threats to civil aviation remain formidable. The *use of planes as weapons* proved to be the most costly form of terror attack to date and still presents and imminent threat—as does the use of shoulder-launched surface-to-air missiles (a.k.a. MANPADS).

Description of Scenario[20]

> Al-Qaeda decides to carry out an attack of similar 9/11 dimensions with an added twist. After three South American civilian aircrafts are shot at by MANPADs as they take off from European airports toward North America, three other civilian aircrafts from different airlines in the United States lose radio contact with ground control and are feared to be hijacked. Shortly after radio contact is lost, the civilian aircrafts crash into crowd-concentrated areas in three different U.S. cities. One of the planes, carrying cargo, crashes into a 70,000-people packed football stadium and a short time after the crash, peculiar debris is noticed in the air.

Threat

INTENTIONS. The motivation to target the aviation industry is as high as ever. Two clear illustrations of this are (a) Richard Reid's attempt to ignite his shoes on board an aircraft almost immediately after 9/11 as security awareness was at its peak and (b) the sophisticated plot on August 10, 2006 to target up to 10 flights with liquid explosives leaving Heathrow Airport for the United States. There are a number of other thwarted plots and indications that show the ever-present motivation to target the aviation industry.

CAPABILITIES. The capabilities of attacking airborne aircrafts both in-flight and by shoulder-launched missiles are evident.

Shoulder-Launched Missiles. The proliferation of the shoulder-launched SAM systems during the Cold War era, ease-of-use, and relative low black market price have made them popular with terrorist and guerilla groups the world over. Yet it was during the war in Afghanistan 1979–1989 that shoulder-launched missiles, referred to at the time as missiles or man-portable air-defense systems (MANPADs), proved their effectiveness against Soviet military helicopters and transport planes. At the time the weapon was considered a defense system, but today it is referred to as a weapon of attack.

From a civilian perspective, there have been over 20 successful missile attacks on a variety of commercial airlines, resulting in over 500 deaths over the past 30 years.[21] Infamous incidents include the two Rhodesian jetliners that were downed by SA-7's in the late 1970s and a Sudan Airways jetliner that was shot down in 1986—a total combined loss of 167 lives.[22] Other countries that have experienced missile attacks on domestic airliners include Angola, the former Soviet republic of Georgia, Bosnia, Armenia, Sri-Lanka, Turkey, Costa Rica, Mozambique, Mauritania, and Somalia.[23]

Recent notable attacks include the November 2002 al-Qaeda failed attempt to down an Israeli Arkia Boeing-757 airliner using a Russian-made Strella missile (the missile was launched prematurely) and the November 2003 attack on a DHL cargo jet, hit by a Russian-made SA-7 or SA-14 missile as it took off from Baghdad's airport (the plane landed back safely).[24] Logic would have it that the threat is relevant to the United States as well.

This type of scenario is usually beyond the scope of airport security systems today; and with an estimated 500,000 operable MANPAD units in the world, simplicity of use, and a reported price tag of as little as $1000 on the international arms black market, it is clear why this particular threat to aviation is imminent.[25]

Although not attacked by missiles, the bombing of two Russian commercial airliners on August 24, 2004 (explosive charges were presumably infiltrated into the cargo section of the planes) should also sound the alarm bells.[26]

At a briefing before the Israeli Parliament's Security and Foreign Affairs Committee, in mid-May 2005, the head of the Israeli Security Agency labeled shoulder-launched missile attacks against civilian aircraft as the top security risk at the time.[27]

Planes as Weapons. Traditionally, most aircraft hijackings have been perpetrated for the purpose of using the passengers as hostages in an effort to either obtain political asylum for the perpetrator or facilitate the release of terrorists being held in prison. Since 1947, 60% of hijackings have been related to refugee issues. In 1968–1969 there was a considerable increase in hijackings: In 1968 there were 27 hijackings and attempted hijackings to Cuba alone; 1969 saw 82 recorded hijacking attempts worldwide—more than twice the total attempts for the entire period of 1947–1967.

Since then, hijackings have declined: In the peak years between 1967 and 1976, there were 385 incidents; in 1977–1986 the total had dropped to 300 incidents and in 1987–1996 it was 212.[28]

As many in the industry viewed the steady decline in hijackings as a clear trend, the 9/11 attacks brought about a new terrorist threat to aviation. An aircraft commandeered by terrorists can be transformed into an ultimate "smart cruise missile" against air and ground targets.

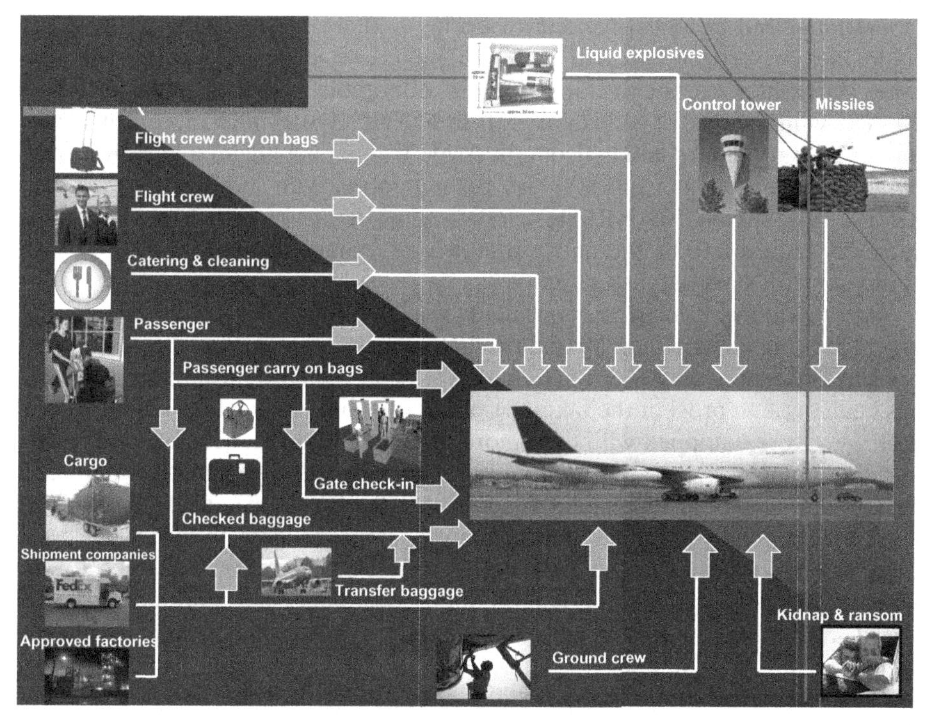

Figure 5.1. Plane vulnerability.

In the 9/11 attacks, the use of hijacked planes as suicide missiles changed the way hijacking was perceived as a security threat. Aircrafts throughout the world have become a form of missile stock pile for terrorists. Available and vulnerable aircrafts are prime game.

The level of threat is high.

Vulnerability

With the level of vulnerability being composed of three primary factors—the *visibility* and *accessibility* of potential perpetrators to the target and the existing security apparatus' *ability to prevent* the attack—Figure 5.1 clearly illustrates the many access points and visibility of an aircraft and the difficulty of preventing attacks.

Vulnerability Points. The high level of vulnerability stems from the relative ease of gaining attack information, constraints on security intelligence, and the vast industries and activities that are intertwined with the aviation industry.

Operational and logistical handling, catering and cleaning, cargo and shipment companies, passengers, ground and in-flight crews, baggage transferring, and other activities make the aviation industry vulnerable.

The vulnerability of civilian aircraft to MANPADs is much more acute than that of military planes and helicopters. Aside from being bigger and slower than military planes, civilian airliners are not equipped with any countermeasure/defense system from MANPADs or other anti-aircraft missiles. Israel's El-Al is currently believed to be the only airline whose jetliners have been or are being equipped with such defense systems.[29]

Most aviation security enhancements since 9/11 concentrated on developing and upgrading airport security and preflight check-in procedures in an effort to prevent illicit smuggling of weapons and explosives onboard the aircraft. But little has been done to prevent the takeover of aircrafts by perpetrators—like those in 9/11, who were equipped with little more than an intense malicious intent.

The main in-flight security reforms were limited to installing reinforced cockpit doors (that are opened a number of times per flight) and an increase of air marshals deployed on designated and very limited flights. In fact, it has been recently reported that of the 28,000 commercial airline flights that take to the skies on an average day in the United States, less than 1% are protected by on-board, armed federal air marshals.[30] The ability to monitor and control in-flight security is still very limited.

Therefore, the vulnerability is still high.

Consequence

The level of damage or consequence of a successful attack is composed of three primary factors: *causalities, economic ramifications, and public morale.*

Without going into a detailed calculation of the wide range of damage such an attack would have, it is enough to state that the *causality, economic, and public morale* consequences would be high based on the 9/11 precedence and more.

Therefore the consequence of a successful attack is high.

Means of Mitigation

In terms of the MANPAD threat, security plans should include (a) increased intelligence capabilities and elevation of constraints, (b) physical security in terms of fences, radars, intrusion detection systems, monitoring and surveillance equipment; (c) trained personnel conducting sweeps, inspections, observation posts and surveillance; (d) public awareness, and (f) installment of anti-missile defense systems on commercial airliners.

As for the pre-flight vetting process and in-flight threat, there are two elements worth noting:

1. One of the main goals of pre-flight security is to verify that passengers with malicious intent are not allowed access to the aircraft. For that end,

passengers are often questioned and luggage is frequently inspected. The main problem with that security process is its efficiency and the time it takes. Today, there are relatively inexpensive pre-flight questioning technologies that expedite and facilitate the process. Such technologies should be implemented in all international and domestic airports.

2. Monitoring and controlling in-flight security and safety has been limited to date. However, today there are operational concepts and technological means that can facilitate the ability to monitor and control irregular events on flight. If such systems were available and installed in the 9/11 airplanes, the tragedy might have been avoided. In order to mitigate the risk of such incidents reoccurring, these concepts and means should be implemented on all international and domestic flights.

Aside from the above, it should be noted that if widespread international security cooperation is not achieved, the efforts of individual nations and airlines may all be futile. Therefore, international legislation, standards, and, most importantly, enforcement is required.

PROSPERITY OR SECURITY IN MARITIME TRADE

The next two scenarios will deal with the maritime threat. With the global economy being a primary target for terrorists in general and al-Qaeda in particular, the maritime industry serves as a leading target. Despite the time elapsed and measures taken to secure mass transportation venues since 9/11, the maritime sector remains extremely vulnerable to terrorism. It is expected that an attempted attack on a significant maritime target will take place.

World trade depends mainly on maritime transport. The United Nations Conference on Trade and Development (UNCTAD) estimates that 5.8 billion tons of goods were traded by sea in 2001, accounting for over 80% of global trade volume. Over 46,000 vessels, servicing nearly 4000 ports throughout the world, carry the bulk of this trade. Great strides have been made in recent years to render this system as open and frictionless as possible, so as to prompt yet greater economic growth. According to UNCTAD reports, world sea-borne trade peaked at 5.89 billion tons in 2002, exceeding the previous record set in 2000.[31] These trade figures have grown and are expected to continue to grow.

The very factors that allow maritime transport to contribute to economic prosperity also leave it uniquely vulnerable to terrorism. The risks are numerous and encompass passenger cruise ships and ferries, container and bulk shipping, and the port facilities themselves. The threats are significant and range from potential physical attacks on ports, vessels, and shipments to document fraud and illicit fundraising for terrorist groups, illicit trafficking in and/or use of weapons, explosives, nonconventional or hazardous materials, and much more. The stakes are extremely high, since any major breakdown in the maritime transport system could cost dearly in terms of lives and could fundamentally cripple global trade.[32]

SCENARIO NO. 3: DIRTY BOMB IN MARITIME CONTAINER

Description of Scenario

With the goal of carrying out an attack of mass disruption and to show terrorist capabilities not seen before, al-Qaeda or an organization of its ilk decides to carry out an attack that entails the use of unconventional weaponry. For that end:

> A container vessel leaves its port of origin in Africa on voyage to a mega port in the United States; it makes two loading stops on the way. As the vessel approaches the mega port, a large ammonium nitrate fertilizer-enhanced explosion occurs in one of the containers. With the explosion, debris is seen from miles away.

The damage done seems substantial but perhaps even worse is the unknown. The authorities detect the immediate physical damage but are not able to verify the exact type of attack or rule out the possibility or extent of radiological contamination.

Threat. As noted, the level of threat is composed of two primary factors: *intention* and *capability* of potential perpetrators to attack.

> **What Is a Dirty Bomb?**
> "Dirty bombs," known also as radiation dispersal devices (RDD), are weapons that use conventional explosives to disperse radioactive materials, thereby augmenting the injury and property damage caused by the explosion. This type of a device lacks the complex nuclear-fission chain reaction that makes a nuclear bomb a weapon of unique devastating proportions. A radiological weapon can come in the form of a conventional explosive such as dynamite, packaged with radioactive material that scatters when the bomb goes off.[33] Essentially, a "dirty bomb" is a conventional bomb designed to disperse radioactive material to cause destruction contamination, along with injury from the radiation produced by the material. An RDD can be almost any size, defined only by the amount of radioactive material and explosives.

INTENTIONS. The main motivation of carrying out such an attack is to cause destruction and mass disruption and to show capabilities not seen before. There is no record of a major dirty bomb attack to this day. However, on May 8, 2002 the United States arrested a U.S. citizen named Jose Padilla (a.k.a. Abdullah al-Mujahir) at Chicago's O'Hare airport after arriving from Pakistan. He was an alleged al-Qaeda terrorist who was arrested for plotting to build and use a dirty bomb[34]—a clear sign that the motivation to use a dirty bomb is high.

Most Probable Perpetrator. A number of terrorist groups are known to have experimented with maritime terrorism. Intelligence reports point out that radicals from Jemaah Islamiah, a group linked to the al-Qaeda network, have been trained in sea-borne guerilla tactics, such as suicide diving and ramming, developed by the Liberation Tigers of Tamil Eelam (LTTE).[35]

In 2000, a small boat filled with explosives rammed into the USS *Cole* in Yemen, and in October 2002 the French-owned supertanker *Limburg* was attacked in a similar fashion in the Persian Gulf region. These incidents, coupled with updated intelligence, indicate that terrorists may be stepping up attacks against shipping. Oil tankers and warships are the only ships under threat; future attacks could target commercial shipping, including cargo ships, cruise liners or ferries, and port facilities practically anywhere in the world.[36]

Al-Qaeda is known to have a maritime military manual that (a) deals with how to attack ships and (b) shows different classes of vessels and indicates how to hit them and how much explosives are needed.[37] In addition, al-Qaeda has a history of nautical attacks. The group's tactics indicate an increase in strikes against shipping and port facilities as part of its push to hit economic targets. Intelligence officials have identified cargo freighters they believe are controlled by al-Qaeda.[38]

CAPABILITIES. The feasible modes of maritime terrorism operation are far-reaching. To mention a few, terrorists could hijack a vessel, or they could register a ship in a "flag of convenience" nation and use it for terrorist activities; they could purchase and make use of a legitimate shipping company and its vessels to carry out acts of terrorism without coming under suspicion. These ships could be loaded with explosives and crashed into other vessels, port facilities, critical infrastructure, or population centers on the coast.

Alternatively, oil tankers or vessels carrying hazardous materials could be used as terrorist weapons. Vessels, major ports, coastal oil depots, power stations, harbors, or bridges could be ideal targets for attacks. Maritime attacks may also involve the use of small underwater craft, such as small submarines or underwater motor-propelled sleds for divers.[39]

But one of the most frightening terrorist threats to maritime security involves terrorists smuggling and/or activating explosives such as dirty bombs in a sovereign country using cargo containers. Such a scenario became less notional five weeks after the 9/11 attacks, when alert Italian security personnel in Gionia Tauro revealed the "Container Bob" incident on October 18, 2001. In that incident, an Egyptian man with Canadian citizenship, nicknamed "Container Bob," was discovered when Port police in the Italian city of Gionia Tauro heard an unusual noise coming from a cargo container. Upon opening it, they discovered a well-dressed man drilling ventilation holes. He was equipped with a bed, a toilet, water supply, satellite phone, laptop computer, cameras, and maps. He also had security passes and an airline mechanic's certificate valid for New York's JFK, Newark, LA International, and O'Hare airports.[40]

The "Container Bob" incident illustrates the threats and vulnerability. The container was chartered by Maersk Sealand's Egyptian office, and the container was loaded in Port Said onto the German-owned, Andrew Weir-chartered, Antigua & Barbuda-flagged 2959 TEU Ipex Emperor. The container was set to be transshipped at Gioia Tauro, carried to Rotterdam before once again being transshipped to its final destination in Canada. There were about 2.5 million other containers handled at Gioia Tauro in 2001; had the stowaway not been attempting to widen the container's ventilation holes when port workers were nearby, the container would probably have passed through unhindered to its final destination.[41] A weapon smuggled in such a container could be detonated upon arrival at the port or at any strategic point along the container's route.

Within the gamut of the nonconventional spectrum that includes chemical, biological, radiological, and nuclear means, the radiological "dirty bomb" seems to be one of the most feasible to apply.[42] In terms of the required capabilities, the know-how required for the construction of a dirty bomb is not much more than the one needed to make a conventional bomb. No special assembly is required: The regular explosive would simply disperse the radioactive material packed into the bomb. It is precisely the relative ease of constructing such weapons that makes them a particularly worrisome threat. Even so, expertise matters. Not all dirty bombs are equally dangerous: The cruder the weapon, the less damage caused. The hard part is acquiring the high-grade radioactive material (e.g., cesium-137, americium-241, strontium-90, iridium-192) and handling of the radiological substance by an attacker without being contaminated, a concern that is obviously obsolete as far as suicide attackers are concerned. The *Washington Post* reported that the Bush administration's consensus view was that Osama bin Laden's al-Qaeda terrorist network probably has stolen radioactive contaminants such as strontium-90 and cesium-137, which could be used to make a dirty bomb.[43]

Nevertheless, the key question remains *capability* and *consequence*. Both aspects are not completely clear, primarily due to the lack of actual case studies.

Whether terrorists could handle, construct, and detonate high-grade radioactive material without fatally injuring themselves first is still a matter for speculation, but the working premise must be that it will be possible in the future.[44]

Therefore the level of threat should be considered high.

Vulnerability

As noted, the level of vulnerability is composed of three primary factors: the *visibility* and *accessibility* of potential perpetrators to the target and the existing security apparatus' *ability to prevent* the attack. Based on those criteria, the maritime industry is very vulnerable. The size, accessibility, and metropolitan location of many port facilities facilitate a free flow of trade and travel, but these factors also make the monitoring and controlling of traffic (in general) and containers through the ports very difficult.

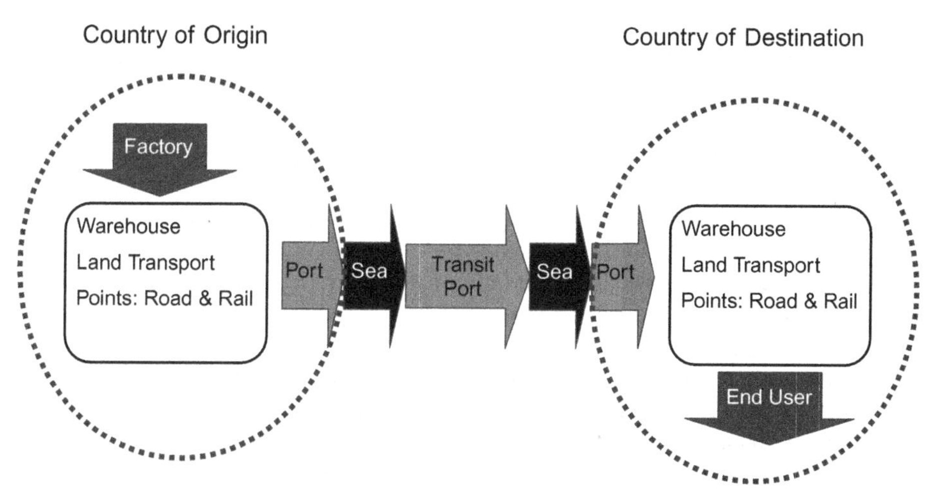

Figure 5.2. Vulnerability points in trade chain. (Source: Phillipe Crist, 2003)

About 10 million containers arrive in U.S. ports from foreign countries each year,[45] with only about 2% of those containers inspected prior to 9/11.[46] The ability to inspect all or most of these containers is still very problematic. This predicament was summed up fittingly by Edward H. Bikey, the chief operating officer of Dubai Ports World, who was seeking to manage port terminals in six major American cities "How do you bring good inspection on the millions and millions of boxes? Nobody has it in the whole world."[47]

Places of vulnerability exist throughout the cargo logistic chain (Figure 5.2). Usually starting from the manufacturer's premises where the container is packed, it is then transported by either road or rail to the port. On the way to the port, containers may be spotted at truck stops or freight yards. Once in port, the container is sent to a staging area before it is usually placed next to the vessel at dock. Even within the port area, a container may be moved around in accordance to port operator and/or customs. After being placed on board a vessel, the container can be removed and transshipped in another port onto another before arriving at its destination port. Here again, the container may be moved several times for customs clearance and into temporary storage areas while waiting to be picked up. Carried by road, rail, or inland waterway to its final destination, the shipment may again transit through several facilities where the container may be packed and unpacked.[48]

Throughout the above-detailed process, people make contact with the containers. Each one of these points is a point of vulnerability.

Therefore, the vulnerability of attack is high.

Consequence

The level of damage or consequence of a successful attack is composed of three key factors: *causalities, economic ramifications, and public morale.*

The exact ramifications of a dirty bomb attack are hard to assess, particularly since such an attack has fortunately yet to occur. Therefore, consequence assessments vary.

One of the main hurdles in confronting "dirty bomb" attacks is the problem of acknowledging that an attack has actually taken place. Radiological materials are not recognizable by the senses, being colorless and odorless. Specialized equipment is required to determine the size of the affected area and whether the level of radioactivity poses an immediate or long-term health hazard. Due to the delayed onset of symptoms in a radiological incident and the role of wind conditions (as in biological weapons), the affected area may vary.[49]

In addition to death and destruction, any such attack using WMD would undoubtedly have a traumatic effect on the national and even global psyche. Such an attack could cause prices to skyrocket and bring America's economy to a halt. There may even be a dominos effect that cascades throughout the world, leading to a deep global recession.

The "dirty bomb" kills or injures through the initial blast of the conventional explosive and by airborne radiation and contamination emitted by the material's particles—hence the term "dirty." Such bombs could vary in size, from a miniature device to something as big as a truck bomb.

The potential capacity for an RDD to cause significant harm is strongly reliant on the type of radioactive material used and the means by which it is dispersed. Other important variables in this "potential impact equation" include location of the explosive device, the population density in the area, and the prevailing weather conditions.

The calculations are complicated. According to an assessment of the Center for Defense Information, the actual harsh physical health threat might be confined to a radius of a few city blocks plus areas under a narrow wind-carried cloud.[50] Others assess that the explosion of a dirty bomb containing one kilogram of plutonium in the center of an average size metropolis could ultimately lead to about 150 cancer cases directly attributable to the blast.[51]

According to a scenario drafted by David Howe of the Homeland Security Council, an RDD bombing in a medium- to large-sized city would result in 180 fatalities and about 270 injured requiring medical care. In addition, up to 20,000 individuals would potentially have detectable superficial radioactive contamination. In the blast, one building and 20 vehicles would be destroyed, and eight other buildings would suffer varying degrees of damage, such as minor structural damage and broken windows. Radioactive contamination would be found inside and outside buildings over an area of approximately 36 blocks.[52]

Others attest that a dirty bomb is not so much a weapon of mass destruction but more a weapon of mass *disruption*, augmenting its conventional potential for physical damage by dispersing radioactive material into the air, carried further away by the wind. In its capacity to cause terror and disruption versus its ability to inflict heavy casualties, the dirty bomb is far superior to conventional explosive devices. Depending on the sophistication of the bomb, wind conditions, and the speed with which the area of the attack was evacuated, the *direct* number of casualties may not be substantially greater than a conventional explosion, though the long-term casualties—mostly as a result of radiation-induced cancer—can be substantial.

At any rate, the consequence of a successful attack is high.

Means of Mitigation

The only way of virtually guaranteeing that a dirty bomb container scenario does not occur is to *efficiently inspect every container and verify that it is not tampered with after inspection.*

Unfortunately, there is currently no efficient manner of actually doing that without bringing international trade to a virtual standstill. There are a number of international security conventions and initiatives that, if properly implemented, could mitigate the risk considerably and prevent an attack.[53] The International Ship and Port Security Code (ISPS Code) developed by the International Maritime Organization is one facilitating initiative and the Container Security Initiative (CSI) is another. The Secure Freight Initiative is a third. The ISPS Code contains security-related requirements for governments, port authorities, and shipping companies, together with a series of guidelines on how to meet these necessities. The CSI allows the U.S. Customs and Border Protection CBP, working with host government Customs Services, to examine high-risk maritime containerized cargo at foreign seaports before they are loaded on board vessels destined for the United States. There are currently 58 foreign ports participating in CSI, accounting for 85% of container traffic bound for the United States.[54]

The initiatives are a step in the right direction but voids are still evident.

The implementation of the initiatives should complement a layered security approach that includes (1) intelligence, (2) the early provision of more and better information and documentation about container contents, (3) activating shippers, all the way up and down the chain, to greater procedural uniformity, fastidiousness, and vigilance, (4) greater control and background screening of those having access to containers and ports, and (5) developing and installing new inspection and tracking technologies, including effective radiation detectors.[55]

Having said all that, the initiatives and security layers might all be for naught, if international agreements and cooperation are not upheld.

To successfully confront terror, it has become widely acknowledged that candid and constant international cooperation and enhanced technological means of detecting are a must. Such cooperation is most vital when attempting to mitigate the maritime risk.

SCENARIO NO. 4: BLOCKING WORLD OIL TRANSIT BY SEA[56]

Ofer Israeli

In the aftermath of the September 11, 2001 terror attacks, substantial efforts and resources were allocated to the security of America's homeland. As a result, the homeland may indeed have become less vulnerable. However, terrorists who want to cripple the U.S. economy need not necessarily attack American territory. Significant damage can result from attacks in international waters, which lack adequate security means.

Description of Scenario

In full understanding of the above-noted premise, Islamic militant organizations, such as al-Qaeda and its ilk initiate a comprehensive attack on the six main oil transit strategic chokepoints, which are straits and channels along widely used global sea routes narrow enough to be blocked and vulnerable to piracy and terrorism—the Strait of Hormuz, the Strait of Malacca, the Suez Canal, the Strait of Bab el-Mandab, the Turkish Straits, and the Panama Canal. The terrorists hijack tankers, sail them into three chokepoints, and blow them up. At the same time, they also use boats as well as one-man submarines, which they crash into additional oil tankers passing through the three other narrower strategic chokepoints.

Oil transit by sea comes to an almost halt for weeks. The psychological effect of the heavy smoke filling the skies for days is colossal. Consequently, the global demand for oil dramatically increases and its global market price sharply and immediately reaches heights never seen before—sparking a three-digit raise in oil price. As a result, the world energy market is crippled; the global economy, and especially the U.S. economy that relies heavily on oil energy, stagnates; the economic damage to the livelihood of the United States and other industrialized countries is far-reaching.

The mentioned strategic chokepoints are of critical importance to U.S. national interests. Blocking any of the six could sharply raise oil rates. The U.S. economy would be directly affected. A simultaneous impasse of a few chokepoints sustained for weeks would cause a worldwide oil shortage and, at least in the short term, place harsh pressure on the global economy. The United States has direct and immediate economic interests in protecting these chokepoints for the secure passage of oil tankers as well as other commercial vessels that sail through them.

Threat. The world's most strategic maritime chokepoints are very vulnerable. Surrounded by volatile and perilous areas, they are almost defenseless against terrorist and pirate attacks. In the same way that al-Qaeda used commercial aircrafts as missiles for targeting the World Trade Center and the Pentagon, oil supertankers could be hijacked and used for blocking the oil transfer through strategic chokepoints. Past attempts launched by terrorist organizations illustrate their ***intent*** as well as their ***capability*** for initiating such attacks.

INTENTIONS. The September 11, 2001 terror attacks against the World Trade Center in New York City were, among other things, motivated by al-Qaeda's desire to cause great damage to America's economy. As the speech of al-Qaeda's leader, Osama bin-Laden, in 2004 shows, the group's goal of forcing America into bankruptcy remains very intense: "We are continuing this policy in bleeding America to the point of bankruptcy. ..."[57] The al-Qaeda organization is well aware that the oil market presents an ideal target for undermining the world economy. Oil supplies are, in al-Qaeda's words, "the provision line and the feeding to the artery of the life of the crusader nation."[58]

The Strait of Hormuz, which leads out of the Persian Gulf, is the world's most important oil chokepoint and has been under expressed terrorist threat since September 11, 2001.[59] The Strait of Malacca, which links the Indian and Pacific Oceans, has become a potential target. Disruption from pirates, including attempted theft and hijackings, are a constant threat to tankers in the Strait of Malacca; Jemaah Islamiyah planned to strike at U.S. naval vessels in the Strait before its members were arrested in Singapore.[60] The Turkish Straits have been under various terrorist threats after September 11, 2001, and the Panama Canal was a suspected terrorist target.[61] A group affiliated with al-Qaeda planned raids on British and American oil tankers passing through the Strait of Gibraltar.[62]

CAPABILITIES. One must also note that a few successful attacks have actually been launched by al-Qaeda and its ilk in the vicinity of several chokepoints. The Strait of Bab el-Mandab is a strategic link between the Indian Ocean and the Mediterranean Sea. It is 18 miles wide, and an estimated 3.3 million BBL/D flow through it daily. A few attacks were launched in the area surrounding the straits: al-Qaeda attempted to ram a boat loaded with explosives into the USS *The Sullivans* in Yemen in January 2000. A tugboat packed with plastic explosives, while anchored in the port of Aden, Yemen, in October 12, 2000, crashed into the USS *Cole*, killing 17 sailors.[63] In October 2002, al-Qaeda badly damaged the French oil supertanker *Limburg* off the coast of Yemen, only 300 miles from the Strait of Bab el-Mandab.[64]

The latest reports of pirate attacks upon vessels in the Gulf of Aden prove the ability of terrorist groups to hijack or blow up oil supertankers: In May 27, 2008, a suspicious speedboat with five persons was noticed proceeding toward a tanker. On May 28, four suspicious high-speed boats tried to approach a tanker. On the same day, four heavily armed pirates in a speedboat attacked and hijacked a general cargo ship under sail. They then sailed the vessel into Somali territorial waters.[65]

Vulnerability

Although alternative energy sources are available, oil remains the leading fuel in the global primary energy mix. Future supplies of oil depend on many variables—such as the political stability of export countries. However, the more vulnerable target, which is of imminent risk, is the world transfer of oil by sea. From total world oil production, amounting to approximately 85 million barrels per day (BBL/D), about one-half, almost 43 BBL/D of oil, was moved by tankers on designated maritime routes: the Strait of Hormuz, 17 million; the Strait of Malacca, 15 million; the Suez Canal (Sumed Pipeline), 4.5 million; the Strait of Bab el-Mandab, 3.3 million; the Turkish Straits (the Bosporus and Dardanelles), 2.4 million; the Panama Canal, 0.5 million.[66]

These six chokepoints are critical targets. By launching a set of coordinated terror attacks upon oil tankers passing through them, the attackers could almost totally cut the sea transfer routes of the crude oil supply into the U.S. and other major industrialized countries in Europe and the Pacific. Sharp reduction of the world oil supply, even for a limited time of a week or two, endangers the global economy, constitutes a great risk to the United States, and emphasizes its susceptibility.

The vulnerability of oil tankers cruising through the six oil transit chokepoints mentioned above is extremely high. Tankers are huge and bold. They are very slow and they lack the ability to rapidly escape from attackers; they limited capability to protect themselves. The six straits mentioned are all very narrow—the narrowest is merely 110 feet wide (the Panama Canal) and the widest is only 21 miles wide (the Strait of Hormuz).

Visibility and Accessibility. Most of the world's oil is shipped through strategic chokepoints located in the world's most terror-infested waters, in which Islamic terror organizations with naval capabilities are widespread and already active. As mentioned by former President Bush, maritime attacks have already been attempted in the Strait of Hormuz and the Strait of Gibraltar.[67] The Strait of Hormuz, which connects the Persian Gulf with the Gulf of Oman and the Arabian Sea, is the world's most important oil chokepoint, with a daily oil flow of 17 BBL/D. It is controlled by Iran, a long time rival of the United States and a home-haven for terrorists at present. A U.S. strike against Iran's nuclear program might very well be met with an Iranian effort to block oil shipments through the Strait of Hormuz; it would be an ideal moment for Iran to cooperate with terrorists in order to block the Strait and to use them as a proxy to harm America's interests.

Other maritime chokepoints are also highly accessible to terror attacks. The Strait of Bab el-Mandab—located between Yemen, Djibouti, and Eritrea, connecting the Red Sea with the Gulf of Aden and the Arabian Sea—is a strategic chokepoint with a daily oil flow of 15 BBL/D toward Europe and the United States. As described above, the straits have been a location for numerous terror and pirate attacks. The Strait of Malacca—located between Indonesia, Malaysia and Singapore and linking the Indian Ocean to the South China Sea and Pacific Ocean—is the key chokepoint in Asia, with a daily oil flow of 15 BBL/D, supplying two of the world's most populous nations, China and Indonesia. The Al-Qaeda-linked Jemaah Islamiyah group has admitted that the group planned to launch attacks on Malacca shipping.[68]

Another point for concern is the increasing sign of collaboration between terrorist groups and piracy activity. Pirate attacks on ships have tripled in the last decade, and piracy figures were up by 20% for the first quarter of 2008.[69] In April 2008, there was a pirate attack on a huge oil tanker off the Somali coast.[70] Vessels transiting the Gulf of Aden between May 19 and June 8, 2008 have reported six attacks, including two hijackings. Pirates fired automatic weapons and rocket-propelled grenades in an attempt to board and hijack vessels.[71]

According to the ICC Commercial Crime Services, the Straits of Malacca, the Gulf of Aden, and the Red Sea are all piracy prone areas.[72] The data presented in Figure 5.3 emphasize the fact that collaboration between piracy and terror would be devastating.

Another Achilles' heel is the chokepoints' narrowness. As presented in Figure 5.4, some of them are extremely narrow and very difficult to maneuver in. It makes the possibility of targeting a supertanker passing through them relatively simple.

In sum, attacking oil tankers as they pass through the six strategic chokepoints would be an almost ideal target in the eyes of militant organizations.

Current Ability to Prevent. The Operation Enduring Freedom, instigated in the wake of the September 11, 2001 terror attacks, polices the seas to fight terror. The task force consists of 13 nations: the United States, France, Saudi Arabia, Denmark, Italy, Spain, Great Britain, Canada, Germany, Bahrain, New Zealand, Pakistan, and Singapore. Nine coalition warships patrol 11,700 miles of coastline, covering 6.2 million square kilometers. However, most pirate attacks occur within territorial waters: In 2005, 276 hijackings and another 24 attempts took place in international waters where the coalition can act. The rest took place close to the coast, where the coalition has no jurisdiction.[73]

At the operative level, "Secure-Ship" is the most recent and effective innovation in the fight against piracy. Secure-Ship is a nonlethal, electrifying fence surrounding the whole ship, which has been specially adapted for maritime use.[74] However, it is only a partial solution for the risk.

Consequence

Simultaneous terror attacks upon oil supertankers passing through the six oil transit chokepoints could create a great environmental disaster and grave economic consequences.

Casualties. The most vulnerable is the Strait of the Bosporus dividing two parts of Istanbul, Turkey. The areas surrounding the other chokepoints mentioned above are extremely isolated areas, which reduces the danger to populations.

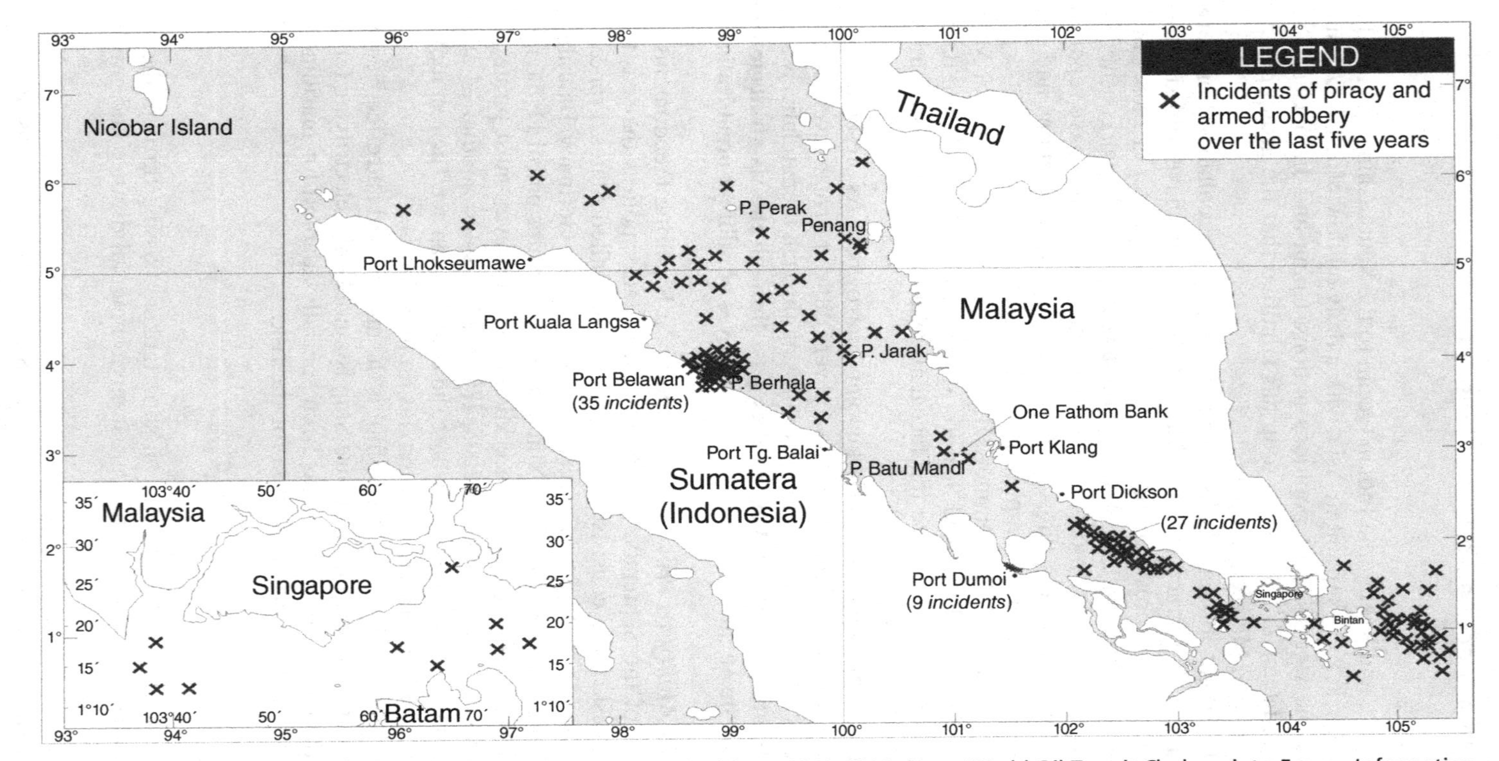

Figure 5.3. The Strait of Malacca: Incidents of piracy and armed robbery, 2000–2005. (From World Oil Transit Chokepoints, Energy Information Administration, U.S. Department of Energy. Available online: http://www.eia.doe.gov/cabs/World_Oil_Transit_Chokepoints/Full.html)

Figure 5.4. Two views of the Strait of the Bosporus. (From World Oil Transit Chokepoints, Energy Information Administration, U.S. Department of Energy. Available online: http://www.eia.doe.gov/cabs/Worle_Oil_Transit_Chokepoints/Full.html)

Treasure. All nations, especially industrialized countries such as the United States, are dependent on oil and its market stability, which links all economies to global market prices. Recent attacks upon oil tankers and oil installations show the vulnerability of oil prices and the colossal effect of such attacks over the world oil stability. The Japanese oil tanker, which was empty when attacked in Middle Eastern waters off the coast of Yemen in April 2008, rattled the nerves of global energy traders.[75] The attack by Nigerian rebels on the Royal Dutch Shell oil pipeline in May 2008, carried out while Shell was trying to repair damage caused by a series of similar attacks in April 2008, forced the firm to cut production and caused world oil prices to reach record highs.[76] If such limited attacks upon oil tankers and oil installations can shock global energy traders, one can imagine the extent of the consequences of a wide-ranging attack throughout the world: blowing up oil tanker's passing the six chokepoints presented above consequently blocking the oil transit to many industrialized countries for weeks. The economic consequences would be horrifying.

Most nations rely on the six narrow strategic chokepoints for import and export of oil and trade. Their economic vitality depends on free access to these sea lanes. U.S. prosperity, in turn, relies on the economic well-being of its trade partners. The United States has a direct and immediate economic interest in protecting free passage through these six chokepoints, since their blockage could immediately and directly disrupt U.S. economy.

The vast majority of the world oil traffic could redirect in a crisis. However, closure of one or more chokepoints would cause a gigantic increase in cargo rates worldwide. Theoretically, blocking shipping routes might not cause great consequences, since alternate routes are available. The closure of the Suez Canal and the Strait of Bab el-Mandab, for example, would divert tankers around the southern tip of Africa, the Cape of Good Hope. Although it adds 6000 miles to transit time, it seems that it could be acceptable—after all, merchant vessels offer one of the cheaper modes of transport. In practice, however, it turns out that blocking these chokepoints would matter a great deal. A great amount of world oil and other merchants' fleets would be required to sail much farther. Closure of the Strait of Hormuz and the Strait of Bab el-Mandab, for instance, would immediately raise cargo rates. It would also disrupt world-shipping markets. Freight rates around the world would be affected, thus adding costs to America's import and export.

The experience with the closure of the Suez Canal seems to indicate that such a disruption might increase freight rates by as much as 500%.[77]

Means of Mitigation

Mitigation of the risk includes short-term tactical as well as long-term strategic means. Other nations, in addition to the United States, have a stake in the free movement of shipping through these six chokepoints. Tactically, the industrialized countries that depend on oil imported through these six vital chokepoints should be motivated to cooperate and share the costs of naval protection. The United States should encourage other nations to share in the costs of the six strategic chokepoints' protection. The United States should work especially closely with countries bordering the strategic chokepoints and facilitate the naval capabilities of those countries; the United States has to help patrol their sea lanes. In that context it should help establish and train national coast guards in the area. Because of the strategic position of the Strait of Bab el-Mandab, for example, both France and the United States have military bases in Djibouti.[78] The United States should also foster consensus in global forums to keep the chokepoints open because world trade in general—and oil trade in particular—requires this.

Strategically, industrialized countries have to diversify their energy sources and not rely solely on imported crude oil. This could also provide a desired solution for the *cross-effect* of consumption of oil by the Western world that later on is used, at least in part, for financing terror states, such as Iran, and sponsoring terror groups, such Hamas and Hezbollah.

Another strategic solution for the problem could be projects designed to bypass some of the dangerous chokepoints. Building a 1200-mile pipeline connecting the Persian Gulf, the main oil exporter to the Mediterranean Sea—from the Saudi coast between its border with Qatar and the Emirates via Jordan to the port of Haifa in Israel—could be a feasible and interesting solution: It would bypass three of the six chokepoints: the Strait of Hormuz, the Strait of Bab el-Mandab, and the Suez Cannal. Aside from the political benefits of such a project, which could be a catalyst to solving the century-old Israeli–Arab conflict, it would sharply reduce the cost of transiting oil, which will lower energy prices.

SCENARIO NO. 5: PC DOOMSDAY[79]

Barak Hudna[80] was a very talented silicon-valley-raised 27-year-old software engineer. He had a well-established job in an extremely important software company. He was a team leader of a group responsible for the last phase of QA (quality assurance). Their task was to make sure that there was no malicious content in a new release of the company operating system software—a very important task indeed. He never tried hard to hide his Muslim beliefs, and his co-workers didn't really care—he was an extremely talented worker and they appreciated his personality and professionalism.

He is the last person to ensure the content of the software. Hudna can actually alter the operating system in such a way that it will shut itself down suddenly and die hard with no one's knowledge.

Three years down the road: For all computers installed with the "Twisgo" version, 75% of all PCs and servers will be shut down and will not recover easily for several days. He could envision the outcome of his plan: dams collapsing, hundreds of air traffic and land traffic accidents, financial collapse, gas pipes bursting. Actually all human activity that relies on his company's PCs and servers will be altered and shut down for minutes, hours, or naturally—the doomsday of the industrial world. Hudna naturally will be the one who ushered in change. However, when the search for a solution begins, Hudna will no longer be around; he will make sure he will add the real flavor to his deeds and will heroically die somehow.

Description of Scenario

Same-day basic malfunction of the all PCs, laptops, and servers which use the same widespread operating system would lead to chaos in the industrial world. Disasters in computer-controlled air traffic, dams, gas, and electricity cause thousands of deaths through out the Western world. Less developed countries that are not as dependent on computers suffer less from what will be remembered as the Doomsday of Computer Blackout.

Reports and testimonies indicate that seconds before global computers shutdown, a message appeared on the screen all of the damaged computers: **"Allahu Akbar"**....

Threat

INTENTIONS. Such an attack may be motivated by the intention to reveal Western weakness and its almost total dependency on technology and science—while showing in a spectacular manner the strength and wisdom of the believers and that of Allah—that could be the ultimate motivation for such an organized attack on the computerized world. Aside from that, it would cause extensive economic damage and human casualties, which is often a strong motivating factor for Jihadist organizations and individuals.

CAPABILITIES. If one person in a relevant sweet spot in an operating software company has knowledge, sophistication, eagerness, and willingness to scarify, he or she can do the job.

Vulnerability

Accessibility. Software companies are not currently considered "critical infrastructure" and therefore are not required, under law, to undertake special security measures. These companies do not use (but are permitted to use) pre-employment security screening, such as background checks or polygraph tests. Moreover, the leading software companies present themselves as multinational, multiethnic enterprise and as such are very sensitive to political correctness. Any existing security measures taken by leading software companies is directed at commercial competition and at "national security." Appling commercial security measures is quite different than "national security" measures. Therefore most software companies do not currently apply significant measures, and they probably won't do so unless compelled to do so by authorities.

Current Ability to Prevent. An insurgence of extremely harmful software into any operating systems is feasible. A skilled person with access to the fundamentals of the software code can carry out such an attack. Software companies currently have little, if any, measures at their disposal to prevent hostile employees from being hired to work and have little, if any, security measures in place. At present there is no realistic implementation of measures that can prevent this scenario from taking place.

Consequence

Casualties. Damage to critical computer systems in electric and water plants, dames, hospitals, SCADA systems (remote control infrastructure systems), air and land traffic etc. can lead to tens of thousands of dead and casualties.

Treasure. Banks may collapse, stock exchange and financial institution computer systems and slow recovery afterwards will lead to loss of billions. Physical damage to infrastructure and property caused by the collapse of the computer systems may lead to the collapse of the insurance systems.

Means of Mitigation

The best way to prevent this from happening is to take the preemptive measures:

1. Declare that major software companies are actually critical software companies.
2. Enforce new security policies on these companies through legislation and dialogue.
3. Implement high-security measures and methodologies used in NSA, CIA, and so on, in the civilian critical software infrastructure companies. These measures should include (a) pre-employment screening and (b) levels of security access to code based on background checks and security classification.
4. Mandate "hostile intent" tests using advanced screening tools available in the market today.

SCENARIO NO. 6: SUICIDE TERRORIST ATTACK ON SUBWAY FOLLOWED BY A SUICIDE TERRORIST ATTACK ON PREMISES OF A LEVEL I TRAUMA CENTER

Shmuel C. Shapira

An attack by a suicide terrorist on a subway or other mode of surface transportation has unfortunately become fairly common. In fact, roughly one-third of all terrorist attacks worldwide target surface transportation systems, with the weapon of choice usually being explosives.[81] Attacks are relatively evenly split between rail system (trains, subways, stations, and rails) and bus systems.[82] If one must choose between rail system or bus system targets, an attack on a subway in the United States seems more likely than on other modes of surface transportation. Examples of terror attacks against surface transportation are numerous. In 1995 and 1996, Algerian terrorists set off several bombs in the Paris subway; IRA bombers have blown up various railway and subway vehicles during their conflict in the United Kingdom; Aum Shinrikyo's Sarin gas attack on Tokyo's subway in 1995 raised the bar by introducing chemical elements into the terrorist arsenal; Palestinians have carried out numerous suicide bombings of buses in Israel; and al-Qaeda proved its train attacking capability in Spain and London. But in the scenario described and analyzed below, we emphasize the risks of a second-wave attack on hospitals that provide dearly needed medical relief.

Description of Scenario[83]

On Monday at 8:20 A.M., a suicide bomber ignited a 10-kg bag of improvised explosives on a Purple Line subway train as it stopped at the 34th Street and 6th Avenue station. Dozens of people are injured, and many suffer major trauma and bystanders report at least 25 fatalities. At 8:55 A.M., an ambulance loaded with 100kg of top industrial grade explosives, driven by a suicide terrorist, explodes at the entrance of the Emergency Department (ED). This is located near the main patio leading to a 750-bed Level I Trauma Center University Hospital located 8 miles away from the site of the first attack. Dozens of staff members, patients, visitors, police officers, and media reporters are killed or injured at the hospital. Major damage and partial collapse occurred in the ED. Windows shattered all over the hospital. The booby-trapped ambulance burned at the ED entrance.

The attack on the medical facility serves as a high-profile event with mass media coverage. Such an event will increase the chaos and fear and will diminish the capabilities to cope with the consequences of the initial Mass Casualty Event (MCE).

Threat

MOTIVATIONS/INTENTIONS. Terrorists seek to destroy, provoke chaos by assaulting infrastructure, and search for soft targets. We are aware of at least three attempts to attack major hospitals in Israel by female suicide bombers.[84] Sequential attacks have occurred and have led to multiple first responder casualties. One of the first examples of sequential attacks took place at the Beit Lid juncture bus station on January 22, 1995.[85]

CAPABILITIES. Suicide belts are relatively easy to construct, and terrorists who are ready to give up their lives while killing others, are in abundance. In Israel, while only 0.5% of terrorist attacks are carried out by suicide terrorists, more than 50% of fatalities and casualties are associated with this mode of attack. Therefore suicide attacks represent a unique model of death distribution.[86] Ambulances have been used by Palestinians to carry terrorists and ammunition and as a mean to bypass security searches.[87]

Vulnerability

Accessibility. Military-quality or home-made explosives in large amounts are easily available. Ambulances are often stolen or are simply fabricated by painting a commercial van.[88]

Current Ability to Prevent. Security at most hospitals in the Western world is minimal and is designed mostly to keep public order and almost not at all to prevent terror attacks. Hospitals are ultrasoft targets. On one hand, during MCE the awareness in hospitals might be enhanced. On the other side the chaotic phase associated with the initial management of MCE and especially the overwhelming flow of ambulances and other rescue vehicles might enhance the penetration of a booby-trapped ambulance.[89]

A abundance of hospital personnel located near the ED entrance makes it a more vulnerable target for attack.

Consequence

Casualties[90]

Initial Event. 37 immediate fatalities; 178 casualties (excluding acute stress disorders, ASD), 43 of whom suffer from major trauma; and 378 others suffer from ASD.

Hospital Attack. 59 immediate fatalities; 212 casualties (excluding ASD), 41 of whom suffer from major trauma; and 196 others suffer from ASD.

Economic Damage[91]

Initial Event. Direct damage, $ 71,000,000; loss of income, $ 110,000,000.

Hospital Attack. Direct damage, $ 115,000,000; loss of income, $ 330,000,000.

Risk level for a suicide attack on subway system and or hospitals is very high.

Means of Mitigation[92]

In terms of mitigating the risk of an attack on a hospital, a number of actions should be implemented as follows:

- Define first responders and hospitals as primary or secondary potential targets for different forms of terror.
- Gather intelligence regarding threats and priorities. Map hospitals and pinpoint vulnerabilities.
- Enhance hospital security by minimizing the number of entrances, positioning of security officers at vulnerable points, erecting fences where feasible, illuminating points at risk, using closed-circuit video cameras, inspecting vehicles and, more specifically periodically changing the location of identification stickers on ambulance windshields, and preventing overcrowding at public locations in hospitals.

- Inscribe relevant standard operation procedures (SOP).
- Train staff and run drills based on different scenarios.
- Upgrade hospital alert level during attack in region.
- Optimize triage in order to avoid overwhelming of a few institutes.
- Construct a contingency plan for MCE victims at alternative hospitals and prepare hospital evacuation plans.

Assessments in Each Jurisdiction. When assessing the risk to a nation, a corporation, or even an individual, the range of risk scenarios are vast. In order to be most efficient in the allocation of efforts and resources, it is of essence to rank the risk level and prioritize the security efforts. Such prioritization requires a qualitative and quantitative assessment and ranking of the various scenarios.

This section limited the assessment to high-risk scenarios on a national or even international level that should all be addressed by the relevant authorities and international organizations as such.

Similar risk assessments should be carried out and adapted to local conditions throughout the country, whereby the responsible security authorities should conduct periodic risk assessments on areas of their jurisdiction. Unlike national risk assessment, local assessments need to pinpoint the targets of risk and therefore need to conduct criticality assessments in an effort to define the regions and assets of highest importance. The criticality assessment is therefore the first step in the context of the risk assessment, in which all the relevant information concerning possible targets within a jurisdiction is gathered.

The main components of risk assessments for each jurisdiction should therefore include:[93]

- *Criticality Assessments.* Identification of the processes and critical areas, activities, and assets that may constitute "high-value" targets and need to be protected.
- *Threat Assessments.* Identification and analysis of the specific threats and scenarios relevant to the jurisdiction as a whole as well as to each critical target.
- *Vulnerability and Consequence Assessments.* The assessment of target vulnerability levels with regard to each credible scenario and a definition of the extent of possible damage that can be anticipated as a result of an effectively executed attack.
- *Risk Prioritization.* Based on the threat, vulnerability, and consequence assessment scores, a rating of scenarios—in descending order of the level of risk—should be carried out. This stage leads to a practical definition of the order of priority and urgency for risks, and it also provides practical directions for security planners.
- *Mitigation Strategies.* After prioritizing the credible "attack" scenarios, essential courses of actions for reducing vulnerability and consequence

levels to specific scenarios within the jurisdiction can be established. This will mitigate the risk level.

The accuracy of the assessed level of risk—and in particular the "threat" component within the risk—is limited to the time the analysis is conducted. Hence, authorities should constantly be threat-minded and adjust the level of threat in accordance to general or specific information regarding an upcoming peril. An escalation of the threat level will also elevate the level of risk compelling authorities to act accordingly.

FUTURE TRENDS

While it may seem that we have already faced most forms and means of terror, analysts often point to the threat of nonconventional agents and the use of chemical, biological, radiological, or even nuclear (CBRN) terrorism as the next imminent threat.

- In the past, the essence of terrorism was to disseminate fear and make a political statement through violence. It was mainly a politically motivated act designed to influence an audience; thus levels of violence were calculated so as to draw attention but not necessarily to be so high as to alienate supporters or trigger overwhelming response from authorities. That continues to be a main theme of "common" or "old-school" terrorism. However, in so-called postmodern or mega-terrorism, motivated by a deranged interpretation of religion, the aim is to maximize the number of casualties and economic damage.[94] For such an end, there seems to be no better method than to carry out an unconventional attack of major proportion.
- As a matter of fact, the use of nonconventional terrorist means is not novel. A number of incidents or plots involving nonstate actors (individuals or organized groups) that made intentional or unintentional use of chemical or biological agents include:
  - *The Weather Underground*, a radical leftist Vietnam-era group that allegedly attempted to steal hazardous material from one of the U.S. Army's biological centers located in Fort Detrick, Maryland (1970).[95]
  - *R.I.S.E.* College students in Chicago, with an extreme environmentalist ideology, plotted to culture deadly biological agents for aerosol release or to contaminate urban water supplies (1972).[96]
  - *Smallpox infection in Yugoslavia (1972).* In early 1972, a 38-year-old Muslim clergyman from Kosovo, named Ibrahim Hoti, undertook the pilgrimage to Mecca. He also visited holy sites in Iraq, where there were known cases of smallpox and where he was probably infected. After he returned home, 175 people were infected, 35 of whom died before vaccination of 19 million people was undertaken to prevent an epidemic.[97]

  - *The Alphabet Bomber.* Muharem Kurbegovic (a.k.a. Isauk Rasin), a Yugoslavian-born terrorist who called himself the "alphabet bomber," sent toxic chemicals through the mail to a Supreme Court justice and threatened to use nerve-gas devices against the Capitol, the president of the United States, and all nine justices of the Supreme Court. He was arrested in August 1974 for a bombing that killed three people at the Los Angeles International Airport; following the arrest, police searching his California home found that he had assembled virtually all the ingredients necessary to construct a nerve-gas bomb (1974).[98]
  - *The Baader Meinhof Gang* stole 53 canisters of mustard gas from a U.S. bunker in West Germany and threatened to use them in Stuttgart (1975).[99]
  - *The Rajneeshis Food poisoning*, a religious cult originating in India, deployed biological agents in Oregon in an effort to manipulate local elections, sickening 751 people in 1984. The incident represents the only bioterrorist attack known to have caused illness, and it illustrates the complexities of employing biological weapons.[100]
  - A chemical leak in a pesticide plant in Bhopal, India, possibly caused by a disgruntled worker, resulted in over 3000 deaths (1984).[101]
  - *Aum Shinkriyo's* sarin nerve gas attack, an attempt to kill three judges, resulted in the deaths of seven civilians in Matsumoto (1994)[102]; and the most famous use of nonconventional weapons at the Tokyo subway station in 1995 killed 12, injured 5000, and transformed the threat of the use of unconventional weapons into a risk to national security.[103]
  - The anthrax attacks or hoaxes almost immediately after the 9/11 attacks were probably the most consequential unconventional attacks on the United States to date and illustrate the public's vulnerability.[104]
- Within the realm of the CBRN threat, it is important to distinguish between the elements and to assess the probability of use and risk of attack.
- Chemical terror incidents are characterized by the rapid onset of medical symptoms (minutes to hours) and usually easily observed marks (colored residue, dead foliage, pungent odor, and dead insect and animal life).
- In the case of biological weapons, the onset of symptoms caused by many of the pathogens and toxins comes hours or even days after an exposure, with typically no palpable characteristic signatures. Because of the delayed symptoms of most biological attacks and the importance of weather conditions, especially winds, in spreading the agents from the epicenter of the attack, as well as due to the movement of infected individuals, especially in a busy metropolitan setting and its transportation systems, the area affected and the number of victims may be greater than initially thought and would entail a costly lapse of time, in terms of detection and initial treatment. A high-grade biological agent dispersed efficiently in large enough quantities could, in some scenarios, prove to be calamitous: One report suggests that well-dispersed anthrax (in ideal weather conditions)

could inflict 20% more casualties than a 12.5-kiloton nuclear bomb,[105] while another report suggests that 110 kg of anthrax could inflict as much damage on a densely populated metropolis as a 1-megaton hydrogen bomb.[106]

- With respect to radiological weapons, the problem of acknowledging that an attack has taken place is similar: Radiological materials are not recognizable by the senses, being colorless and odorless. Specialized equipment is required to determine the size of the affected area and whether the level of radioactivity poses an immediate or long-term health hazard. Due to the delayed onset of symptoms in a radiological incident and the role of wind conditions (as in biological weapons), the affected area may be very large.[107]
- In the nuclear realm, ever since the breakup of the former Soviet Union and the deteriorating level of control, regulation, and monitoring of nuclear facilities (weapon depots and reactors), there has been ample cause for concern. The alarm has only grown nuclear capability being acquired by unstable regimes, like the one in Pakistan and the relentless drive for weapons of mass murder by the terror-sponsoring regime in Iran.
- In terms of prioritizing the CBRN risk, it would seem that *the probability (Threat X Vulnerability) of a radiological, chemical, and even biological attack (in that order) is higher than a nuclear terrorist attack*, due to the currently assessed capabilities of potential perpetrators.

The above cases and several others that were not noted illustrate (a) the motivation of individuals and organizations to acquire lethal agents and (b) the fact that the inability of individuals and organizations to cause mass destruction is mainly due to lack of military-grade capabilities and our good fortune.

Those hurdles may be overcome by terrorist organizations if their sponsoring states, such as Iran and Syria, acquire nuclear and other mass destruction capabilities.

The risk is imminent.

In light of the risk, U.S. Undersecretary of State for Arms Control, John Rood, recently stated in a news conference that "Combating nuclear terrorism is especially important today" and that "Regrettably we continue to see indications in the United States from information we collect of the very terrorist groups we are most concerned about making concerted efforts to acquire nuclear capabilities with the express intent to use them against our peoples."[108]

ENDNOTES

1. Planning Scenarios, The Homeland Security Council David Howe, Senior Director for Response and Planning July 2004.
2. Interview with Col. (retired.) Lior Lotan, former director of International Institute for Counter-Terrorism in Herzlyia, June 1, 2008.
3. We have limited ourselves to six high-risk scenarios, but that does not imply by any means that this is a closed list. To the contrary. Furthermore, this book has focused

primarily on suicide terrorism; however, the high-risk scenarios analyzed in this section are not limited to that mode of attack. It should also be noted that this risk assessment section is a generic one. An in-depth analysis and assessment can and should be done after obtaining specific information and conducting on-ground surveys of the assessed target.

4. Our risk methodology is based on computing the *threat*, *vulnerability*, and *consequence* levels of each scenario. For further details on risk methodology, refer to "Assessing the Risk," by Ophir Falk and Lior Lotan, published in *Law Enforcement Executive Forum*, **7**(7), 2007.
5. The National Strategy for the Physical Protection of Critical Infrastructures and Key Assets, The White House, Washington, DC, February 2003, p. 52.
6. In order to understand the capabilities and intentions, it is important to review past case studies of successful and thwarted attacks in an effort to forecast future trends. Dr. Joshua Sinai has detailed seven indicators of "How to Forecast and Preempt al-Qaeda's Catastrophic Terrorist Warfare." According to Dr. Sinai, these indicators are: (1) previous terrorist attacks, failed attacks or plots not yet executed, which serve as blueprints for intentions and future targeting; (2) a terrorist group's modus operandi, especially tactics; (3) use of particular types of weapons and devices that a terrorist group perceives will achieve its objectives; (4) the objectives of a group's state sponsor; (5) the geographic factor; 6) historical dates of particular significance to terrorist groups; and (7) triggers that propel a group to launch attacks in a revenge mode as quickly as possible ahead of a previous timeline. For further details, refer to "How to Forecast and Preempt al-Qaeda's Catastrophic Terrorist Warfare," by Joshua Sinai, Ph.D.
7. Excerpts from a speech given by Osama bin Laden, October 31, 2004, http://english.aljazeera.net/News/archive/archive?ArchiveId=7387
8. These and other statements illustrate the motivation to carry out attacks of major economic consequence and are relevant for this and other economic related scenarios appearing in this chapter.
9. Alaska Clean-up Could Take Years, http://news.bbc.co.uk/1/hi/world/americas/1584553.stm, *BBC News*, October 7, 2001 (retrieved July 1, 2008).
10. Dr. Joshua Sinai, The Terrorist Threat Against the Gas and Oil Sectors: A Comprehensive Threat Assessment, presented at the EAPC/PfP Workshop Critical Infrastructure Protection and Civil Emergency Planning, Interdependencies and Vulnerabilities of Energy, Transportation and Communication, Zurich, Switzerland, September 22–24, 2005.
11. Simon Henderson, Al-Qaeda Attack on Abqaiq: The Vulnerability of Saudi Oil, published by The Washington Institute for Near East Policy, February 28, 2006, http://www.washingtoninstitute.org/templateC05.php?CID=2446
12. Country Profile: Iraq August 2006, Library of Congress—Federal Research division, http://lcweb2.loc.gov/frd/cs/profiles/Iraq.pdf
13. For further analysis concerning cyber pipeline threat feasibility, see *Security Guidelines for the Petroleum Industry*, 3rd ed. Washington, DC: American Petroleum Institute, April 2005.
14. Dr. Joshua Sinai, The Terrorist Threat Against the Gas and Oil Sectors: A Comprehensive Threat Assessment, presented at the EAPC/PfP Workshop Critical Infrastructure Protection and Civil Emergency Planning, Interdependencies

and Vulnerabilities of Energy, Transportation and Communication, Zurich, Switzerland, September 22–24, 2005.

15. *Ibid.* Further analysis on cyber type attacks is discussed in Scenario no. 5 of this chapter
16. *Ibid.*
17. Alaska Clean-up Could Take Years, http://news.bbc.co.uk/1/hi/world/americas/1584553.stm, *BBC News*, October 7, 2001 (retrieved July 1, 2008).
18. Trans-Alaska Pipeline Shot by a Drunk, http://solcomhouse.com/trans.htm
19. *Ibid*, and The National Strategy for The Physical Protection of Critical Infrastructures and Key Assets, The White House, Washington, DC, February 2003.
20. The formulation and analysis of this scenario was facilitated by Mr. Shmuel Sasson. Mr. Sasson served in the ISA for over 20 years in senior security and protection positions. He is the former head of El Al security division, a position he held between the years 2003 and 2006.
21. http://www.jinsa.org/articles/articles.html/function/view/categoryid/1701/documentid/2884/history/3,2360,655,1701,2884
22. http://www.washtimes.com/upi-breaking/20050111-062545-1800r.htm
23. http://www.janes.com/security/international_security/news/jir021128_1_n.shtml
24. http://www.cnn.com/2003/WORLD/meast/11/25/sprj.irq.missile.tape/index.html http://www.dw-world.de/dw/article/0,,1039411,00.html http://www.usatoday.com/news/world/iraq/2003-11-22-iraq_x.htm
25. James Bevan, Big Issues, Big Problems MANPADS, *Small Arms Survey*, 2004, p. 83. http://www.smallarmssurvey.org/files/sas/publications/year_b_pdf/2004/2004SASCh3_full_en.pdf
26. http://www.globalsecurity.org/security/library/news/2006/09/sec-060908-rferl02.htm
27. Maariv daily newspaper, May 18, 2005.
28. T. Dugdale-Pointon, 14, June 2005, *Hijacking*, http://www.historyofwar.org/articles/concepts_hijacking.html
29. http://www.cnn.com/2004/WORLD/meast/05/24/air.defense/
30. Drew Griffin, Kathleen Johnston, and Todd Schwarzschild, Sources: Air Marshals Missing from Almost All Flights http://edition.cnn.com/2008/TRAVEL/03/25/siu.air.marshals/index.html March 25, 2008.
31. Review of Maritime Transport, 2003, *Report by the UNCTAD Secretariat,* United Nations Conference on Trade and Development, Geneva, http://www.unctad.org/en/docs/rmt2003_en.pdf
32. Ophir Falk and Yaron Schwartz, The Maritime Threat, http://www.ict.org.il/articles/articledet.cfm?articleid=532, April 25, 2005, translated version, Piraten under gruner Flagge (German), published in *International Politik,* November 2005, pp. 28–31.
33. *Ibid.*
34. US Foils Terror Attack, *BBC News*, Tuesday, June 11, 2002, http://news.bbc.co.uk/1/hi/world/americas/2036705.stm
35. Experts says Islamic Militants Trained for Sea Attacks, *Reuters*, January 22, 2003.
36. International Shipping Vehicles Vulnerable to Terrorist Attacks, FBI Warns, Gregory Katz, *Dallas Morning News*, December 1, 2002.
37. *Ibid.*

38. Fifteen Freighters Believed to Be Linked to Al Qaeda, by John Mintz, *Washington Post*, December 31, 2002.
39. Possible Use of Scuba Divers to Conduct Terrorist Attacks, Information Bulletin 02-006, National Infrastructure Protection Center, May 23, 2002, http://www.nipc.gov/publications/infobulletins/2002/ib02-006.htm
40. Phillipe Crist, Security in Maritime Transport: Risk Factors and Economic Impact, Maritime Transport Committee, Directorate for Science, Technology and Industry, Organization for Economic Cooperation and Development, Paris, July 2003. p. 9.
41. Ibid.
42. Yaron Schwartz and Ophir Falk, *Chemical, Biological, Radiological, Nuclear Terror—Assessing the Threat,* The International Policy Institute for Counter Terrorism at the Interdisciplinary Center, Herzlyia, Israel, May 15, 2003, http://www.ict.org.il/articles/articledet.cfm?articleid=487
43. Dafna Linzer, Attack with Dirty Bomb More Likely, Officials Say, *The Washington Post*, December 29, 2004.
44. Donald Rumsfeld, U.S. Secretary of Defense, was quoted as saying "It is inevitable that terrorists will obtain weapons of mass destruction and that they will use them against us." John Arquilla, The Forever War, *San Francisco Chronicle*, January 9, 2005, p. 6.
45. Stephen Cohen, Boom Boxes: Containers and Terrorism, *Protecting the Nation's Seaports: Balancing Security and Cost*, Library of Congress, 2006.
46. Phillipe Crist, Security in Maritime Transport: Risk Factors and Economic Impact, Maritime Transport Committee, Directorate for Science, Technology and Industry, Organization for Economic Cooperation and Development, Paris, July 2003. p. 8.
47. *New York Times,* February 26, 2006.
48. Phillipe Crist, Security in Maritime Transport: Risk Factors and Economic Impact, Maritime Transport Committee, Directorate for Science, Technology and Industry, Organization for Economic Cooperation and Development, Paris, July 2003. p. 24.
49. Tara O'Toole, *Biological Weapons: National Security Threat & Public Health Emergency,* CSIS, 2000.
50. Nuclear Attack a Real, if Remote, Possibility, by Brad Knickerbocker, *Christian Science Monitor*, October 30, 2001, http://www.cdi.org/terrorism/nuclearcsm.cfm
51. See *Scientific American,* November 2002 issue and Federation of American Scientists' website: www.fas.org ("Radiological Bomb" features).
52. Planning Scenarios, The Homeland Security Council David Howe, Senior Director for Response and Planning July 2004, 11-1.
53. This section is limited to a discussion on prevention and deterrence. Issues of emergency management and response are beyond the section scope.
54. http://www.dhs.gov/xprevprot/programs/gc_1165872287564.shtm
55. Stephen Cohen, Boom Boxes: Containers and Terrorism, *Protecting the Nation's Seaports: Balancing Security and Cost*, Library of Congress, 2006, p. 108.
56. This scenario was formulated and analyzed by Dr. Ofer Israeli. Dr. Israeli is an international expert in the theories of International Relations. He has published extensively and lectures at the University of Haifa on these issues.
57. Osama bin Laden, as quoted in *CNN.com, World* (November 1, 2004). Available online: http://www.cnn.com/2004/WORLD/meast/11/01/binladen.tape/

58. Gal Luft and Anne Korin, Pirates and Terrorists: Yo Ho Ho, Infidel, *International Herald Tribune*, October 29, 2004. Available online: http://www.iht.com/articles/2004/10/29/edluft_ed3_.php
59. World Oil Transit Chokepoints, Energy Information Administration, U.S. Department of Energy. Available online: http://www.eia.doe.gov/cabs/World_Oil_Transit_Chokepoints/Full.html
60. John Brandon, Asian shipping at risk: Terrorism on the High Seas, *International Herald Tribune*, June 5, 2003. Available online: http://www.iht.com/articles/2003/06/05/edbrandon_ed3_.php
61. World Oil Transit Chokepoints, Energy Information Administration, U.S. Department of Energy. Available online: http://www.eia.doe.gov/cabs/World_Oil_Transit_Chokepoints/Full.html
62. Gal Luft and Anne Korin, Terror's Next Target, *The Journal of International Security Affairs,* December 2003. Available online: http://www.iags.org/n053004a.htm. Also see: Gal Luft and Anne Korin, Pirates and Terrorists: Yo Ho Ho, Infidel, *International Herald Tribune*, October 29, 2004. Available online: http://www.iht.com/articles/2004/10/29/edluft_ed3_.php
63. L. Steven Myers, Failed Plan to Bomb a U.S. Ship Is Reported, *The New York Times*, November 10, 2000. Available online: http://www.nytimes.com/2000/11/10/world/10KUWA.html?ex=1213848000&en=3148bf9faed0ebfc&ei=5070
64. S. Craig Smith, Fire on French Tanker Off Yemen Raises Terrorism Fears, *The New York Times*, October 7, 2002. Available online: http://query.nytimes.com/gst/fullpage.html?res=9906E5D8123BF934A35753C1A9649C8B63&sec=&spon=&pagewanted=all
65. Weekly Piracy Report, *ICC Commercial Crime Services,* May 27–June 2, 2008. Available online: http://www.icc-ccs.org/prc/piracyreport.php
66. All data for the year 2007, from the World Oil Transit Chokepoints, Energy Information Administration, U.S. Department of Energy. Available online: http://www.eia.doe.gov/cabs/World_Oil_Transit_Chokepoints/Full.html
67. State of the Nation, *The White House*, January 28, 2003. Available online: http://www.whitehouse.gov/news/releases/2003/01/20030128-19.html
68. Gal Luft and Anne Korin, Pirates and Terrorists: Yo Ho Ho, Infidel, *International Herald Tribune*, October 29, 2004. Available online: http://www.iht.com/articles/2004/10/29/edluft_ed3_.php
69. International Maritime Bureau (IMB). Available online: http://www.icc-ccs.org/prc/overview.php
70. Agence France-Press, Pirates Attack Oil Tanker Off Somalia: Malaysian Watchdog, *Hindustan Times* (April 21, 2008). Available online: http://www.hindustantimes.com/StoryPage/StoryPage.aspx?id=673b63e8-876d-494a-916d-84972ea7d2e5&ParentID=5b411340-4b08-47a1-943e-dbe1614286ac&&Headline=Pirates+attack+oil+tanker+off+Somalia
71. International Maritime Bureau (IMB). Available online: http://www.icc-ccs.org/main/all_piracy_al.php
72. Weekly Piracy Report, ICC Commercial Crime Services, May 27–June 2, 2008. Available online: http://www.icc-ccs.org/prc/piracyreport.php
73. The Associated Press, "Warships fight silent war against terror threat on high seas," *International Herald Tribune*, March 19, 2007. Available online: http://www.iht.com/articles/ap/2007/03/19/africa/AF-FEA-GEN-East-Africa-Terrorism.php

74. Weekly Piracy Report, ICC Commercial Crime Services, May 27–June 2, 2008. Available online: http://www.icc-ccs.org/prc/piracyreport.php

75. Martin Fackler, Oil Market Rattled by Attack on Tanker, *The New York Times*, April 22, 2008. Available online: http://www.nytimes.com/2008/04/22/world/asia/22tanker.html

76. Nick Tattersall, Nigeria pipeline attack forces cut in oil output, *International Herald Tribune* (May 26, 2008). Available online: http://www.iht.com/articles/reuters/2008/05/26/europe/OUKWD-UK-NIGERIA-DELTA-ATTACK.php

77. H. John Noer, Southeast Asian Chokepoints: Keeping Sea Lines of Communication Open, *Institute for National Strategic Studies* 98 (December 1996). Available online: http://www.ndu.edu/inss/strforum/SF_98/forum98.html

78. Jeffrey Gettleman, Little Djibouti Thinks Big in Trade, *International Herald Tribune*, May 30, 2008, p. 2.

79. This scenario was formulated and analyzed by an international counterterrorism expert that served as a chief scenario formulator for IDF intelligence and is currently the Chief Executive Officer of an advanced technological company, recently listed on the NASDAQ. The company developed an innovative product to detect malice intent of terrorist suspects.

80. The names and terms appearing in the scenario are notional and do not intend to offend anyone.

81. On the Road to Transportation Security, Institute for Security Technology Studies and Dartmouth College, February 2003, p. 21, http://www.ists.dartmouth.edu/analysis/trans.pdf

82. *Ibid.*

83. This scenario was formulated and analyzed by Professor Shmuel C. Shapira. MD MPH. Professor Shapira is the Deputy Director General of Hadassah University Hospital in Jerusalem, Director of Hebrew University Hadassah School of Public Health and is an international counter-terrorism expert. Professor Shapira has decades of first-hand experience with "terror medicine" and provided immediate medical response to attacks of terror on many occasions.

84. http://www.mfa.gov.il/MFA/Terrorism-+Obstacle+to+Peace/Terrorism+and+Islamic+Fundamentalism-/Attack+by+female+suicide+bomber+thwarted+at+Erez+crossing+20-Jun-2005.htm?DisplayMode=print (accessed June 9, 2008).

85. http://www.intelligence.org.il/Eng/sib/3_05/pji.htm (accessed June 9, 2008).

86. S. C. Shapira et al. Mortality in Terrorist Attacks: A Unique Model of Temporal Death Distribution, *World J of Surgery*, **30,** 1–8, 2006.

87. http://www.mfa.gov.il/MFA/MFAArchive/2000_2009/2003/12/The+Palestinian+use+of+ambulances+and+medical+mate.htm (accessed June 10, 2008).

88. S. C. Shapira and L. Cole, Terror Medicine: Birth of a Discipline. *Journal of Homeland Security Emergency Management* **3**, 1–8, 2006.

89. Suspicious Activity Involving Emergency Services and Hospitals, Indiana Department of Homeland Security, Indiana Intelligence Center, June 2008.

90. The estimated casualty rate is based on the following sources: N. Sheffy, Y. P. Mintz, A. I. Rivkind, and S. C. Shapira, Terror Related Injuries: A Comparison of Gunshot Wounds versus Injuries Induced by Secondary Explosive Fragments, *Journal of the American College of Surgeons* **203**, 297–303, 2006. S. C. Shapira, R. Adatto-Levi, M. Avitzour, A.I. Rivkind, I. Gershtenstein, and Y. Mintz, 2006. Mortality in terrorist

attacks: a unique model of temporal death distribution. *World Journal of Surgery* **30**, 1–7, 2006.

91. The estimate of direct and indirect economic consequence is based on the calculation of an attack on a large-scale hospital that treats approximately 10,000 in house patients daily.
92. S. C. Shapira and S. Mor-Yosef, Terror Politics and Medicine—the role of Leadership, *Studies in Conflicts Terrorism* **27**, 65–71, 2003.
93. Further detail on each component can be found in Ophir Falk and Lior Lotan, Assessing the Risk, *Law Enforcement Executive Forum* **7**(7), 2007.
94. Walter Lacquer, Postmodern Terrorism, *Foreign Affairs*, September/October 1996; and Yaron Schwartz and Ophir Falk, *Chemical, Biological, Radiological, Nuclear Terror—Assessing the Threat*, The International Policy Institute for Counter Terrorism at the Interdisciplinary Center, Herzlyia, Israel, http://www.ict.org.il/articles/articledet.cfm?articleid=487 May 15, 2003.
95. Jonathan B. Tucker ed., *Toxic Terror—Assessing Terrorist Use of Chemical and Biological Weapons*, Cambridge, MA: MIT Press, 2000.
96. *Ibid.*
97. Anthony H. Cordesman, *The Risks and Effects of Indirect, Covert, Terrorist, and Extremist Attacks with Weapons of Mass Destruction*, CSIS, 2001, p. 57 and http://www8.georgetown.edu/centers/cndls/applications/posterTool/index.cfm?fuseaction=poster.display&posterID=881
98. Jeffrey D. Simon, in *Toxic Terror—Assessing Terrorist Use of Chemical and Biological Weapons*, Jonathan B. Tucker, ed., Cambridge, MA: MIT Press, 2000, Chapter 5.
99. David Claridge, in *Toxic Terror—Assessing Terrorist Use of Chemical and Biological Weapons*, Jonathan B. Tucker, ed., Cambridge, MA: MIT Press, 2000, Chapter 6.
100. W. Seth Carus, *Toxic Terror—Assessing Terrorist Use of Chemical and Biological Weapons*, Jonathan B. Tucker, ed., Cambridge, MA: MIT Press, 2000, Chapter 8.
101. Anthony H. Cordesman, *The Risks and Effects of Indirect, Covert, Terrorist, and Extremist Attacks with Weapons of Mass Destruction*, CSIS, 2001, p. 19.
102. Richard A. Falkenrath, Robert D. Newman, and Bradley A. Thayer, *America's Achilles' Heel: Nuclear, Biological and Chemical Terrorism and Covert Attack*, Cambridge, MA: MIT Press, 2000.
103. Richard A. Falkenrath, Robert D. Newman, and Bradley A. Thayer, *America's Achilles' Heel Nuclear, Biological and Chemical Terrorism and Covert Attack*, Cambridge, MA: MIT Press, 2000.
104. The 2001 Outbreak: New Hoaxes and Public Anxiety, http://www.adl.org/learn/Anthrax/NewHoaxes.asp?xpicked=1&item=newHoaxes (Last accessed June 19, 2008)
105. Anthony Cordesman, *The Risks and Effects of Indirect, Covert, Terrorist and Extremist Attacks with Weapons of Mass Destruction*, CSIS, September 2000.
106. Tara O'Toole, *Biological Weapons: National Security Threat & Public Health Emergency*, CSIS, 2000.
107. Yaron Schwartz and Ophir Falk, *Chemical, Biological, Radiological, Nuclear Terror—Assessing the Threat,* The International Policy Institute for Counter Terrorism at the Interdisciplinary Center, Herzlyia, Israel, http://www.ict.org.il/articles/articledet.cfm?articleid=487 May 15 2003.
108. Statement given at an international conference held in Spain on June 18, 2008, http://www.nti.org/d_newswire/issues/2008_6_18.html

6

METHODS FOR CONFRONTING SUICIDE TERROR

William Cooper

It is said that if you know your enemies and know yourself, you will not be imperiled in a hundred battles; if you do not know your enemies but do know yourself, you will win one and lose one; if you do not know your enemies nor yourself, you will be imperiled in every single battle.

—General Sun Tzu, 6th Century b.c., *The Art of War*

The FBI and Department of Homeland Security agree that the probability of improvised explosive devices is an ongoing and increasing threat to the United States. Such devices, both human- and vehicle-borne, have continued to evolve and create significant damage and destruction in Iraq and are likely to travel to America. These devices, with suicidal attackers, are a simplified version of a "smart bomb," which can be guided directly to its intended target or can change its path at any given moment to another target. These suicide bombers can select targets at will and can deliver themselves at a time of their choosing. For first responders, both stopping them before they happen and responding to them after the detonation present significant challenges. This chapter is designed to articulate the threat and to provide responders with basic tools to interdict the threat, or react to it after it happens.

Suicide Terror: Understanding and Confronting the Threat. Edited by Falk and Morgenstern

Before the problem can be addressed, there must be an understanding not only what it is, but what the consequences of failing to deal with it are. Americans, particularly first responders, are all acutely aware of the United States being attacked on 9/11, along with the significant loss of life in the first responder community. One of the most important questions first responders must ask is whether or not they are prepared for another, possibly more substantial, attack. In many cases they are not, believing they will not be a target, or have more important, immediate needs to address, or may simply not recognize the true nature of those who wish us harm. An exploration of the initial step in identifying the true nature of the threat is in order.

Most Americans believe that more attacks are coming—that is, that it is a matter of when, not if. On the other hand, there are few who truly understand the extent to which those who wish the United States harm are prepared to go. Virtually everyone is familiar with Osama bin Laden and his ongoing series of threats, but many believe he is isolated and unable to carry them out. There are other examples of terrorist threats, most of which draw little attention. The reality of the threats is that since 9/11 there have been more than 10,000 terrorist attacks, many in the Middle East and other parts of the globe not in close proximity to America. In addition, direct threats have frequently been made toward the United States, any of which are possible to carry out. The intent of terrorism, besides to casualties and economics, is to create psychological effects that go beyond the attack itself. The attacks on 9/11, years later, continue to have that impact.

The unclassified version of the July 2007 National Intelligence Estimate states the following:

> "We judge the US Homeland will face a persistent and evolving terrorist threat over the next three years. The main threat comes from Islamic terrorist groups and cells, especially al-Qaeda, driven by their undiminished intent to attack the Homeland and a continued effort by these terrorist groups to adapt and improve their capabilities.
>
> al-Qaeda can and will remain the most serious terrorist threat to the Homeland, as its central leadership continues to continues to plan high impact plots, which pushing others in extremist Sunni communities to mimic its efforts and to supplement its capabilities. al-Qaeda will intensify its efforts to put operatives here.
>
> We assess that al-Qaeda will continue to enhance its capabilities to attack the Homeland through greater cooperation with regional terrorist groups.
>
> We assess that al-Qaeda's Homeland plotting is likely to continue to focus on prominent political, economic, and infrastructure targets with the goal of producing mass casualties, visually dramatic destruction, significant economic aftershocks, and/or fear among the US population. The group is proficient with conventional small arms and improvised explosive devices, and is innovative in creating new capabilities and overcoming security obstacles.
>
> We assess that al-Qaeda will continue to try to acquire and employ chemical, biological, radiological, or nuclear material in attacks and would not hesitate to use them if it develops what it deems is sufficient capability.

> We assess Lebanese Hizballah, which has conducted anti-US attacks outside the United States in the past, may be more likely to consider attacking the Homeland over the next three years if it perceives the United States as posing a direct threat to the group or Iran.
>
> We assess that the spread of radical—especially Salafi—Internet sites, increasingly aggressive anti-US rhetoric and actions, and the growing number of radical, self-generating cells in Western countries indicate that the radical and violent segment of the West's Muslim population is expanding, including in the United States.
>
> We assess that other, non-Muslim terrorist groups—often referred to as "single issue" groups by the FBI—probably will conduct attacks over the next three years given their violent histories.
>
> We assess that globalization trends and recent technological advances will continue to enable even small numbers of alienated people to find and connect with one another, justify and intensify their anger, and mobilize resources to attack—all without requiring a centralized terrorist organization, training camp, or leader."[1]

Examples of such threats include recent statements not only by terrorist group leaders, but by state leaders:

Iranian President Mahmoud Ahmadinejad:

"To those who doubt, to those who ask is it possible, or those who do not believe, I say accomplishment of a world without America and Israel is both possible and feasible."

Iranian Supreme Leader Ayatollah Ali Khamenei:

"The world of Islam has been mobilized against America for the past 25 years. The people call, 'Death to America.' Who used to say 'Death to America?' who, besides the Islamic Republic and the Iranian people used to say this? Today everyone says this."

Hassan Abbassi, Revolutionary Guards Intelligence Advisor to the Iranian President:

"We have a strategy drawn up for the destruction of Anglo-Saxon civilization … we must make use of everything we have at hand to strike at this front by means of our suicide operations or by means of our missiles. There are 29 sensitive sites in the US and in the West. We have already spied on these sites, and we know how we are going to attack them."

Iraqi Ayatollah Ahmad Husseni:

"If the objective circumstances materialize, and subjective there are soldiers, weapons, and money—even if this means using biological, chemical, and bacterial weapons—we will conquer the world, so that 'There is no God but Allah, and Muhammad is the Prophet of Allah' will be triumphant over the domes of Moscow, Washington, and Paris."

Venezuelan President Hugo Chavez:

Chavez has called on Iran to "save the human race, let's finish off the US empire."

Osama bin Laden:

"Kill the Americans and their allies, civilian and military. It is an individual duty for every Muslim who can do it in any country in which it is possible to do it."

Abu Ubeid al Qurashi:

"al-Qaeda takes pride in that, on 11 September it destroyed elements of America's strategic defense, which the former USSR and every other hostile state could not harm. In addition to the destruction, al-Qaeda has dealt America the most severe blow ever to their morale. The best means of bringing about a psychological defeat is to attack a place where the enemy feels safe and secure."

Ayman al-Zawahiri:

"The need to inflict the maximum casualties against the opponent, for this is the language understood by the West, no matter how much time and effort such operations take."

al-Qaeda Manual:

"Islamic governments have never and will never be established through peaceful solutions and cooperative councils. They are established as they (always) have been by pen and gun by word and bullet by tongue and teeth."

Having read these few of the many examples of the threats facing the United States, the simplest question for those who may doubt the intent is to ask, "What part of we're going to kill you doesn't the American population understand"? Literally tens of thousands of people have been killed in these types of attacks and many thousands more injured. Of note is that the largest single terrorist event in history occurred in the United States, and it comes with promises of more to follow.

Stratfor (Strategic Forecasting) stated in its July 25, 2007 report that

> As long as the ideology of jihadism exists and jihadists embrace the philosophy of attacking 'the far enemy,' they will pose a threat on US soil. There are likely homegrown and transnational jihadists in the United States right now plotting attacks. We believe the United States is long overdue for an attack.[2]

It is therefore imperative that first responders not only know and understand the threat, but understand the philosophies, strategies, and tactics employed to carry them out. It is also essential that responders understand that they have been targeted themselves as they react to a terrorist attack. First responders (and all Americans as well) must understand the enemy before the enemy can be defeated.

First responders should take particular note that a number of terrorist attacks have been interdicted in the United States prior to and since 9/11, further ratifying the true threat. While there has been some information released, the useful details of these potentially damaging attacks have not largely been shared. Responders should be aware that some of these attacks were prevented either by a member of law enforcement or by an alert member of the community reporting a suspicious activity. Examples of interdicted attacks include:

- **1999: Ahmed Ressam—the "Millennium Bomber."** Ressam was attempting to transport explosives into the United States to explode at Los Angeles International Airport to mark the millennium. He was arrested due to the observance of U.S. Customs officers.
- **2001: Richard Reid—the "Shoe Bomber."** Reid attempted to blow up a trans-Atlantic commercial jet by detonating explosives in his shoe. He was stopped due to an alert flight attendant.
- **2002: Jose Padilla.** Padilla was arrested returning to the United States after training with alleged intent of blowing up apartment buildings using natural gas, as well as allegedly intending to detonate "dirty bombs" in the United States.
- **2003: Majid Khan.** Khan allegedly intended to target U.S. bridges and gas stations (including underground tanks), and to poison water reservoirs.
- **2003: Iyman Faris.** Faris was involved in determining the probability of downing the Brooklyn Bridge and derailing passenger trains.
- **2004: Herald Square, NY.** Plot to use suicide bombers to collapse Manhattan Mall. Attackers had previously discussed NYPD precincts on Staten Island and the Verrazano Bridge.
- **2004: Dhiren Barot.** Barot and his team plotted attacks both in the United States and the United Kingdom. Targets in the United States included the International Monetary Fund, World Bank, New York Stock Exchange, Citigroup Headquarters, and the Prudential Building. Attacks involved the use of vehicles containing explosives.
- **2005: Shayk Shahaab Murshid.** Targeted U.S. military facilities, the Israeli Consulate, and Jewish synagogues in Los Angeles. Interdicted due to local law enforcement action.
- **2006: "Miami Seven."** Targeted Sears Tower in Chicago and conducted surveillance of Miami Police Department, Federal Justice Building, FBI Building, courthouses, and the federal Detention center.
- **2006: Derrick Shareef.** Targeted a mall in Rockford, Illinois at Christmas time.
- **2006: Khaled Sheik Mohammad.** At Guantanamo Bay he stated that he had planned more than 30 attacks, including several in the United States.
- **2007: Fort Dix Plot.** Six members of local group targeted Fort Dix military base for an attack. They intended to enter the base and kill as many soldiers as possible before escaping. Interdicted due to an observant citizen.

- **2007: JFK Airport Plot.** Four members of local group intended to blow up fuel deport and fuel lines for JFK airport, causing massive damage and many casualties.

As may be seen, there actually have been attempts to attack the Homeland on multiple occasions, and first responders should be aware of the methodologies used in planning and preparing to execute the attacks. It also provides significant evidence that the Homeland will be attacked again, perhaps on multiple occasions, using multiple strategies. Understanding the attackers provides responders with the ability to examine possibilities and probabilities to form strategies to try and stop the attack, or respond appropriately if it actually does occur.

It is simply a matter of understanding who the attackers are and what and how they think and believe. *Time* magazine printed an article on June 26, 2005 entitled "Inside the Mind of a Suicide Bomber." The article involved the interview of a suicide bomber waiting for the call to conduct his attack. He is quoted here to demonstrably show the seriousness of the threat of such incidents coming to the United States. "First I will ask Allah to bless my mission with a high rate of casualties among the Americans. The most important thing is that he should let me kill many Americans."[3] Further evidence of terrorist activities in the United States may be found in the identification of terror cells in Virginia, New Jersey, Oregon, New York, and two cities in California. These do not include the cells or groups discussed elsewhere in this chapter. It is clear that the threat is here and, as time passes, potentially becomes more imminent. First responders must take note and prepare accordingly.

DETECTION OF KEY TERRORIST ACTIVITIES

With this in mind, the next step is to examine how to detect key terrorist activities. What are those things a terrorist or group is doing that may assist law enforcement in detecting their actions? How does a terrorist attack occur, and can it be stopped? One of the best systems to utilize, and one being used across the country, is entitled the *Seven Steps of Terrorism*. History has shown that the planning, preparation, and execution of an attack typically follow these seven steps. Through an understanding of these steps, interdiction of an attack is possible during any one of them.

As previously stated, it is well known and documented that there are terrorist sleeper cells in the United States, from Florida to Washington State, New York to California, Illinois to Texas. It is unreasonable to expect or believe that there are not others being planned or developed.

Law enforcement continues to develop information- and intelligence-sharing capabilities that has significantly enhanced their ability to identify and track terrorist members of these cells. With these abilities, and the growing cooperation of the public, opportunities to recognize potential targets, potential terrorists, and likely methods of attack is increasing. Through education of the people, they are learning to combat fear and apprehension through action, by emphasizing safety, understanding, and preparation. One major element is for us to understand what the terrorists have done and how they carry out their plans. Historically, they are meticulous planners, and they seem to be in no hurry to launch an attack unless and until they are ready. They spend considerable time and effort gathering information in order to maximize the effect of an attack.

The weapon of choice tends to be explosives, frequently by means of suicide bombers. They use IEDs (improvised explosive devices) and VBIEDs (vehicle-borne improvised explosive devices) as the most common explosive approaches. Their knowledge and use of these devices is growing through experiences in Iraq and other countries. Terrorists use a methodological approach for planning and executing attacks. The Department of Homeland Security prepared and distributed an informational guide entitled *The Seven Signs of Terrorism*; the guide is an accurate representation of the strategies employed by terrorists. These Seven Signs offer an opportunity to look for unusual circumstances and report them to law enforcement. The benefit to all Americans is that at any one of these steps, a potential attack can be prevented.

Sign 1: Surveillance. Prior to an attack, often weeks or months before, terrorists will conduct surveillance and scouting operations. Such methods include photography, videotaping, diagramming, mapping, measuring, and observing our security, or other out-of-the-ordinary practices. These acts have occurred on our ferry systems in New York and Washington State, the JFK airport, power grids, dams, buildings (Sears Tower in Chicago), bridges (Brooklyn and George Washington in New York), tunnels (Lincoln and Holland in New York), subway systems, rail, and many others.

What to look for:

- Suspicious people taking video or photos, diagramming, or measuring in areas not normal; they may be there on multiple occurrences, or may be different people collecting the same information; staying in position for longer than normal times.
- The use of technology such as cell phone cameras and mini-cameras.
- Taking photos or video of security; timing security, measuring distances from police or fire stations to the target.
- People in possession of maps of critical infrastructure, highlighted in key areas.
- The use of communications equipment where not normally used—cell phones, walkie-talkies.
- The use of GPS units where it seems unusual.

Sign 2: Elicitation. Terrorists will attempt to gather and obtain information about a place, person, or operation that is more likely than not a critical infrastructure, either public or private. They may ask questions and make inquiries, and they may obtain plans or blueprints, or other information, much of it from the Internet.

What to look for:

- People asking questions, unusual or normal, looking to get information. Pay particular attention to questions about security, access to facilities or information systems, and delivery schedules.
- Attempts to access information via computer—blueprints, plans, schedules, anything to do with strengths or weaknesses. Information in the media has been of interest.

Sign 3: Tests of Security. Terrorists will test security/law enforcement by entering or attempting to enter secured or essential facilities or locations, and/or they will time responses and routes of response. The locations of police, fire, and emergency medical services will be identified, and routes of response will be located; timing of response likely will happen as well, usually through a false alarm.

What to look for:

- People attempting to enter secured or forbidden areas. If contacted by security while doing so, they will usually have a plausible story.
- Attempts to move prohibited materials through security to determine if they will be detected and what the response will be.
- False alarm or false report of an incident to test response times, deployment, and numbers of responders.
- Testing of alarm systems to determine reaction and timing.
- Unattended packages or briefcases to see what the reaction will be.

Sign 4: Acquiring Supplies. Terrorists will purchase weapons and ammunition, explosives or the components of explosives, chemicals, equipment, or military or law enforcement identification and uniforms to allow easier access into areas.

What to look for:

- People who buy excessive amounts of dangerous chemicals or components that can be used in constructing explosive devices—an example would be fertilizer (ammonium nitrate).
- People buying weapons and ammunition at unusual levels.
- People who are not in law enforcement or security buying (or stealing) uniforms, badges, and identification cards.
- People buying used emergency vehicles that may be used in an effort to access the target without raising suspicion.
- People attempting to obtain access cards to facilities.
- Thefts of weapons, especially military-grade weapons.

Sign 5: Suspicious People. Look for people who are out of the ordinary, who do not belong, or whose actions are out of the ordinary. Appearance, position, and actions may each be indicators. It may be as simple as knowing it when you see it. Profile their behaviors, not the people. Terrorists are not all of one race, color, or gender.

What to look for:

- People whose actions are not ordinary or normal.
- People who seem out of place.
- People not dressed for the weather or location.
- People trying to avoid detection or letting others see who they are.
- People who are evasive when spoken to or confronted.
- Carrying materials not suitable to the location.
- People attempting to hide.
- Reports of weapons practice in out of the way areas.

Sign 6: Dry Runs. Prior to the actual attack, the terrorists will conduct one or more dry runs to look for flaws or unanticipated problems. This is a critical time when they may be identified or caught. There are many examples of this occurring. The 9/11 hijackers flew the same planes they ultimately hijacked on multiple occasions to scout them out. The terrorists plotting to hijack 10 airliners out of the United Kingdom were involved in dry runs. Virtually all major attacks, before they occur, will have had at least one or more dry runs. Terrorists will conduct a test of their plan to look for problems they may overcome before the actual operation.

What to look for:

- Tests of the system: security, response times, reactions to the test.
- Suspicious people or suspicious actions—if they seem out of the ordinary they probably are.

Sign 7: Deployment of Assets. This is immediately prior to the attack and is the last opportunity to stop the attack. The terrorists will move into their predetermined positions just prior to the attack. Once in place and the timing is appropriate, they will attack the target. The history of attacks demonstrates that the terrorists may not always assemble in one location. These are the activities where the terrorists stage for the attack.

What to look for:

- People or vehicles in unusual positions or places.
- People dressed in clothing not aligned with weather or location.
- Unusual number of people or vehicles (or vehicle types) in vicinity of possible attack site.

While these indicators and signs are not a complete list, they give the reader an idea of what to look for and report, or respond to.

THE SUICIDE BOMBER

Suicide terror, also known as martyrdom operations, involves an "individual who makes great sacrifices or suffers much in order to further a belief, cause, or principle" (www.answers.com). Adam Fosson, terrorism analyst, writes in his paper "The Globalization of Martyrdom: Cause and Effect" (July 2007) that in recruitment of suicide bombers, considerable planning is involved. He writes that "One extremely crucial way these individuals are indoctrinated, trained, and recruited for martyrdom operations worldwide is through the Internet. It exploits the humiliation and anger sensed by many Muslims, while offering them an opportunity to make a difference."[4] Fosson further writes that "Both religion and the humiliation of life under occupation are often key motives for suicide bombers seeking a better life in paradise."[5] Fosson's extensive research concluded, for example, that in Palestine, "the idea of a child wanting to become a martyr is nearly the same as an American child wanting to visit Disneyland."[6]

Fosson concludes his paper with an extremely insightful couple of statements, one which first responders should pay particular attention to as they plan and prepare. He states, "As time goes by, the only way to anticipate a possible martyrdom operation is to essentially use your imagination. The statement that the United States had a *"failure of imagination"* prior to September 11, 2001 is pertinent here. The United States did not believe that an enemy such as al-Qaida, based thousands of miles away, could do harm to our homeland."[7]

DETECTION OF TERRORIST RECRUITMENT IN THE COMMUNITY

Recruitment of potential terrorist operatives, including suicide bombers, is an ongoing and global set of actions that terrorist organizations are engaged in. One of the more frequently discussed reasons for recruits to join is the anger with the United States for the war in Iraq.

The key for first responders to dealing with terrorist recruitment in the community is twofold:

1. Understand where and how recruitment is occurring;
2. Disrupt recruitment methods.

Terrorist operations require personnel, money, and materials, as well as places from which to work and to live. If responders know how recruitment works for terrorists, they find themselves in a better position to interdict it. The basis for recruitment involves several elements, each of which presents an opportunity to identify and react. These elements include where terrorist groups recruit, how they recruit, who they target, and why those people are targeted.

Where they recruit:

1. The Internet—there are currently more than 5000 sites on the Internet of a Jihadist nature. Particular segments of the population are identified and targeted—some accept and some do not. For those invited or who may seek out the site, there may be an expectation of what the content is.
2. The use of videos and CDs. There has been an ongoing use of fairly high-quality videos produced by al-Qaeda and others that contain considerable anti-American rhetoric, designed, in part, to add to the number of people being recruited globally, including within the United States itself.
3. Mosques may have the reputation for being "radical." For those attending, there may be an understanding of what would be expected to be said. For some attending these mosques, they may be seeking certain behaviors or philosophies they can align with. In some mosques there is no shortage of ever escalating anti-American campaigning.
4. Law enforcement should look for ritualistic behaviors as recruiting gets underway. Such behaviors include "hazing rituals and group identity building exercises or, in the case of al-Qaeda, validation of commitment to its principles through the recruit's demonstrated knowledge of radical Islam and the use of violence to achieve its goals. These techniques can result in radically polarized and altered attitudes among those who successfully navigate them, usually along the lines desired by the recruiting group."[8]
5. Another way to target select members of a population is to infiltrate them. The infiltrator will recruit from within the group. Law enforcement should be aware of such groups in their communities which may be specifically targeted.
6. Colleges and universities—there are a number of professors and students at major U.S. universities who subscribe to Jihadist philosophies. Terror candidates or operatives have entered the country using student visas. In some cases, these philosophies are openly taught.

What they look for to recruit (Rand Report):

1. A high level of current distress or dissatisfaction (emotional, physical, or both)
2. Cultural disillusionment in a frustrated seeker (i.e., unfulfilled idealism)
3. Lack of an intrinsic religious belief system or value system
4. Some dysfunctionality in family system (i.e., family and kin exert "weak gravity")
5. Some dependent personality tendencies (e.g., suggestibility, low tolerance, or ambiguity)

For example, potential shoe bomber Richard Reid had four of the five traits listed above. When discussing anti-American sentiments, the "recruitment pitch is simple: American policies are directly responsible for Muslim misery, all over the world."[9] Who is being recruited has changed considerably. Previously, suicide terrorists were younger Middle Eastern men. Now, terrorist organizations are recruiting men and women, various races and cultures, older and younger, and in some extreme cases they have used or will use small children to attack a target. Given the diverse society of America, intervention goals include identifying the recruiters and potential people targeted for recruiting. It is important that law enforcement have a good working relationship with its community, certainly sufficient to trust that potential recruiting centers, efforts, and recruit possibilities are reported.

Jails and prisons are one of the more common sources where terrorist recruits are sought and make commitments to terrorist philosophies and acts. The majority of those incarcerated possess some, if not all, the characteristics previously discussed. John S. Pistole, Assistant Director of the FBI's Counterterrorism Division, testified before Congress in October 2003 that U.S. correctional facilities are a viable venue for radicalization and recruitment. "Recruitment of inmates within the prison system will continue to be a problem for correctional institutions throughout the country. Inmates are often ostracized by, abandoned by, or isolated from their family and friends, leaving them susceptible to recruitment. Membership in the various radical groups offer inmates protection, positions of influence, and a network they can correspond with both inside and outside of prison."[10]

The al-Qaeda Manual itself describes the characteristics of recruits they seek. While the list of traits is fairly extensive, parts of it are included here to provide law enforcement (and ultimately members of the community to be sought for assistance) the tools necessary to assist in identifying potential terrorist suicide operatives. This chapter makes extensive use of the al-Qaeda Manual to provide the first responder the philosophies and strategies employed by the terrorists. The manual's second chapter is quoted here.

Necessary Qualifications for the Organization's Members:

1. The member of the organization must be Muslim.
2. The member must be committed to the organization's ideals.
3. Maturity—the requirements of military work are numerous, and a minor cannot perform them. The nature of hard and continuous work in dangerous conditions requires a great deal of psychological, mental, and intellectual fitness.
4. Sacrifice—the member must be willing to do the work and undergo martyrdom for the purpose of achieving the goal and establishing the religion of majestic Allah on earth.
5. Listening and obedience—this is known today as discipline. It is expressed by how the member obeys the orders given to him.
6. Keeping secrets and concealing information—this secrecy should be used even with the closest people, for deceiving the enemies is not easy.

7. Free of illness—the military organization's member must fulfill this important requirement.
8. Patience—the member should have plenty of patience for enduring afflictions if he is overcome by the enemies. He should be patient in performing the work, even if it takes a long time.
9. Tranquility and unflappability—the member should have a calm personality that allows him to endure psychological traumas such as those involving bloodshed, murder, arrest, imprisonment, and reverse psychological traumas such as killing one or all of his organization's comrades. He should be able to carry out the work.
10. Intelligence and insight.
11. Caution and prudence.
12. Truthfulness and counsel.
13. Ability to observe and analyze.
14. Ability to act, change positions, and conceal oneself.[11]

IDENTIFYING SAFE HOUSES AND PLANNING CENTERS

The al-Qaeda manual provides first responders with its blueprint for the selection and use of safe houses. This information is useful for law enforcement investigations and patrol operations. It is also use for fire and EMS in the event that they respond to a fire or other incident at one of these locations. Each will have some awareness of the dangers and threats involved. The manual's fourth lesson is extensively quoted here.

Security Precautions Related to Apartments:

1. Choosing the apartment carefully as far as the location and the size for the work necessary (meetings storage, arms, fugitives, work preparation).
2. It is preferable to rent apartments on the ground floor to facilitate escape.
3. Prepare secret locations in the apartment for securing documents, records, arms, and other important items.
4. Prepare ways of vacating the apartment in case of a surprise attack (stands, wooden ladders).
5. Under no circumstances should anyone know about the apartment other than those using them.
6. Provide the necessary cover for the people who frequent the apartment (students, workers, employees, etc.).
7. Avoid seclusion and isolation from the population and refrain from going to the apartment at suspicious times.
8. It is preferable to rent these apartments using false names, appropriate cover, and non-Muslim appearance.

9. A single brother should not rent more than one apartment in the same area, from the same agent, or using the same rental office.
10. Care should be exercised not to rent apartments that are known to the security apparatus such as those used for immoral or prior Jihad activities.
11. Avoid police stations and government buildings. Apartments should not be rented near these places.
12. When renting these apartments, one should avoid isolated or deserted locations so the enemy would not be able to catch those living there easily.
13. It is preferable to rent apartments in newly developed areas where people do not know one another. Usually, in older quarters people know one another and strangers are easily identified, especially since these quarters have many informers.
14. Ensure that there has been no surveillance prior to the members entering the apartment.
15. Agreement among those living in the apartment on special ways of knocking on the door and special signs prior to entry into the building's main gate to indicate to those who wish to enter that the place is safe and not being monitored. Such signs include hanging out a towel, opening a curtain, placing a cushion in a special way, etc.
16. If there is a telephone in the apartment, calls should be answered in an agreed-upon manner among those who use the apartment. That would prevent mistakes that would otherwise lead to revealing the names and nature of the occupants.
17. For apartments, replace the locks and keys with new ones. As for the other entities (camps, shops, mosques), appropriate security precautions should be taken, depending on the entity's importance and role in the work.
18. Apartments used for undercover work should not be visible from higher apartments in order not to expose the nature of the work.
19. In a newer apartment, avoid talking loud because prefabricated ceilings and walls used in the apartments do not have the same thickness as those in old ones.
20. It is necessary to have at hand documents supporting the undercover member. In the case of a physician, there should be an actual medical diploma, membership in the medical union, the government permit, and the rest of the routine procedures used in that country.
21. The cover should blend well with the environment. For example, select a doctor's clinic in an area where there are clinics, or in a location suitable for it.
22. The cover of those who frequent the location should match the cover of that location. For example, a common laborer should not enter a fancy hotel because that would be suspicious and draw attention.[12]

First responders, in looking for or entering a bomb-making lab, should be aware of chemicals, explosives, and other bomb-making materials (pipes, electronic components, and computers—particularly laptops). These computers usually prove invaluable in terms of intelligence. Responders also should be aware of various chemical odors.

A case example involves the 1995 Operation Bojinka plots in the Philippines. The plot involved the assassination of the Pope while in the Philippines, the destruction of as many as a dozen commercial airliners over the Pacific Ocean. The plot was discovered when a fire started in the terrorists' apartment in Manila. When responders arrived, they found a large number of items of evidentiary and intelligence value. They recovered items associated with the plot to kill the Pope, including maps of the papal motorcade, pictures of the Pope, rosaries, and clothing priests wear.

In addition, evidence of a plot to destroy airliners was present. Chemicals in the apartment included various acids, ammonium nitrate, nitroglycerin, cylinders, fuses, and chemistry equipment. Completed pipe bombs were found, Casio watches to be used as explosive devices were present, a manual on building liquid-based bombs was discovered, and a number of falsified passports were also collected.

The most valuable piece of evidence in the apartment was a laptop computer, containing a wealth of intelligence. Present on the computer were the schedules of the flights to be destroyed, times of explosions aboard each aircraft, the names of numerous people associated with the group, as well as photographs, and a number of documents regarding anti-U.S. and anti-Israel sentiments. Responders should be well aware of the possibility of such evidence any time they enter a possible safe house. They should also be aware the terrorists will go to extensive lengths to protect these locations.

TERRORIST MEANS OF COMMUNICATION

Terrorist organizations use meticulous planning in all aspects of their operations. One of the key elements is their ability to communicate with one another. With the number of terror attacks worldwide, the United States employed a number of technologies to intercept communications among the groups and cells. As a consequence, the terrorist organizations adapted and used other technologies, including the Internet. Jihadist forums on the Internet are extensive and contain many hundreds of files on Jihadist philosophies, strategies, and methodologies. Examining these files provides law enforcement and private sector security a wealth of information and data on which to base prevention, response, and mitigation strategies. In addition, considerable training materials are shared across hundreds of Internet sites, which are disbanded when training is completed. Given that following 9/11 the United States destroyed the training camps in Afghanistan, the Internet has become the vehicle for providing extensive levels and types of training. The terrorist can learn from the comfort of his home.

U.S. efforts were successful in obtaining additional information, so again the terrorists adjusted their operations. Their communications strategy is clearly articulated in the al-Qaeda Manual, and much of that strategy is included here so law enforcement and community members understand the basics behind how information is passed among the groups. The key here is to understand that the terrorists are continually adapting and adjusting their methodologies to evade interdiction efforts. The communications portion of the manual is included here. The Fifth Lesson in the al-Qaeda Manual is provided in part.

Communication Means

The Military Organization in any Islamic group can, with its modest capabilities, use the following means:

1. The telephone
2. Meeting in-person
3. Messenger
4. Letters
5. Some modern devices, such as the facsimile and wireless [communication]

Communication may be within the county, state, or even the country, in which case it is called local communication. When it extends expanded between countries, it is then called international communication.

Secret Communication Is Limited to the Following Types:

1. *Common Communication.* It is a communication between two members of the Organization without being monitored by the security apparatus opposing the Organization. The common communication should be done under a certain cover and after inspecting the surveillance situation [by the enemy].
2. *Standby Communication.* This replaces common communication when one of the two parties is unable to communicate with the other for some reason.
3. *Alarm Communication.* This is used when the opposing security apparatus discovers an undercover activity or some undercover members. Based on this communication, the activity is stopped for a while, all matters related to the activity are abandoned, and the Organization's members are hidden from the security personnel.

Method of Communication Among Members of the Organization

1. Communication about undercover activity should be done using a good cover; it should also be quick, explicit, and pertinent. That is just for talking only.
2. Prior to contacting his members, the commander of the cell should agree with each of them separately (the cell members should never meet all in

one place and should not know one another) on a manner and means of communication with each other. Likewise, the chief of the organization should [use a similar technique] with the branch commanders.

Cell or cluster methods should be adopted by the Organization. It should be composed of many cells whose members do not know one another, so that if a cell member is caught the other cells would not be affected, and work would proceed normally.

3. A higher-ranking commander determines the type and method of communication with lower-ranking leaders.

First Means: The Telephone

Because of significant technological advances, security measures for monitoring the telephone and broadcasting equipment have increased. Monitoring may be done by installing a secondary line or wireless broadcasting device on a telephone that relays the calls to a remote location. ... That is why the Organization takes security measures among its members who use this means of communication (the telephone).

1. Communication should be carried out from public places. One should select telephones that are less suspicious to the security apparatus and are more difficult to monitor. It is preferable to use telephones in booths and on main streets.
2. Conversation should be coded or in general terms so as not to alert the person monitoring [the telephone].
3. Periodically examining the telephone wire and the receiver.
4. Telephone numbers should be memorized and not recorded. If the brother has to write them, he should do so using a code so they do not appear as telephone numbers (figures from a shopping list, etc.).
5. The telephone caller and person called should mention some words or sentences prior to bringing up the intended subject. The brother who is calling may misdial one of the digits and actually call someone else. The person called may claim that the call is for him, and the calling brother may start telling him work-related issues and reveal many things because of a minor error.
6. In telephone conversations about undercover work, the voice should be changed and distorted.
7. When feasible, it is preferable to change telephone lines to allow direct access to local and international calls. That and proper cover facilitate communications and provide security protection not available when the central telephone station in the presence of many employees is used.
8. When a telephone [line] is identified [by the security apparatus], the command and all parties who were using it should be notified as soon as possible in order to take appropriate measures.

9. When the command is certain that a particular telephone [line] is being monitored, it can exploit it by providing information that misleads the enemy and benefits the work plan.
10. If the Organization manages to obtain jamming devices, it should use them immediately.

Second Means: Meeting in-Person

This is direct communication between the commander and a member of the Organization. During the meeting the following are accomplished:

1. Information exchange
2. Giving orders and instructions
3. Financing
4. Member follow-up

Stages of the In-Person Meeting:

A. Before the meeting
B. The meeting [itself]
C. After the meeting

A. *Before the Meeting.* The following measures should be taken:

1. Designate the meeting location
2. Find a proper cover for the meeting
3. Specify the meeting date and time
4. Define special signals between those who meet

1. *Identify the meeting location.*
If the meeting location is stationary, the following matters should be observed:

i. The location should be far from police stations and security centers.
ii. Ease of transportation to the location.
iii. Select the location prior to the meeting and learning all its details.
iv. If the meeting location is an apartment, it should not be the first one, but one somewhere in the middle.
v. The availability of many roads leading to the meeting location. That would provide easy escape in case the location ware raided by security personnel.
vi. The location should not be under suspicion (by the security [apparatus]).
vii. The apartment where the meeting takes place should be on the ground floor, to facilitate escape.

viii. The ability to detect any surveillance from that location.
ix. When public transportation is used, one should alight at some distance from the meeting location and continue on foot. In the case of a private vehicle, one should park it far away or in a secure place so as to be able to maneuver it quickly at any time.

If the meeting location is not stationary, the following matters should be observed:

i. The meeting location should be at the intersection of a large number of main and side streets to facilitate entry, exit, and escape.
ii. The meeting location (such as a coffee shop) should not have members that might be dealing with the security apparatus.
iii. The meeting should not be held in a crowded place because that would allow the security personnel to hide and monitor those who meet.
iv. It is imperative to agree on an alternative location for the meeting in case meeting in the first is unfeasible. That holds whether the meeting place is stationary or not.

Those who meet in-person should do the following:

i. Verify the security situation of the location before the meeting.
ii. Ensure that there are no security personnel behind them or at the meeting place.
iii. [Do] not head to the location directly.
iv. Clothing and appearance should be appropriate for the meeting location.
v. Verify that private documents carried by the brother have appropriate cover.
vi. Prior to the meeting, design a security plan that specifies what the security personnel would be told in case the location were raided by them, and what [the brothers] would resort to in dealing with the security personnel (fleeing, driving back, ...).

2. *Find a proper cover for the meeting.*

i. [The cover] should blend well with the nature of the location.
ii. In case they raid the place, the security personnel should believe the cover.
iii. [The cover] should not arouse the curiosity of those present.
iv. [The cover] should match the person's appearance and his financial and educational background.
v. [The cover] should have documents that support it.

vi. Provide reasons for the two parties meeting (for example, one of the two parties should have proof that he is an architect. The other should have documents as proof that he is a land owner. The architect has produced a construction plan for the land).

3. *Specify the meeting date and time.*

i. Specify the hour of the meeting as well as the date.
ii. Specify the time of both parties' arrival and the time of the first party's departure.
iii. Specify how long the meeting will last.
iv. Specify an alternative date and time.
v. [Do] not allow a long period of time between making the meeting arrangements and the meeting itself.

4. *Designate special signals between those who meet.*

If the two individuals meeting know one another's shape and appearance, it is sufficient to use a single safety sign. [In that case] the sitting and arriving individuals inform each other that there is no enemy surveillance. The sign may be keys, beads, a newspaper, or a scarf. The two parties would agree on moving it in a special way so as not to attract the attention of those present.

If the two individuals do not know one another, they should do the following:

a. The initial sign for becoming acquainted may be that both of them wear a certain type of clothing or carry a certain item. These signs should be appropriate for the place, [should be] easily identified, and [should] meet the purpose. The initial sign for becoming acquainted does not [fully] identify one person by another. It does that at a rate of 30%.
b. Safety Signal: It is given by the individual sitting in the meeting location to inform the second individual that the place is safe. The second person would reply through signals to inform the first that he is not being monitored. The signals are agreed upon previously and should not cause suspicion.
c. A second signal for getting acquainted is one in which the arriving person uses while sitting down. That signal may be a certain clause, a word, a sentence, or a gesture agreed upon previously, and should not cause suspicion for those who hear it or see it.

B. *The Stage of the Meeting* [itself]. The following measures should be taken:

1. Caution during the meeting.
2. Not acting unnaturally during the meeting in order not to raise suspicion.

3. Not talking with either loud or very low voices ([should be] moderate).
4. Not writing anything that has to do with the meeting.
5. Agreeing on a security plan in case the enemy raids the location.

C. *After the Meeting.* The following measures should be taken:

1. Not departing together, but each one separately.
2. Not heading directly to the main road but through secondary ones.
3. Not leaving anything in the meeting place that might indicate the identity or nature of those who met.

Meeting in-person has disadvantages, such as:

1. Allowing the enemy to capture those who are meeting.
2. Allowing them [the enemy] to take pictures of those who are meeting, record their conversation, and gather evidence against them.
3. Revealing the appearance of the commander to the other person. However, that may be avoided by taking the previously mentioned measures such as disguising himself well and changing his appearance (glasses, wig, etc.).

Third Means: The Messenger

This is an intermediary between the sender and the receiver. The messenger should possess all characteristics mentioned in the first chapter regarding the Military Organization's member.

These are the security measures that a messenger should take:

1. Knowledge of the person to whom he will deliver the message.
2. Agreement on special signals, exact date, and specific time.
3. Selecting a public street or place that does not raise suspicion.
4. Going through a secondary road that does not have check points.
5. Using public transportation (train, bus, ...) and disembarking before the main station. Likewise, embarking should not be done at the main station either, where there are a lot of security personnel and informants.
6. Complete knowledge of the location to which he is going.

Fourth Means: Letters

This means (letters) may be used as a method of communication between members and the Organization, provided that the following security measures are taken:

1. It is forbidden to write any secret information in the letter. If one must do so, the writing should be done in general terms.

2. The letter should not be mailed from a post office close to the sender's residence, but from a distant one.
3. The letter should not be sent directly to the receiver's address but to an inconspicuous location where there are many workers from your country. Afterwards, the letter will be forwarded to the intended receiver. (This is regarding the overseas-bound letter.)
4. The sender's name and address on the envelope should be fictitious. In case the letters and their contents are discovered, the security apparatus would not be able to determine his [the sender's] name and address.
5. The envelope should not be transparent so as to reveal the letter inside.
6. The enclosed pages should not be many, so as not to raise suspicion.
7. The receiver's address should be written clearly so that the letter would not be returned.
8. Paying the post office box fees should not be forgotten.

Fifth Means: Facsimile and Wireless

Considering its modest capabilities and the pursuit by the security apparatus of its members and forces, the Islamic Military Organization cannot obtain theses devices. In case the Organization is able to obtain them, firm security measures should be taken to secure communication between the members in the country and the command outside. These measures are:

1. The duration of transmission should not exceed five minutes in order to prevent the enemy from pinpointing the device location.
2. The device should be placed in a location with high wireless frequency, such as close to a TV station, embassies, and consulates in order to prevent the enemy from identifying its location.
3. The brother, using the wireless device to contact his command outside the country, should disguise his voice.
4. The time of communication should be carefully specified.
5. The frequency should be changed from time to time.
6. The device should be frequently moved from one location to another.
7. Do not reveal your location to the entity for which you report.
8. The conversation should be in general terms so as not to raise suspicion.[13]

In terms of other communication methods, this time directed at the United States, videotapes from high-ranking terrorist leaders are distributed not infrequently. These videos contain messages directed at influencing Americans in their thinking and their politics, and they may be used to communicate with other terrorist groups in code.

The Internet is also used by terrorists to convey their messages. For example, on May 14, 2007, Abu Kandahar and Roslan al-Shami used the Internet to send a five-part message to Americans. In the message they discussed a number of near-simultaneous attacks in various cities using nuclear weapons hidden in trucks. They named the cities of New York, Los Angeles, and a city in Florida, suspected to be Orlando (due to its proximity to the space center). At the approximate same time, they would initiate attacks in other cities, including Seattle, Washington, DC, and oil cities in Texas. The writers further discussed the economic damage and number of casualties. While appearing to be serious in nature, these types of communications typically are designed to create fear rather than be an actual attack. What is important, though, is to understand that terrorists, particularly al-Qaeda, are known to state what their intentions are prior to actually doing the attack.

IDENTIFYING TERRORIST TRANSPORTATION

In their ongoing efforts to plan and execute attacks, terrorist groups utilize a variety of vehicles. There have been numerous reports of stolen cars, trucks, and heavy tractor–trailer combinations. While there is no direct nexus to terrorists stealing these vehicles, they are associated with the potential types of attacks they'd like to carry out. In addition, law enforcement, fire, and EMS should be acutely aware of the potential theft of emergency vehicles, as well as first-responder uniforms and equipment. Much information has been shared regarding these possibilities, so the operational security undertaken by first responders typically is not a large-scale issue. What is more important is that law enforcement and other first-response organizations identify and share such information with private sector agencies that have similar operations. Law enforcement should also, as time and ability exist, assist the private sector, where necessary, with the development of their own operational security.

With this in mind, an examination of historic terrorist attacks provides first responders with the information on the strategies and methodologies to understand what types of attacks may be in the future and the potential to stop them before they occur. Within recent memory, attacks include:

- *Oklahoma City Murrah Federal Building.* Domestic terrorists rented a Ryder truck, filled it with explosives, and detonated the vehicle bomb in front of the building.
- *1993 World Trade Center Attack.* International terrorists rented a Ryder truck, filled it with explosives, and detonated it after parking it in the underground garage.
- *Iraq.* Vehicle bombs have evolved significantly, from the use of passenger cars, to small trucks, to larger trucks, to cement mixers. In addition to the terrorists learning how much explosive to deploy in what type of vehicle for what type of operation to achieve maximum effect, they are now learning how to deploy chemicals in the explosives. There have been a number of attacks using chlorine with the explosives. While the successes desired

> have not been achieved, additional attacks are used to learn how to properly manage chemicals with explosives. The United States should expect this type of attack within the foreseeable future.

Various reports provide information that while the application of chlorine bombs in Iraq has not been as successful as desired, terrorists continually learn and adapt; therefore the attackers will continue to develop and refine their devices to maximize the amount and placement of explosives and chemicals and to achieve a much higher number of casualties and damage.

The deployment of these vehicle bombs to determine how to perfect them is invaluable training for the terrorists. It reinforces the premise that they are a determined and deadly foe. In the United States, more than 13 million tons of chlorine are used every year. To deploy these devices in the United States, first responders and citizens alike should be aware of their historic use of stolen or modified emergency vehicles, driven by terrorists dressed as emergency responders.

Again, the referral to the al-Qaeda Training Manual provides invaluable intelligence for first responders. Their Fifth Lesson regarding transportation includes the following:

Transportation Means:

The members of the Organization may move from one location to another using one of the following means:

a. Public transportation
b. Private transportation

Security Measures That Should Be Observed in Public Transportation

1. One should select public transportation that is not subject to frequent checking along the way, such as crowded trains or public buses.
2. Boarding should be done at a secondary station, as main stations undergo more careful surveillance. Likewise, embarkment should not be done at main stations.
3. The cover should match the general appearance (tourist bus, first-class train, second-class train, etc.).
4. The existence of documents supporting the cover.
5. Placing important luggage among the passengers' luggage without identifying the one who placed it. If it is discovered, its owner would not be arrested. In trains, it [the luggage] should be placed in a different car than that of its owner.
6. The brother traveling on a "special mission" should not get involved in religious issues (advocating good and denouncing evil) or day-to-day matters (seat reservation, ...).

7. The brother traveling on a mission should not arrive in the [destination] country at night because then travelers are few, and there are [search] parties and check points along the way.
8. When cabs are used, conversation of any kind should not be started with the driver because many cab drivers work for the security apparatus.
9. The brother should exercise extreme caution and apply all security measures to the members.

Security Measures that Should be Observed in Private Transportation:

Private transportation includes: cars, motorcycles

A. *Cars and motorcycles used in overt activity*:

1. One should possess the proper permit and not violate traffic rules in order to avoid trouble with the police.
2. The location of the vehicle should be secure so that the security apparatus would not confiscate it.
3. The vehicle make and model should be appropriate for the brother's cover.
4. The vehicle should not be used in special military operations unless the Organization has no other choice.

B. *Cars and motorcycles used in covert activity*:

1. Attention should be given to permits and [obeying] the traffic rules in order to avoid trouble and reveal their actual mission.
2. The vehicle should not be left in suspicious places (deserts, mountains, etc.). If it must be, then the work should be performed at suitable times when no one would keep close watch or follow it.
3. The vehicle should be purchased using forged documents so that getting to its owners would be prevented once it is discovered.
4. For the sake of continuity, have only one brother in charge of selling.
5. While parking somewhere, one should be in a position to move quickly and flee in case of danger.
6. The car or motorcycle color should be changed before the operation and returned to the original after the operation.
7. The license plate number and county name should be falsified. Further, the digits should be numerous in order to prevent anyone from spotting and memorizing it.
8. The operation vehicle should not be taken to large gasoline stations so that it would not be detected by the security apparatus.[14]

Terrorists obtain vehicles through a variety of means. Some may buy them, use them, and discard or sell them. Others may steal them. First responders should be aware of thefts of vehicles, particularly emergency vehicles like police cars, fire aid vehicles, and ambulances. The more common way that terrorists obtain vehicles is to rent or lease them. The first World Trade Center bombing in 1993 involved a rented van. One of the terrorists was identified when he returned to the rental company to get his $400 payment back. The Oklahoma City bombing was done using a rented truck. Other plots, interdicted before becoming operational, would have used limousines to attack financial institutions on the east coast. Quite simply, terrorists will use whatever vehicle for an attack that is necessary to complete it. Land, sea, and air transportation will be used as they see fit.

IDENTIFYING TERRORIST FINANCING

Once again, the al-Qaeda Manual provides excellent guidance on how terrorists finance an operation. Money laundering, charities, robberies, burglaries, and other seemingly legitimate operations are commonly used. One example used more recently has been amassing and selling cigarettes for profit. This author was recently involved in the investigation of a terrorist organization laundering money after raising millions of dollars through a charity cosponsored by a major transnational company.

Financial Security Precautions:

1. Dividing operational funds into two parts: One part is to be invested in projects that offer financial return, and the other is to be saved and not spent except during operations.
2. Not placing operational funds (all) in one place.
3. Not telling the Organization members about the location of the funds.
4. Having proper protection while carrying large amounts of money.
5. Leaving the money with nonmembers and spending it as needed.[15]

IDENTIFYING PAPER FALSIFICATION

Terrorists commonly use forged identification with various aliases or iterations of their names. There may be more than a couple of dozen variations on their name, any of which may have identification associated with it. The iterations may involve interchanging last name with first, or variations of middle names or titles. The al-Qaeda Manual, Third Lesson, provides an excellent overview of their methodologies and security precautions.

Forged Documents (Identity Cards, Record Books, Passports):

The following security precautions should be taken:

1. Keep the passport in a safe place so it would not be seized by the security apparatus, and the brother it belongs to would have to negotiate its return (I'll give you your passport if you give me information).
2. All documents of the undercover brother, such as identify cards and passport, should be falsified.
3. When the undercover brother is traveling with a certain identity card or passport, he should know all pertinent information such as the name, profession, and place of residence.
4. The brother who has special work status (commander, communication link...) should have more than one identity card and passport. He should learn the content of each, the nature of the indicated profession, and the dialect of the residence area listed in the document.
5. The photograph of the brother in these documents should be without a beard. It is preferable that the brother's public photograph on these documents be also without a beard. If he already has one document showing a photograph with a beard, he should replace it.
6. When using an identity document in different names, no more than one such document should be carried at one time.
7. The validity of falsified travel documents should always be confirmed.
8. All falsification matters should be carried out through the command and not haphazardly (procedure control).
9. Married brothers should not add their wives to their passports.
10. When a brother is carrying the forged passport of a certain country, he should not travel to that country. It is easy to detect forgery at the airport, and the dialect of the brother is different from that of the people from that country.[16]

LAND ATTACK CHARACTERISTICS

This chapter has discussed extensively tactics and methods used by terrorists to plan and execute attacks. Land-based attacks will typically include suicide bombers using vehicle-borne improvised explosive devices (VBIEDS): chemical, bacteriological, or radiological. They likely will apply the extensive lessons learned in Iraq to attacks in the United States. First responders should follow the type, scale, and methods of those attacks.

The preferred methodology for land-based attacks by terrorists is the use of explosives. VBIEDS are the majority of instruments, and terrorist groups have become experts at building, deploying, and detonating them to maximum effect. Their talents include building devices that are formed and shaped specifically to target military vehicles, and they have practiced attacking law enforcement in their training. The second most prominent explosive device is that used by the suicide bomber, wearing a vest containing explosives and other materials/projectiles.

These devices are extremely dangerous because they are essentially self-guided devices capable of being delivered exactly where the attacker wants it to go. Many thousands of innocent people have been killed by these devices, and the successes and evolution of tactics and the devices themselves have continued to escalate.

Among the previously mentioned characteristics, law enforcement may look for the following traits for an impending suicide bombing:

- The bomber may utilize their final time preparing a will—usually a videotape.
- Contacting family members to say good-bye.
- Eliminating any evidence related to the attack or their identity.
- Preparing a list of the 70 names their attack will assure a place in heaven.

In addition, the suicide bomber, as the attack nears, will likely be seen substantially less as he prepares his final time. There is the possibility that other suicide attackers may gather in one location just prior to the attack. Finally, many attackers will appear to be either increasingly nervous or much happier and at peace as the attack nears.

When dealing with land and sea characteristics, the majority of tactics and strategies are similar, with the exception of the type of vehicle used. For land attacks using a VBIED, much research has shown that the typical car used as a bomb may contain between 500 and 1000 pounds of explosives. This will produce a large-scale explosive, capable of fatalities within several hundred feet and damage or injury up to blocks away. Responders should look for cars that look weighed down with heavy loads, have wiring inside the passenger compartment, or contain large boxes or bags of materials.

Trucks and vans will hold considerably more explosives, and the same characteristics will apply. Terrorists, however, have evolved and continue to evolve these vehicle bombs to maximize their success in approaching and detonating the bomb. Responders should take great care in approaching such a vehicle, and they should take equal care in responding to an after the fact explosion—looking always for second, third, or fourth devices in the area designed to attack them.

In addition, first responders should be aware that terrorists, in their pre-operational planning, have likely tried to identify locations that responders will use as command posts and staging areas. When setting up in one of these locations, the location should be carefully scouted and high-level operational security should be maintained.

SEA ATTACK CHARACTERISTICS

The difference with a waterborne attack is that its targets and approaches are limited to vessels and water. The same tactics and strategies will be used, with the exception that responders need to be aware of far less visible attacks coming from underwater. Primary targets include ferry systems and cruise ships. Container ships may be used to smuggle materials and personnel into the country.

A definition of maritime terrorism has been crafted by the Council for Security Cooperation in the Asia Pacific Working Group:

> ... the undertaking of terrorist acts and activities within the maritime environment, using or against vessels or fixed platforms at sea or in port, or against any one of their passengers or personnel, against coastal facilities or settlements, including tourist resorts, port areas, and port towns or cities.

The sophistication, expense, and training to carry out maritime terrorism necessitates considerable overhead. It would require terrorist organizations to acquire appropriate vessels, mariner skills, and specialist weapons/explosive capabilities.[17] Abdul al-Rahim al-Nashiri, one of the responsible parties for the USS *Cole* bombing, used extensive pre-operational intelligence to identify and attack the ship. His agents rented apartments overlooking the harbor and surveilled numerous Navy ships to determine when the ships would arrive, how long they would be in the harbor refueling, and when and how to attack them. Their plans followed the seven signs of terrorism, the attack characteristics being essentially the same as those of a land or air attack.

The one failure in the original attack was that they did not conduct a dry run. In their case, the original target was the USS *The Sullivans*. The same surveillance and other steps were followed; but when the USS *The Sullivans* steamed in to the harbor, they loaded their boat with explosives; and on putting it in the water, it sunk because it was too heavy with explosives. The USS *The Sullivans* left before they could resurrect the boat. Understanding that other American warships would enter the harbor, they started the cycle again and ultimately attacked the ship, killing and injuring dozens of Americans.

Several months following the attack on the USS *Cole*, the terrorists launched an attack on a very large crude oil freighter, the MV Limburg. Although the attack did not result in large numbers of fatalities, it did cause sufficient damage to the ship to release 90,000 barrels of crude oil into the sea. The same tactics were utilized in this attack. Because of these attacks, ships—in particular, military ships—increased security substantially, making them harder to attack. As a consequence, terrorists shifted their emphasis to softer targets such as ferries and cruise ships. A large ferry in the Philippines was attacked with explosives inside the boat, resulting in sinking the craft with loss of life. As mentioned previously, the ferry systems in the United States have been extensively scouted, and the impact of an attack on any of them would result in considerable impacts.

AIR ATTACK CHARACTERISTICS

Although America itself and its interests were subjected to numbers of terrorist attacks, the country as a whole did not truly awaken to the nature of the threat until 9/11. The use of commercial aircraft as flying explosive devices, to be taken to targets of their choosing, produced extensive security modifications to commercial air service. Several other air-based attacks have been interdicted, clearly showing the terrorist's ongoing interest in using aircraft to initiate attacks. The original hijackings involved small numbers of terrorists who aggressively took control of the planes, killing passengers or flight attendants to get the attention of the passengers. Once done, they led the passengers to believe they would be returning to airports to negotiate demands. Given that most hijackings led to essentially peaceful conclusions, the passengers complied with demands of the hijackers, believing that they would survive by doing so, and all were killed when the planes impacted the buildings. One plane did not reach its target after the passengers learned of the attacks and tried to re-take the plane.

In subsequent attempts to use airliners, explosives were smuggled aboard; or, in some cases, tests of security on airliners involved a dozen or more potential hijackers. Characteristics are continually evolving and law enforcement must interact closely with civilian passengers, encouraging immediate reporting when suspicious activities occur.

THE PUBLIC AND THE BATTLE AGAINST TERRORISM

During the 1980s and 1990s, the American public was largely unaware of, and unconcerned about, international terrorism. Although American interests overseas had been frequently targeted and numbers of Americans had been killed, little attention was paid because it hadn't happened here. Americans believed that because of our distance from those countries, we were not seriously at risk. In 1993 the World Trade Center was bombed, resulting in fatalities and many injuries. Most Americans watching the news viewed the pictures of black smoke coming out of the garage and occupants of the buildings being escorted out, some with soot and other injuries. No pictures were shown of the considerable damage to the structures or that there was a near collapse of the towers, which the attack was designed to do. It was identified as a terrorist attack, but little attention was paid to it. In the eight years following that attack, more Americans and more American interests were killed and damaged in ongoing attacks against our interests abroad.

On 9/11, citizens of the United States watched the attacks in real time, watched some 3000 of their friends, neighbors, and family members, as well as the first responders who tried to save them, all die. Because of those attacks, Americans changed the way they think and operate. First responders also changed operations by adding intelligence capability, and interaction with one another went to new levels.

Awareness became a critical element of safety and prevention. The Seven Signs of Terrorism is reaching more people, and cooperation with law enforcement has resulted in stopping a number of attacks. Because of this, along with the military destroying terrorist training camps overseas, terrorists are finding it increasingly difficult to complete attacks with the ease they previously experienced.

More and more citizens are getting involved in the battle against terrorism, and the successes are mounting. The partnership between law enforcement and their communities is growing faster as a result of the attacks on our country. Citizens are reporting suspicious activities and have provided law enforcement with the ability to actually stop pending terrorist attacks within the United States. They are developing new methodologies to assist first responders and citizens alike to identify terrorists and criminals in their communities.

One of the more effective programs that first responders are teaching their communities is FEMA's Community Emergency Response Teams (CERT). This program is designed to prepare communities and individuals for both natural and man-made disasters. The premise is that with a disaster, government resources—police, fire, EMS—will essentially be overwhelmed with activities and emergencies for a period of time. The program partners these government services with their communities to build a level of preparedness and individual response far beyond that of the norm. Many communities across the United States have been trained and others are lined up to receive it. This, in turn, provides first responders with substantially fewer nonemergency issues and allows them to focus on critical problems.

An example of an individual who has taken the initiative to help her community and first responders alike is a resident of a city in Washington State. This person lives in a residential neighborhood and for years that community had, as friends, a foreign family. They visited together, their children played together, and attended school together. After hearing a terrorism presentation she approached this author with a question and concern. While in her front yard, the neighbor opened his garage door and she observed the garage to be completely full of cases of cigarettes. She felt this was of concern as she had heard about money laundering and cigarettes. The name and address of the family was obtained and forwarded to the authorities to check. As it turned out, the father was on a terror watch list and the case proved successful. It is through the interaction of such people with their police that a number of terrorist attacks have already been stopped in the United States. These efforts should be continued.

LEGISLATION IN THE BATTLE AGAINST TERRORISM

Prior to 1993, there were few laws that specifically addressed terrorist attacks. Now, there is an array of federal statutes addressing terrorism, including laws introduced for the use of weapons of mass destruction. Recent legislation was also passed protecting people who report suspicious activities; this as a consequence of airline passengers and flight crew reporting suspicious behaviors aboard a domestic U.S. flight. The passengers and crew were sued for doing so, and legislation was passed to protect against this type of litigation.

Other laws enacted since 9/11 include the Financial Terrorism Act of 2001, which deals with investigating terrorism funding and freezing assets of terrorist groups. Laws increasing security (via the Transportation Security Administration) have been successfully implemented; and the often used and discussed U.S. Patriot Act, expanding law enforcement powers, was established. Additional laws regarding intelligence were added also.

NEGOTIATION TACTICS TO USE IN INCIDENTS OF TERROR

One of the best examples involving negotiations with terrorists is the seizure of a school in Beslan, Russia. Prior cases, including aircraft hijackings, the 1972 Olympics, and the Russian theater incident, clearly show that traditional negotiation techniques used by law enforcement may not be successful in a terrorist takeover event. As an example, the Beslan school incident is described here; and again, the al-Qaeda Manual provides guidance to the terrorist philosophy regarding such events.

The al-Qaeda Manual states: "Islam does not coincide or make a truce with unbelief, but rather confronts it. The confrontation Islam calls for with these godless and apostate regimes does not know Socratic debates, Platonic ideals, nor Aristoltelian diplomacy. But it knows the dialogue of bullets, the ideals of assassination, bombing and destruction, and the diplomacy of the cannon and machine gun."[18]

This language, and the large number of ongoing threats made against the United States, lends itself to the premise that terrorists are serious about what they say, particularly since they have been repeatedly carrying out those threats.

A case in point involves the terrorist seizure of a school in Beslan, Russia. On September 1, 2004, 30 heavily armed terrorists took control of a school, taking hundreds of school children and adults hostage. The situation continued for nearly three days and resulted in the deaths of 344 civilians, 172 of them children, with hundreds more being injured. The initial attack included more than 1300 hostages. They were herded into the school's gymnasium and kept seated there while the terrorist strung explosives all around them. During this time the terrorists killed some 20 male hostages—the biggest males and those who appeared to be in authority positions—in efforts to reduce potential risk to themselves.

During the second day the terrorists released 26 women and small children to negotiators, which was taken as a sign of progress by the negotiators. On the third day, an explosion occurred—there is disagreement on who did it—and gunfire followed. Government forces initiated action against the school, and the terrorists activated bombs in the gym, destroying it and many of the hostages; some of the terrorists were killed as well.

The investigation of this incident revealed that the terrorist, after taking the hostages and setting up the explosives around them, systematically executed anyone they believed was a potential threat—as mentioned, large males or those perceived to be leaders or in leadership positions. Once done, they began to barricade the gym against the attack they knew would come. They created corridors in the school to the gym that the troops would be required to use to get in—specifically designed to maximize the casualties of the troops. It was also believed that the hostage takers did not intend to survive and intended that the hostages would be killed as well.

A second example involves a terrorist event in Moscow, Russia in 2002. Forty terrorists seized the theater with some 850 customers inside. The demands of the terrorists were unreasonable and after two days Russian Special Forces entered the theater after reports the hostages were being killed. Similar tactics were employed in this incident as in Beslan. Most of the terrorists were killed, as were more than 129 hostages.

Presuming these reports to be true, it may effectively negate the use of traditional law enforcement methodologies for dealing with hostage situations involving terrorists. If the terrorists never intend to release the hostages and force law enforcement to engage them, it will be on their terms, in a manner they choose, in a situation they have designed. The advantages are theirs. Law enforcement must address such circumstances through their policy and procedures. For example, if such an event were to occur at an American school, law enforcement may need to discuss an immediate response tactic, similar to the active shooter event. In essence, the longer the incident continues, the less likely there will be a successful outcome. Departmental use of force and tactical policies should be addressed and modified to add such incidents and how the department intends to respond. Mutual aid agreements with other agencies should also be modified to include any such policy changes.

RULES OF ENGAGEMENT

Also when dealing with departmental policy and procedure, the rules of engagement should also be included. The events officers may encounter will challenge the department's policies, the law, and certainly traditional police methods. For example, assume that the department has a Fourth of July parade and that there has been intelligence that such gatherings may be a suicide bomb terrorist target. In response the department adds additional officers, a more visible presence, and what plans they believe will be deployed if an attack occurs. During the parade, one of the officers, while walking through the crowd, observes a suspicious person, meeting the appearance of a potential suicide bomber. All the characteristics are present. The individual is standing in the middle of the crowd. The officer needs to do something. What should be done—really?

Should the officer confront the person? If so, will the person detonate the explosives, killing bystanders and the officer? Should the officer engage with force, including deadly force? What has the person actually done, other than *appear* to be a potential bomber? What if he turns out not to be? What should the officer do? These discussions and others about the tactics and strategies employed by terrorists must occur now, and policy and procedure must be developed to address them that is both reasonable and defendable. It is the terrorist's strategy to undermine public confidence in the ability of the authorities to protect and defend citizens. The Israelis have been dealing with this problem for years. Their advice to law enforcement is to teach officers what to do the moment of an attack or the attempt. Prevention begins with the officer on the street, who is forced to make the instant life-or-death decisions affecting citizens nearby. Rigorous training is required for identifying a potential bomber, confronting the suspect, the mitigation of the situation, and preservation of the crime scene whether the bomb detonates or not. The officer must have the authority to take action without waiting for supervisor approval.[19]

RESPONSE TO SUICIDE/HOMICIDE BOMBERS

The police, the military, and intelligence agencies can take steps that work from the outside in, beginning far in time and distance from a potential attack and ending at the moment and the site of an actual attack. Although the importance of these steps is widely recognized, they have been implemented only unevenly across the United States. An article in *Atlantic Monthly* in 2003 provides first responders with a set of general guidelines to consider. They are included here:

- Understand the terrorists' operational environment. Know their *modus operandi* and targeting patterns. Suicide bombers are rarely lone outlaws; they are preceded by long logistical trails. Focus not just on suspected bombers but on the infrastructure required to launch and sustain suicide-bombing campaigns. This is the essential spadework. It will be for naught, however, if concerted efforts are not made to circulate this information quickly and systematically among federal, state, and local authorities.
- Develop strong, confidence-building ties with the communities from which terrorists are most likely to come, and mount communications campaigns to eradicate support from these communities. The most effective and useful intelligence comes from places where terrorists conceal themselves and seek to establish and hide their infrastructure. Law-enforcement officers should actively encourage and cultivate cooperation in a nonthreatening way.
- Encourage businesses from which terrorists can obtain bomb-making components to alert authorities if they notice large purchases of, for example, ammonium nitrate fertilizer; pipes, batteries, and wires; or chemicals commonly used to fabricate explosives. Information about customers who simply inquire about any of these materials can also be extremely useful to the police.
- Force terrorists to pay more attention to their own organizational security than to planning and carrying out attacks. The greatest benefit is in disrupting pre-attack operations. Given the highly fluid, international threat the United States faces, counterterrorism units, dedicated to identifying and

targeting the intelligence-gathering and reconnaissance activities of terrorist organizations, should be established here within existing law-enforcement agencies. These units should be especially aware of places where organizations frequently recruit new members and the bombers themselves, such as community centers, social clubs, schools, and religious institutions.

- Make sure ordinary materials don't become shrapnel. Some steps to build up physical defenses were taken after 9/11—reinforcing park benches, erecting Jersey barriers around vulnerable buildings, and the like. More are needed, such as ensuring that windows on buses and subway cars are shatterproof, and that seats and other accoutrements are not easily dislodged or splintered. Israel has had to learn to examine every element of its public infrastructure. Israeli buses and bus shelters are austere for a reason.
- Teach law-enforcement personnel what to do at the moment of an attack or an attempt. Prevention comes first from the cop on the beat, who will be forced to make instant life-and-death decisions affecting those nearby. Rigorous training is needed for identifying a potential suicide bomber, confronting a suspect, and responding and securing the area around the attack site in the event of an explosion. Is the officer authorized to take action on sighting a suspected bomber, or must a supervisor or special unit be called first? Policies and procedures must be established. In the aftermath of a blast the police must determine whether emergency medical crews and firefighters may enter the site; concerns about a follow-up attack can dictate that first responders be held back until the area is secured. The ability to make such lightning determinations requires training—and, tragically, experience. We can learn from foreign countries with long experience of suicide bombings, such as Israel and Sri Lanka, and also from our own responses in the past to other types of terrorist attacks.[20]

PATROL LEVEL RESPONSE

This chapter has conducted extensive discussions of the problems first responders likely will encounter. There are additional considerations that law enforcement needs to understand and apply in order to enhance their own chances for survivability, as well as begin to address the wide array of issues they will confront. "Most law enforcement experts agree that a patrol officer is the most likely person to identify and potentially confront a suicide bomber. We must train patrol officers in the most unthinkable scenarios they have ever faced."[21]

- Suicide bombers, especially on foot, will very likely have a surveillance team in the vicinity. Their role is multifaceted. First, they are present to assure the attack is carried out. They may have the capability to remotely detonate the device if the bomber hesitates or changes his mind. Second, they are there to ensure that law enforcement does not interdict the attack.

They will make efforts to eliminate officers to protect their bombers. Third, they will be present to attack first responders who arrive after the fact.

- The same characteristics will apply to an attack on land, sea, or air, so anyone trying to stop an attack or respond to one must pay close attention to the surroundings and people in the area, as well as be aware of possible devices.
- First responders should have considerable understanding of the scale and magnitude of a bombing, be it a vehicle-borne device or a suicide bomber. Responders should expect an extensive and disturbing scene and should be prepared to react accordingly.
- In responding to an attack, officers, fire fighters, and emergency medical personnel should carefully consider whether they want to proceed immediately into the scene. While there will be damage and casualties, entering the scene without assessing possibilities or the landscape for potential secondary attackers will place the responders at considerable additional risk.
- Also, while responding to an event of any scale, responders historically will place a command post to direct response actions from. Those responsible for setting the location for the command post, staging areas, and other critical locations must consider that the terrorists have likely scouted those sites as well and may have been set up to destroy responders in those locations as well.
- Responders should have a fundamental understanding of the effects of various sizes and types of explosive or other devices and should deploy accordingly to maximize safety. Minimum safe distances should be known by all responders. They should also know that if a device is present, it can be detonated remotely.
- Responders must consider the possibility of chemical, biological, or radiological materials disseminated by an explosion. If an explosion occurs, consider immediate questions, such as which way the wind is blowing, and respond accordingly.
- Responders must consider the probability of bloodborne pathogens at any such scene. Terrorists have used attackers infected with various diseases. Projectiles in the explosive device are often coated with pathogens or such material as rat poison (an anti-coagulant) to increase casualty rates.
- Responders should be trained in advance in the tactics, strategies, and philosophies of the terrorists. They should understand what they may confront and should train in advance. Responding to an actual attack is not the time to train.
- Law enforcement, fire, and emergency medical services should plan and train together. Other local agencies, internal and external, should also be included.

OPERATIONAL PHILOSOPHY

First responders need to take serious note of the ongoing threats being made against the United States. The country is involved in a war with an enemy unlike any in the history of America—one that is clearly and directly telling America what it is going to do and is doing it. To ignore such threats is to create enormous risk. "Americans need to recognize that we underestimated bin Laden's motivation, complexity, and determination. The United States has never had an enemy who has more clearly, calmly, and articulately expressed his hatred for America and his intention to destroy our country by war or die trying. For five years in media interviews, public statements, and letters to the press, bin Laden told us that he meant to defeat the United States and that he would attack—and urge others to attack—U.S. military and civilian targets both in the United States and abroad. In response, the United States never seemed to take bin Laden too seriously, let alone accept the fact that our nation was in the path of real danger."[22] The author also states that "bin Laden unambiguously pledges to use weapons of mass destruction."[23]

The evidence of a forthcoming attack cannot be expressed more clearly, and first responders must be aware of the type and scale of potential attacks and must be prepared to conduct preventive operations, as well as prepare to respond appropriately.

"The United States will remain the Jihad's primary target. In targeting the United States, al-Qaeda will kill as many Americans as possible in as many attacks as it can carefully prepare and execute. Al-Qaeda is clearly building up to the point where it will use a chemical–biological–radiological–nuclear (CBRN) weapon."[24]

Although law enforcement, fire, and emergency medical services have many other important and clear duties, terrorism should be included among the most critical issues. Intelligence development and sharing is absolutely essential, and most law enforcement agencies are involved. In partnering with the community, all completed educational programs such as CERT provide more eyes and ears for responders, further enhancing information gathering. This, in turn, provides the information necessary to target hardening, driving potential attackers to softer targets.

Also critical are (a) development of policy as discussed earlier in this chapter and (b) training aligned with that policy. Every member of the departments should clearly understand what it is and how to apply it. Additional training, such as tabletop exercises, should occur on a regularly scheduled basis and should address the types of attacks the jurisdiction may encounter.

Another critical discussion that must take place among first responders involves the decisions of whether to enter the scene of a terrorist attack at all—or until it is rendered potentially safe from additional attacks directed at the responders. These conversations involve the principle of acceptable losses—in other words, serious decisions regarding who may live or die. Once the incident occurs, there will be numbers of innocent victims injured and in need of immediate attention at or near the scene. Officers, fire fighters, and emergency medical personnel, trained to respond to help, now are faced with a contradiction to that training.

Knowing that terrorists are specifically targeting first responders, if they enter the scene to help the victims, there is a probability that they will be attacked and possibly killed by secondary devices or people. If they don't enter the scene, but instead attempt to isolate and contain it, while waiting for specially trained experts to enter and clear the scene, some of the victims in the scene may not survive. The argument on the one hand is that responders are trained to react to a crisis and assist those victims, to save lives and mitigate further problems. On the other hand, if they do react in this manner and some or all are killed or neutralized, they have lost the ability to assist in either case. Departments must engage in these conversations and arrive at policies, procedures, and protocols that are in the best interests of their communities and their responders. All these department may be sure that the terrorists have done exactly that planning, knowing what the responses historically have been. Making these decisions is neither simple nor easy, but they must be described before an event happens.

Finally, agreements with neighboring agencies, including all responder organizations, should be reviewed and updated to include response to these types of incidents. All the agencies should agree in principle with response plans. Suicide terror is evolving as an art and in its sophistication. First responders can do no less in identifying such problems, preparing for them and learning how to interdict them before they happen.

Bob Mahoney, a retired 24-year veteran of the FBI with experience in multiple major terrorism incidents, including having been present in the World Trade Center on 9/11, makes a compelling statement for consideration. "In an age where it is realized that destroying hijacked commercial airliners may be necessary, where vaccines for pandemic disease may need to be rationed, and weapons of mass destruction might be used in our communities, the traditional military concept of 'acceptable losses' being visited upon the civilian population becomes an important issue for discussion, policy, planning, and operations of governmental entities at all levels. It is an issue with multiple dynamics and of singularly significant consequences that has yet to receive the level of discourse it requires." Mahoney is definitely an expert to be heard. In addition to his extensive work at the World Trade Center on and after 9/11, he was appointed General Manager for Security Programs for the Port Authority of New York and New Jersey and was a core member for the Lower Manhattan Counterterrorism Advisory Team, tasked with developing the Master Security Plan for the new World Trade Center. Here is a resource with the knowledge and actual experiences for first responders to hear.

ENDNOTES

1. The Terrorist Threat to the US Homeland, July 2007, National Intelligence Estimate, www.dni.gov
2. Fred Burton and Scott Stewart, Al Qaeda and the Strategic Threat to the US Homeland, July 25, 2007, Stratfor, Austin, TX.

3. Bobby Ghosh, Inside the Mind of a Suicide Bomber, *Time Magazine*, June 26, 2005.
4. Adam Fosson, The Globalization of Martyrdom: Cause and Effect, July 2007, Research thesis paper Tiffin Online University.
5. *Ibid.*
6. *Ibid.*
7. *Ibid.*
8. Scott Gerwehr and Sara Daly, Al Qaeda: Terrorist Recruitment and Selection, 2002, Rand Corporation, San Francisco.
9. Don Van Natta and Desmond B. Ondon, *New York Times*, March 16, 2003.
10. John S. Pistole, FBI Assistant Director of Counterterrorism, Congressional Testimony, October 2003.
11. Al-Qaeda Training Manual, Lesson 2.
12. Al-Qaeda Training Manual, Lesson 4.
13. Al-Qaeda Training Manual, Lesson 5.
14. Al-Qaeda Training Manual, Lesson 5.
15. Al-Qaeda Training Manual, Lesson 3.
16. Al-Qaeda Training Manual, Lesson 3.
17. Akiva J. Lorenz, Al Qaeda's Maritime Threat, 2007, Maritime Terrorism Research Center, www.maritimeterrorism.com.
18. Al-Qaeda Training Manual, Introduction. First published by the International Institute of Counter-Terrorism, www.ict.org.il
19. *Response to Suicide/Homicide Bomber Policies and Training*, Tactical Security Network, Inc., Sparks, NV, 2006.
20. Bruce Hoffman, The Logic of Suicide Terrorism, *Atlantic Monthly*, June 2003.
21. Police Executive Research Forum, *Patrol-Level Response to a Suicide Bomb Threat: Guidelines for Consideration*, April 2007.
22. Anonymous, *Through Our Enemies' Eyes*, Brassey's, Inc., Dulles, VA, 2002, p. xi.
23. *Ibid.*
24. *Ibid.*

7

MEDICAL MANAGEMENT OF SUICIDE TERRORISM

Shmuel C. Shapira and Leonard A. Cole

Terrorist attacks, especially by suicide bombers, commonly cause many deaths and injuries. The willingness of a person to die in order to kill others makes him (or her) an unusually efficient weapon. A suicide killer not only can place himself in the midst of a target area, but can suddenly change positions by a few feet, the timing of an attack by a few seconds, or even leave the scene entirely in anticipation of causing greater damage on another occasion.[1]

During the second "Intifada"—the Palestinian uprising against Israelis—between 2000 and 2006, some 150 Palestinian bombers blew themselves up in Israeli restaurants, shopping malls, buses, and other places of public accommodation. These suicide attacks represented only 1% of all terror assaults against Israelis during this period, yet they accounted for nearly half the 1100-odd people who died at the hands of terrorists. Similarly, about half the 6500 individuals who were injured from terror attacks during this period were victims of suicide bombings.[2]

Difficult as this period was, the repeated attacks provided opportunities for Israeli medical responders to develop more effective practices to rescue and treat victims. In the words of one analyst, the Israeli experience has shown that "despite the significant death toll inflicted by suicide attacks initially, with the proper

Suicide Terror: Understanding and Confronting the Threat. Edited by Falk and Morgenstern

attention, focus, preparations, and training, this threat can be effectively countered."[3] The validity of this observation is underscored by a review of Israeli medical management of suicide attacks. Israeli practices under these circumstances have become part of the emergent field of terror medicine, and they can inform preparedness and response efforts in other countries as well.

At the outset it is important to understand the distinctive challenges posed by terrorist attacks, especially in the form of suicide bombings. Typically, a bomber wears a belt containing several pounds of explosives packed with nails, bolts, and other metal particles that will be dispersed at great speed and maximize the probability of injury. By detonating himself amid crowds of people, the bomber ensures a variety of damage including blunt trauma, penetration wounds, blast effects, crush, burns, bone fractures, and inhalation injuries. This combination, which under ordinary circumstances is rarely seen in a single individual, may apply to dozens of victims of a terror attack. Appreciation of the challenge begins with an understanding of the nature and characteristics of an explosion.

EXPLOSIVES

A conventional explosion (one not involving chemical, biological, or radiological agents) takes place as a result of a sudden release of energy, which produces a localized increase in pressure and temperature. The rapid combustion may be categorized as low-order or high-order. Low-order explosives, such as gunpowder and Molotov cocktails, do not produce high pressure; although if detonated in a confined area like a room or a bus, the containment can exacerbate the effects. But detonation of high-order explosives such as TNT, dynamite, and Semtex commonly induces very high positive pressure, which exceed ambient atmospheric pressure of 760 mmHg (millimeters of mercury) at sea level. High-order explosives are further categorized as stable or unstable, depending on their tendency to detonate spontaneously.[4] These characteristics may also affect the magnitude of injuries.

Blast waves are caused by the intense overpressurization impulse after detonation of a high-order explosive. The very brief (microseconds) production of positive pressure in a blast wave is followed by a sudden vacuum. The resultant negative pressure wave, whereby the pressure is lower than normal atmospheric pressure, is of longer duration and lower magnitude (Figure 7.1). These rapid changes in pressure also produce the blast wind, which is a forced super-heated airflow. A change in pressure of 0.25 psi (pounds per square inch) is comparable to a wind velocity of 200 km/h (kilometers per hour), equivalent to hurricane force.

The close proximity of a victim to the blast epicenter may result in total body disruption, especially if in a closed space where waves bounce from the walls and other fixed objects. Thus, explosions in restaurants, buses, and other contained facilities cause many more casualties—severe injuries and deaths—than in open spaces.[5] Blast injuries, especially to the lungs, can produce high mortality rates as well, and require exceptionally skillful medical intervention.

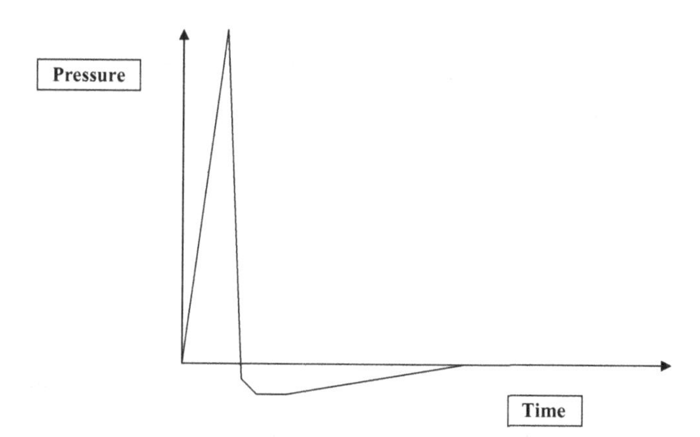

Figure 7.1. Pressure wave after blast.

The mechanism of blast injury may be divided broadly into four categories, from primary to quaternary.[6] Primary includes direct effects from the blast wave, which may rupture gas-filled body structures like the lungs, gastrointestinal tract, and middle ear.

The blast wind (as distinguished from the blast wave) can then cause the hurling of victims, flying debris, shattering of glass, and uprooting of bus seats, depending on the proximity and intensity of the explosion. Extremely high positive-pressure explosions have reportedly reached magnitudes of several million psi. But even a blast inducing an "overpressure" of 130 psi can cause 50% mortality, that is, LD_{50} (lethal dose for 50% of those affected). The closer the victim is to the core of the explosion, the stronger the shock front and the more likely severe injury to the lungs, ears, and gastrointestinal tract. In effect, the rapid change in pressure causes a mini explosion in the body tissues.[7]

Secondary mechanisms of blast injury are flying debris and bomb fragments that can cause penetration wounds. The debris may include the nails, screws, ball bearings, and bolts that have been packed into suicide bomb belts and other conveyers of explosives. These particles become asymmetric projectiles as opposed to symmetric ammunition, like bullets. The irregularly shaped shrapnel not only causes multiple penetrating injuries, but may tear wads of tissue, major blood vessels, or vital organs resulting in life-threatening injury.

Victims of suicide attacks typically suffer from multiple penetrating and exit wounds. But the ballistics of these fragments—low-velocity materials against resistant body tissue—often means they lack sufficient energy to exit the body. In order to identify wound tracks and analyze damage, multiple X-rays and CT (computerized tomography) scans are necessary.[8]

A terrorist attack in 2002 at a shopping mall in Hadera, Israel, prompted new concerns when a bone fragment from the suicide bomber was found embedded in a 31-year-old woman. The fragment tested positive for hepatitis B.[9] The incident raised additional medical management concerns, resulting in a requirement

TABLE 7.1. Injuries Associated with Explosions

Mechanisms of Injury	Cause of Injury	Patterns of Injuries
Primary	Blast	Lungs Gastrointestinal tract Ears Rare: brain, heart
Secondary	Shrapnel Asymmetric projectiles	Penetrating wounds
Tertiary	Blast wind	Blunt wounds
Quaternary	Heat Inhalation Ruins	Burns Inhalational injuries Crush injuries

See also: Mechanisms Associated with Blast Injury, U.S. Centers for Disease Control and Prevention, http://www.cdc.gov/masstrauma/preparedness/primer.pdf.

that all victims of suicide bombers receive a vaccination for hepatitis B.[10] This policy is another illustration of a distinctive response prompted by suicide terrorism.

Tertiary mechanisms of blast injury principally arise from the blowing effect of the blast wind. The immense force can cause fractures, amputation of limbs, and brain injury. Finally, quaternary mechanisms refer to all injuries and illnesses not covered by the first three categories, including burns, crush injuries, asthma, and other breathing problems from dust, smoke, or toxic fumes.

The mechanisms of blast injury are summarized in Table 7.1.

EFFECTS AND MANAGEMENT OF SUICIDE TERRORISM

Mass casualty medical management that is associated with suicide attacks may overlap management procedures during other disasters. Still, the singular flexibility of location and timing available to a suicide bomber can enhance the scale of damage. The consequent medical response—from the mechanics of rapid evacuation to treatment of large numbers with multi-varied injuries—is unique to terrorist attacks especially by suicide bombers.[11] Repeated exercises and debriefings for responders and health workers have enhanced their abilities to manage terror attacks, mitigate injury levels, and strengthen steps toward rehabilitation. Indeed, this information should be incorporated into the curricula for physicians, medical students, nurses, and paramedics.

Israeli efficiency is attributable to frequent drills as well as the experience gained by responders from actual events. In addition, a policy of "scoop and run" means that victims are moved quickly to a hospital rather than provided treatment onsite. Only in the most extreme life-threatening cases, such as (a) blockage of an airway or (b) excessive bleeding, is treatment administered before placement of a victim in an ambulance.

TABLE 7.2. Injury Anatomical Distribution

Body Region	Percentage
Head, face, and neck	96.5
Chest	39
Abdomen	47
Upper extremities	42
Lower extremities	48
Spine	9

Assessments of the effects of suicide bombings in Israel during the "Intifada" offer the following findings:

1. The mortality rate, defined as the number of dead divided by the total number of casualties (injured plus dead), was 13%.
2. Among the fatalities, 87% died immediately onsite and another 6% died within 24 hours of the attack.
3. Twenty-seven percent of casualties were injured in one body region, 31% in two, and 41% in three or more body regions. The anatomical distribution of injuries is recorded in Table 7.2.[12]
4. Most attacks occurred in urban areas and evacuation procedures became increasingly efficient in the course of the "Intifada."[13,14] By 2005, about 60% of injured victims were arriving at a hospital within 30 minutes of an attack, and nearly 90% within one hour. This compares favorably with the emergency medical response to the 2005 terror bombings in London, when nearly an hour elapsed from the time of the initial explosion to the time the first victim received hospital care.[15]

Beyond explosives, the use of nonconventional agents—chemical, biological, radiological, and/or nuclear—by suicide terrorists could pose even greater threats. The potential for massive damage from a chemical attack was demonstrated in 1995 when the Japanese cult Aum Shinrikyu released sarin nerve agent in the Tokyo subway. Twelve people died and more than 1000 became ill. The sarin had been improperly refined; but had it been a more pure product, casualties could have been far greater.[16] In the case of a biological weapon, especially a microorganism that causes contagious diseases, every infected person could become a biological bomb. Someone infected with a virulent form of the smallpox virus or plague bacterium could infect others through close-quarter contact. Similarly, a person willing to risk his own life to kill others could disperse radioactive materials through aerosol devices while wandering unprotected through an unsuspecting crowd.

Hundreds of individuals have sacrificed themselves as suicide bombers, and doubtless many others would be willing to do so using nonconventional weapons.

TABLE 7.3. Checklist for Operations During Weapons of Mass Destruction (WMD) Terror Attack

- Detect the event
- Identify the agent by:
  a. Clinical signs
  b. Lab tests
- Operate containment measures
- Decontaminate
- Use antidotes and supportive measures

A key reason for the infrequent use of these weapons is the challenge to produce or procure them, but also the variability of their effectiveness. Unlike explosives, which are easy to acquire and whose controlled detonation will almost surely injure and kill, the effects of an aerosolized chemical, biological, or radiological agent are less predictable. A bacterial or viral agent, for example, must be highly virulent, exposed individuals must be susceptible, and wind and other environmental conditions must be conducive for effective dispersion.[17] Still, the potential for these weapons to cause great damage remains real. The medical community, first responders in particular, must be prepared for the worst. This means training with appropriate technical and protective equipment to accomplish the following:

1. Confirm the presence of a potentially harmful chemical, biological, or radiological agent.
2. Identify the particular agent.
3. Contain as necessary the area and population at risk.
4. Protect the responders and healthcare workers.
5. Decontaminate the affected people.
6. Provide appropriate treatment to victims.

See Table 7.3.

A mass casualty event can put a strain on available resources including manpower, number of ambulances, available operating theaters, and more. Large-scale terror events involving more than 500 physically injured patients, or lower numbers with special injuries—a few dozen burn victims, for example—could greatly stress a medical system even in countries with advanced capabilities.[18] Just one case of smallpox, a disease that in 1980 had been eradicated, might stress the system by provoking a massive demand for vaccinations and the need for population containment. Thus, mass casualty incident management can present logistical as well as medical challenges. In this regard, preparedness for mass-casualty terrorism can fall within the bounds of an "all-hazards" model that seeks to identify things that commonly occur in many types of large-scale events, such as emergency warning and the need for mass evacuation.[19]

Israeli experience has demonstrated that mortality rates and permanent disability can be mitigated through effective communication, coordination, and teamwork. This can be achieved through preparation and drills and, most importantly, the personal commitment of relevant authorities and respondents.[20]

PREPARING FOR SUICIDE TERROR MASS CASUALTY INCIDENTS

The central medical challenge in dealing with a mass casualty incident is to provide appropriate care in the face of scarce resources. Israeli experience has shown the needs under these circumstances to be acute in the realms of trained staff, ambulances, operating theaters, intensive care units, hospital beds, medical imaging devices, and ventilators.[21]

The first stage in preparing for mass casualty suicide terrorism is to define the threat. Health-care providers should be familiar with the management protocols for various injuries from a variety of weapons that could be used in a suicide attack. They also need to understand the unique concerns presented by security issues such as threats to first responders from sequential attack scenarios at the same location, or by suicide assaults on hospitals. Additionally, a list of standard operating procedures (SOP) should be available. Optimally, generic SOP should be produced by the national government and adopted as a framework at lower levels throughout the health-care system. Procedures that are specific to a facility or area should be written by the local emergency medical services and hospital authorities. Detailed SOPs are most effective when they contain brief user-friendly checklists. See Tables 7.4 and 7.5.

Equipment and material for the management of mass casualty suicide terrorism are available in most trauma centers. But backup supplies should be maintained elsewhere in each hospital and at EMS dispatch centers. Additional stockpiles should be stored in regional or nationally designated locations. Other important preparations include planned traffic patterns for ambulances to rapidly enter and exit hospital grounds and a communications network to quickly summon additional staff.[22]

Knowledge should be provided through formal instruction to all potential participants including responders, nurses, physicians, and other hospital personnel. Tutoring of individual groups through multimedia and in-person presentations should be catered to the specialty roles of each group. Additional presentations should be available to multidisciplinary audiences with an emphasis on enhancing intergroup communications and teamwork skills.

Perhaps the most effective means of imparting knowledge toward preparedness is by participation in drills. Engagement in a mock attack will enhance a participant's sense of his role and the network activity during a real event. Drills offer opportunities for rapid decision-making and exposing vulnerabilities that can be corrected before an actual attack. They usually take one of three forms. The first is a table-top exercise, during which participants assume various roles—physician, nurse, incident commander, safety officer—and verbally respond as a terror scenario unfolds. Thus, a triage commander at the (theoretical) site of an

TABLE 7.4. Emergency Medical Service (EMS) Checklist for Terror-Related Mass Casualty Event (MCE)

Dispatch Center

- Confirm the data.
- Gather information: What happened (nature of event), where, estimated number of casualties, hazards.
- Send teams (Use: alarms, radio, pagers).
- Report—police, fire brigades, headquarters, hospitals, regional dispatch center.
- Keep dedicated radio channel for event management.
- Document resources and needs.

Scene Commander [First Paramedic or Emergency Medical Technician (EMT)]

- Take command.
- Assess hazards.
- Report: What happened (nature of event), where, estimated number of casualties, hazards.
- Assess needs and report.
- Report to police commander. Assign access and evacuation routes.
- Assign triage officer.
- Supervise medical treatment.
- Supervise evacuation, in accordance with dispatch center guidelines.

TABLE 7.5. Hospital Administrator Checklist for Mass Casualty Event (MCE) Management

- Confirm, gather data
- Call for extra medical and paramedical staff
- Notify ER, OR, X ray, Blood Bank
- Assign triage officer
- Decide whether decontamination is needed
- Decide whether to open extra ERs
- Open control station
- Open public information center

attack would have to determine when and where to send victims beyond the nearest hospital to avoid overloading any single facility. That determination would have consequences for other participants who then need to make decisions concerning ambulance traffic, coordination with police, and preparations at receiving hospitals. A tabletop drill is relatively inexpensive, but the number of people who can feasibly participate is limited to about 20. Moreover, while an effective means to evaluate decision-making, it offers little help in assessing actual movement during an event.

For that, a full-scale drill would be necessary during which participants handle live "casualties." In this case, mock victims can be made up with red and black coloring and powders to simulate blood and injury, and they can act as if in pain and distress. Ideally, these victims should include medical professionals who could provide skilled feedback. Such an effort allows for evaluation of not only decision-making, but also systems and processes. The main drawback is the far greater expense and diversion of resources that are required.

A third and more recent approach to simulating attacks involves modeling through computer programs. An observer/participant is able to make choices as a disaster scenario unfolds on the computer screen. While yet to be used as extensively as the other type of drills, disaster simulation is likely to gain currency as another useful tool.[23]

After each drill, whether tabletop, full-scale, or computer simulation, a debriefing meeting with the participants is necessary. A review of the event and lessons learned may lead to revisions of the SOP, which should be implemented immediately.[24]

MASS CASUALTY INCIDENT MANAGEMENT

Pre-Hospital Management, Short Term

Suicide terror attacks are usually short-term incidents. The pre-hospital phase of attending to victims should last less than 60 minutes. To achieve this efficiency, considerable preparedness and organization are necessary. When participants in the response efforts function as an efficient system, real time operation will seem almost automatic. The first senior paramedic or emergency medical technician at the scene is to assume command of the emergency medical services (EMS) arriving teams. Commanders and medical managers should follow checklists (Table 7.4). The initial phase is usually too brief to warrant establishing a formal command post. It is essential, however, that rescue teams and hospitals be able to communicate with each other on compatible radio frequencies. EMS, police, and fire brigades should also operate on interoperable radio frequencies with appropriate backup systems.[25]

The incident commander is typically the senior police officer at the scene. Operating in close proximity to him, and to help coordinate efforts, are the senior EMS and fire brigade officials. The police commander has the overall responsibility to:

1. Identify hazards.
2. Obtain rapid clearance for entry to the scene.
3. Secure access and evacuation routes for ambulances.
4. Maintain public order.

Explicit designation of these assignments has proved highly effective during Israeli responses to terror attacks. In comparison, jurisdiction in the United States is less certain since incident leadership varies there from one community to another. When the World Trade towers were hit in 2001, the New York City police and fire departments each considered themselves operationally autonomous.[26] To the consternation of the fire department, in 2005, the mayor of New York assigned the lead role during a biological or chemical attack to the police.[27]

Beyond typically causing many victims, a terror attack poses challenges not ordinarily encountered in other large-casualty incidents. The initial assault may be followed at the same location by additional suicide bombers. Or, more explosions might later occur from materials that did not detonate at first. Moverover, toxic materials may contaminate the area in a nonconventional attack. Thus, fear of sequential explosions or environmental hazards at the same location could deter responders from quickly entering an area. Action protocols now have paramedics approaching only after a bomb squad announces clearance of onsite hazards. But the delay increases the risk of losing lives and worsening injuries. A compromise approach would call for the rapid entry into the zone of attack, along with a quick evacuation of victims to a safer area where lifesaving care could be initiated.

Optimal primary triage includes three essential ingredients: matching victims with hospitals suited to their needs (for example, sending someone with a head injury to a hospital with a neurosurgery department), dispatching victims in the shortest possible time, and distributing them so as not to overload any single hospital. Still, the demands of the moment might sometimes curb these goals. Thus, in the event of a shortage of ambulances, more patients might be sent to the nearest hospital to cut the ambulances' turn-around time.

Unlike in ordinary circumstances, during a mass-casualty event every patient might not receive the best possible medical care. Triage in such a situation shifts the ideal from assuring optimal care for each patient to saving as many lives as possible. If resources are limited, priority care would be given to severely injured patients whose chances of survival seem promising. Care could be denied to those whose chances of survival are doubtful and who would require a disproportionate and lengthy amount of medical attention. See Figure 7.2.

Pre-Hospital Management, Prolonged Term

Although suicide bombings and pre-hospital responses are usually brief, under some circumstances these events could last for many hours. A series of bombings in one area, or the taking of hostages by terrorists who threatens to blow them up, could result in a prolonged incident. Prolongation might also be caused by efforts to extricate victims who are buried under rubble. Indeed, some patients who are wedged under debris may need treatment of exposed areas before they can be removed from the site. Thus, in addition to the preferred policy of "scoop and run" in which victims are quickly moved from the attack scene, responders need a backup organizational plan for extended incidents. See Figure 7.3.

For organizational purposes, the event focal point should be surrounded by two presumed concentric circles. The area in the inner circle is limited to personnel necessary for immediate response. These individuals are specially trained and equipped to act in threat-related incidents. The outer circle should consist of

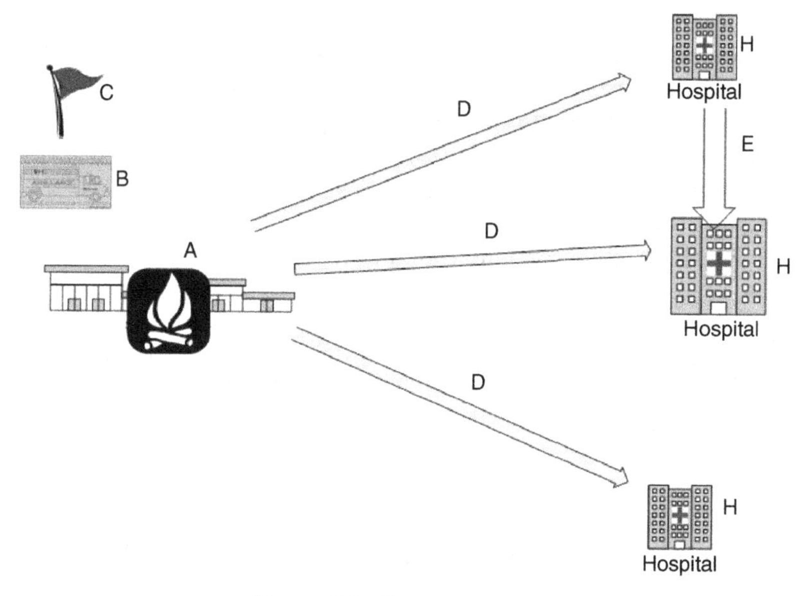

Figure 7.2. Short-term event.

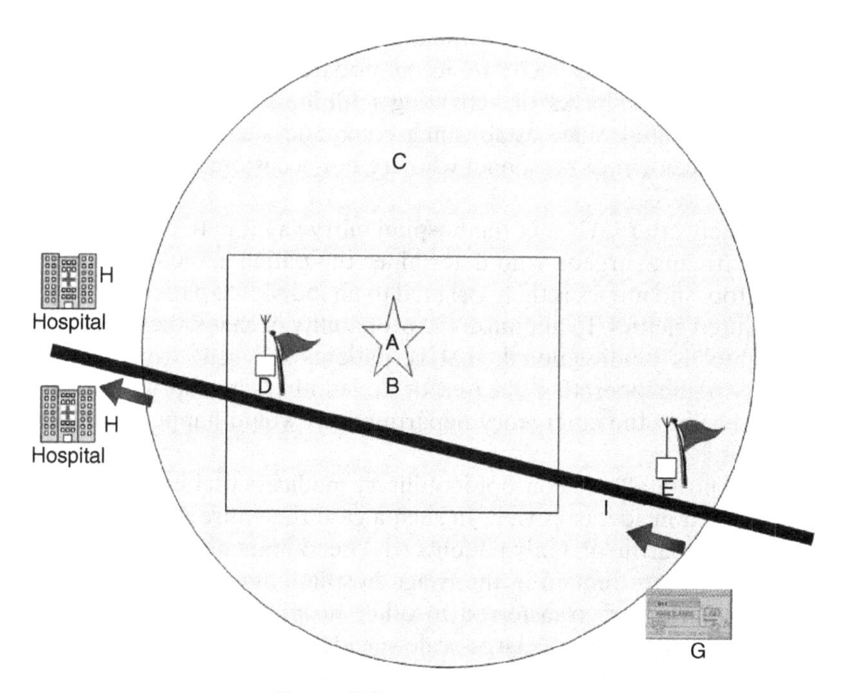

Figure 7.3. Long-term event.

support teams and additional medical staff and equipment. Ambulances are stationed adjacent to the outer circle perimeter. Two command stations are placed near each other, one in each circle. The time taken to extricate victims or to negotiate with the perpetrators should also be used to:

1. Optimize availability of medical staff, ambulances, and helicopters.
2. Review the overall medical plan.
3. Establish effective communication channels with hospitals.
4. Brief responders and medical teams.

Finally, hospital personnel should not ordinarily be sent to an incident site. Unlike emergency responders, they are trained to function principally in a hospital environment and not at a terror scene, where they could hinder rescue efforts.[28]

HOSPITAL MANAGEMENT OF SUICIDE TERRORISM

Hospitals operate in accordance with SOP and checklists (Table 7.5). Limited command and communication centers would have only a few top administrators including the hospital director. Continuous communication should be maintained with EMS to understand the nature and magnitude of the event and the anticipated number of patients. An early management decision must be made about expanding the hospital's capacity to accommodate a large influx of emergency patients. But the drawbacks of activating additional emergency departments should be appreciated, since establishing command sites at multiple locations would mean placement of personnel who are not accustomed to the improvised locations.

Immediately after arrival at the hospital entryway, a patient is triaged by an experienced trauma surgeon who determines the patient's medical priorities. A team of doctors and nurses is then assigned to an individual patient, or to a group of mildly injured victims. To minimize the probability of chaos, the flow of patients in the hospital is unidirectional; that is, patients are sent from the imaging department to the operating theater or to another appropriate department rather than back to the emergency department as would happen under normal circumstances.[29]

A "triage hospital" is a concept in military medicine that can apply to mega-terrorism-related incidents as well. In such a case the entire facility functions as an emergency department. Only patients who need immediate surgical intervention or intensive care remain in the triage hospital. Stable patients, even those with severe injuries, are transferred to other hospitals. This concept could be important to management of a large-scale suicide attack. A hospital near the site of the attack would serve as the triage facility; and after initial evaluation and stabilization, some victims may be evacuated by ambulance or helicopter to other hospitals. Upon arrival, further evaluation and treatment might reveal and correct misjudgments made during primary triage.[30]

The role of hospital security personnel during a suicide terror event includes maintaining public order and providing safety for the hospital, for example, by checking for booby-trapped ambulances. In Israel a special police unit is assigned to the hospital after a mass casualty incident. This unit assists the security personnel to keep access roads clear and to maintain order.

A public information center should also be established in the hospital to respond to phone calls or in-person inquiries from worried relatives and friends. Dedicated phone lines should be part of the information center to avoid overwhelming the hospital switchboard. These phone numbers should be made available to the public through the media. Moreover, spokespersons should offer regular public briefings. The challenge is to maintain a balance between (a) transparency and public relations and (b) medical needs and patients' rights.[31]

As is the case with drills, mass-casualty suicide incidents should quickly be followed by a debriefing of key participants.[32] Administrators, medical directors, EMS directors, police representatives, and others should engage in unsparing assessments, and modifications as needed should be immediately incorporated into the SOP and checklists.

CONCLUSION

Preventing suicide bombings and other forms of terrorism should be the goal of all civilized society. But no less important is to mitigate the consequence of an attack. Medical experts can serve a key role in reducing the impact of suicide terrorism by effectively treating the injured and comforting families and friends. The medical community should be prepared to manage suicide terrorism from the pre-hospital stage through the acute phase of hospitalization, and then during a victim's rehabilitation. Beyond treatment of those directly targeted by a terrorist attack, psychological support is often necessary for responders and others who have been traumatized by repeated rescue and care-giving efforts.

During the many years of terror attacks, Israeli medical workers developed highly efficient methods to rescue, treat, and rehabilitate victims. Their performance has demonstrated that proper management of suicide terrorism can save lives and lessen the impact of injuries. Israeli approaches stand both as lessons for other societies and as key elements in the national resilience of the Jewish state.

ENDNOTES

1. B. Ganor, The Rationality of the Islamic Radical Suicide Attack Phenomenon, International Institute for Counterterrorism. Herzlyia, Israel, March 21, 2007. Accessed November 6, 2007, http://ict.org.il/apage/printv/11290.php

2. Suicide Bombing Terrorism, Intelligence and Terrorism Information Center at the Center for Special Studies, January 2006. Accessed November 6, 2007, http://www.intelligence.org.il/eng/eng_n/pdf/suicide_terrorism_ae.pdf
3. B. Hoffman, *Inside Terrorism*. New York: Columbia University Press, 2006, p. 167.
4. J. Thurman, *Practical Bomb Scene Investigation*. New York: Taylor & Francis, 2006, pp. 5–20.
5. D. Leibovici, O. N. Gofrit, M. Stein, S. C. Shapira, Y. Noga, R. J. Heruti, and J. Shemer, Blast Injuries in Bus Versus Open Air Bombings: A Comparative Study of Injuries in Survivors of Open-Air Versus Confined Space Explosions, *Journal of Trauma* **41**, 1030–1035, 1996.
6. Explosions and Blast Injuries: A Primer for Clinicians, U.S. Centers for Disease Control and Prevention. Accessed October 15, 2007, http://www.cdc.gov/masstrauma/preparedness/primer.pdf
7. D. Leibovici, O. N. Gofrit, and S. C. Shapira, Eardrum Perforation in Explosion Survivors—Is It a Marker of Pulmonary Blast Injury? *Annals of Emergency Medicine* **34**, 168–172, 1999.
8. J. Sosna, T. Sella, D. Shaham, S. C. Shapira, A. Rivkind, A. I. Bloom, and E. Libson, Facing the New Aspects of Terrorism. A Radiologist's Perspective Based on Experience in Israel. *Radiology* **237**, 28–36, 2005.
9. I. Braverman, D. Wexler, and M. Oren, A Novel Mode of Infection with Hepatitis B: Penetrating Bone Fragments Due to the Explosion of a Suicide Bomber. *Israel Medical Association Journal* **4**, 528–529, 2002.
10. Israel Ministry of Health, *Guidelines*. Jerusalem, 2001. (Hebrew)
11. S. C. Shapira, and L. A. Cole, Terror Medicine: Birth of a Discipline, *Journal of Homeland Security and Emergency Management* **3**, 2006 www.bepress.com/jhsem/vol3/1ss2/9
12. Hadassah University Hospital, Trauma Registry 2001–2005. Jerusalem, 2007. (Hebrew)
13. N. Sheffy, Y. Mintz, A. I. Rivkind, and S. C. Shapira, Terror Related Injuries: A Comparison of Gunshot wounds versus explosives' secondary fragments induced injuries. *Journal of the American College of Surgeons* **203**, 297–303, 2006.
14. S. C. Shapira, R. Adatto-Levi, M. Avitzour, A. I. Rivkind, I. Gertsenshtein, and Y. Mintz, Mortality in Terrorist Attacks: A Unique Model of Temporal Death Distribution, *World Journal of Surgery* **30**, 1–8, 2006.
15. L. A. Cole, *Terror: How Israel Has Coped and What America Can Learn*. Bloomington, IN: Indiana University Press, 2007, p. 142.
16. U.S. Senate Subcommittee on Investigations, Hearing on Global Proliferation of Weapons of Mass Destruction, 104th Congress. Minority Staff Statement, "A Case Study on the Aum Shinrikyo." October 31, 1995, p. 52.
17. L. A. Cole, *The Eleventh Plague: The Politics of Biological and Chemical Warfare*. New York: W. H. Freeman, 1998.
18. Israel Ministry of Health. Report of the Committee on Mega-Terror. Jerusalem, 2004. (Hebrew)
19. W. L. Waugh, Terrorism and the All-Hazards Model. Presented at IDS Emergency Management On-Line Conference, June 28–July 16, 2004. Accessed October 15, 2007, http://72.14.205.104/search?q=cache:ZmikwCpskRkJ:training.fema.gov/emiweb/downloads/Waugh%2520-Terrorism%2520and%2520Planning.doc+all-hazards&hl=en&ct=clnk&cd=3&gl=us

20. S. C. Shapira, and S. Mor-Yosef, Terror Politics and Medicine—The Role of Leadership, *Studies in Conflicts and Terrorism* **27**, 65–71, 2004.
21. M. Avitzour, M. Libergal, J. Assaf, J. Adler, S. Beyth, R. Mosheiff, A. Rubin, Z. Feigenberg, R. Gofin, and S. C. Shapira, A Multicasualty Event: Out-of-Hospital and in-Hospital Organizational aspects, *Academic Emergency Medicine* **11**, 1102–1104, 2004.
22. S. C. Shapira, and S. Mor-Yosef, Applying Lessons from Medical Management of Conventional Terror to Responding to Weapons of Mass Destruction Terror: The Experience of a Tertiary University Hospital. *Studies in Conflicts and Terrorism* **26**, 379–385, 2003.
23. A. Hirshberg, and K. L. Mattox, Modeling and Simulation in Terror Medicine, in S. C. Shapira, J. Hammond, and L. A. Cole, eds. *Essentials of Terror Medicine*. New York: Springer, 2009, pp. 79–94.
24. O. N. Gofrit, D. Leibovici, J. Shemer, A. Henig, and S. C. Shapira, The Efficacy of Integrating "Smart Simulated Casualties" in Hospital Disaster Drills. *Prehospital and Disaster Medicine* **12**, 26–30, 1997.
25. M. Avitzour, M. Libergal, J. Assaf, J. Adler, S. Beyth, R. Mosheiff, A. Rubin, Z. Feigenberg, R. Gofin, and S. C. Shapira, A Multicasualty Event: Out-of-Hospital and In-Hospital Organizational Aspects, *Academic Emergency Medicine* **11**, 1102–1104, 2004.
26. The 9/11 Commission Report. *Final Report of the National Commission on Terrorist Attacks Upon the United States*. New York: W.W. Norton, 2004, pp. 284–285.
27. Michelle O'Donnell, New Terror Plan Angers Fire Department. *New York Times*, April 22, 2005, A-1.
28. M. Avitzour, M. Libergal, J. Assaf, J. Adler, S. Beyth, R. Mosheiff, A. Rubin, Z. Feigenberg, R. Gofin, and S. C. Shapira, A Multicasualty Event: Out-of-Hospital and In-Hospital Organizational Aspects, *Academic Emergency Medicine* **11**, 1102–1104, 2004.
29. S. C. Shapira, S. Penchas, and S. Mor-Yosef, Practical Points of Terror Victims' Medical Management, *Journal of Emergency Management* **4**, 47–50, 2006.
30. D. Leibovici, O. N. Gofrit, R. J. Heruti, S. C. Shapira, J. Shemer, and M. Stein, Interhospital Patient Transfer—A Quality Improvement Indicator for Prehospital Triage, *The American Journal of Emergency Medicine* **15**, 341–344, 1997.
31. S. C. Shapira, and S. Mor-Yosef, Applying Lessons from Medical Management of Conventional Terror to Responding to Weapons of Mass Destruction Terror: The Experience of a Tertiary University Hospital, *Studies in Conflicts and Terrorism* **26**, 379–385, 2003.
32. O. N. Gofrit, J. Shemer, D. Leibovici, B. Modan, and S. C. Shapira, Quaternary Prevention: A New Look at an Old Challenge, *Israel Medical Association Journal* **2**, 498–500, 2000.

INDEX

Printed in the United States
By Bookmasters